Statistical Methods
for Environmental
& Agricultural
Sciences

Second Edition

Statistical Methods

for Environmental

& Agricultural

Sciences

Second Edition

A. Reza Hoshmand
University of Hawaii
West Oahu

CRC Press
Taylor & Francis Group
Boca Raton London New York

CRC Press is an imprint of the
Taylor & Francis Group, an **informa** business

CRC Press
Taylor & Francis Group
6000 Broken Sound Parkway NW, Suite 300
Boca Raton, FL 3487-2742

First issued in paperback 2020

ISBN-13: 978-0-367-57943-2 (pbk)
ISBN-13: 978-0-8493-3152-7 (hbk)

**Visit the Taylor & Francis Web site at
http://www.taylorandfrancis.com**

**and the CRC Press Web site at
http://www.crcpress.com**

Library of Congress Cataloging-in-Publication Data

Hoshmand, A. Reza.
Statistical methods for environmental and agricultural sciences / by
A. Reza Hoshmand. — 2nd ed.
    p.   cm.
Rev. ed. of: Statistical methods for agricultural sciences. c1988.
    Includes bibliographical references and index.
    ISBN 0-8493-3152-8 (alk. paper)
    1. Environmental sciences—Statistical methods. 2. Agriculture—Statistical methods.
    I. Hoshmand, A. Reza. Statistical methods for agricultural sciences. II. Title.
GE45.S73H67 1997
628'.07'27—dc21

                                                                    97-80
                                                                    CIP

Library of Congress Card Number 97-80

Cover Designer: Dawn Boyd

About the Author

A. Reza Hoshmand is Professor of International Business at the University of Hawaii–West Oahu. A native of Kabul, Afghanistan, he holds a Ph.D. in Agriculture and Resource Economics from the University of Maryland. Prior to his current position, he was Professor of Agricultural Business Management and Associate Dean of the College of Agriculture at California State Polytechnic University, Pomona.

Professor Hoshmand has taught courses in statistical analysis and quantitative methods and received a teaching excellence award in 1989. His research interests are in economic development issues and international trade. Hoshmand has published numerous articles in academic journals, and his latest book, *Experimental Research Designs for Agricultural and Natural Sciences*, was published by CRC Press in 1994. He continues to be involved in projects sponsored by the U.S. Agency for International Development.

To my wife Lisa and
my children Anthony and Andrea

Contents

Preface

Mathematical symbols and notations

General mathematical symbols

Part I: Descriptive statistics in environmental and agricultural sciences

Part II: Inferential statistics in environmental and agricultural sciences

Part VI: Analysis of change over time

Appendices

Preface

The first edition of this book has been used extensively in classrooms in the U.S. and countries around the world to teach statistical methods in agricultural and natural sciences. I am grateful to many colleagues, students, and other users who have offered valuable comments to make the second edition a better medium for prospective users of statistics and to increase its pedagogical effectiveness. To this end, I have added new exercises and revised some of the previous ones to provide a broad range of practical and relevant cases in the use of statistics in environmental and agricultural sciences. Many of the exercises are drawn from current professional journals. Theoretical concepts are included to provide a better understanding of techniques used in analyzing data in environmental and agricultural sciences. For more intensive theoretical discussions, the reader is referred to other sources that are presented at the end of each chapter in this text.

As was the case with the first edition, this text is intended as an applied undergraduate text for students of environmental and agricultural sciences. However, the text can also be used by graduate students and other practitioners who have little or no statistical training. Although the examples used are geared toward environmental and agricultural sciences, the text is by no means limited to those areas.

The major changes in this edition are the inclusion of the application of statistical methods in environmental sciences. Each chapter of the book has examples that are relevant to students of agriculture as well as students and practitioners of environmental sciences. Chapter 12 on regression has been revised to include nonlinear equations. Furthermore, each chapter ends with a case study giving the reader the opportunity to evaluate the relevance of statistics in environmental and agricultural sciences. I believe that these changes, along with the original features of this text, make the second edition a useful tool in understanding how to apply and analyze statistical concepts.

The major features of the text, intended to enhance both interest and mastery of the concepts in environmental and agricultural statistics, remain the same. First, emphasis is placed on practical application of statistics. The text gives examples in varied fields of environmental and agricultural sciences to show the relevance of statistical techniques in these fields. Second, each chapter begins with a set of learning objectives that orients the student to the learning criteria of that specific chapter. Third, the chapters are organized into various subsections, each representing a unit of mastery based on incremental learning. Fourth, each chapter includes a summary highlighting the main points of the chapter. Finally, each chapter concludes with a case study and review exercises. As supplementary teaching materials, the instructor's manual includes lecture outlines, transparency materials for all the tables and figures in the text, and solutions to the end-of-the-chapter review questions.

The text's simple, nonmathematical language requires only familiarity with elementary algebra and mathematical notations. A glossary of symbols is provided.

Part I contains an introduction to how statistical concepts apply to different fields of environmental and agricultural sciences, followed by descriptive measures of central tendency and variability. Part II covers probability and sampling concepts as used in inferential statistics. Essential techniques of estimation are discussed as well. Part III presents parametric methods in hypothesis testing, which include research designs of special interest to students of environmental and agricultural sciences. Part IV discusses a number of nonparametric techniques. Part V introduces the tests of association and prediction. Finally, analysis of change over time is discussed in Part VI.

In the Appendices, a number of statistical tables are provided that are reprinted from other works. The mathematical derivations of these tables are beyond the scope of this text, and the reader should consult Fisher and Yates, *Statistical Tables for Biological, Agricultural, and Medical Research* (1963).

I am indebted to many associates and students who have contributed to revising this edition. My special thanks to my wife Lisa for providing editorial assistance and constant encouragement. I am grateful to Mr. Tim Pletscher and the editorial staff of CRC Press for their assistance and cooperation in the preparation of this book.

A. Reza Hoshmand
University of Hawaii–West Oahu
January 1997

Mathematical Symbols and Notations

X	Value of an observation	$P(\overline{A})$	Probability that event A does not occur
$\overline{x}$	Grand (arithmetic mean of all observations	$P(A \mid B)$	Probability of A given B (conditional probability)
X_{ij}	Value of an observation in the ith row and jth column	$P(A \cap B)$	Probability that both A and B occur (joint probability)
$\overline{X}_j$	Arithmetic mean of the jth column of observations	$P(A \cup B)$	Probability that A or B or both occur
$\overline{X}_w$	Weighted arithmetic mean		
$\overline{X}, \overline{x}$	Arithmetic mean of a sample	$z = \dfrac{x - \mu}{\sigma}$	Standard score; deviation of value of an observation from the arithmetic mean of a distribution expressed in multiples of the standard deviation
x	Number of successes in a sample size of n		
N	Number of observations in a population		
n	Number of observations in a sample	$\mu_{\overline{x}}$	Mean of the sampling distribution of $\overline{x}$
Σ	Sum of	$\sigma_{\overline{x}}$	Standard error of the mean, that is, standard deviation of sampling distribution of the mean
$\displaystyle\sum_{i=1}^{n}$	Sum of the terms that follow, from $i = 1$ to $i = n$		
$\displaystyle\sum_{x}$	Sum of the terms that follow, for all values of x	$_nP_r$	Permutation of n things r at a time
$\displaystyle\sum_{j}\sum_{i}$	Summation of terms that follow, first over all values of i, then over all values of j	$\binom{n}{r}$	Combination of n things r at a time
		σ	Standard deviation of a population
μ	Arithmetic mean of a population	π(pi)	(1) Proportion of a population having a particular characteristic (2) Probability of a trial success in a Bernoulli process
f	Number of observations (frequency) in a class interval of a frequency distribution		
$P(A)$	Probability that event A occurs	π_0	Value of the population proportion assumed under the null hypothesis

$\pi_1, \pi_2, \pi_3...$ Values of population proportions in testing for equality of several population proportions

p Sample proportion (estimator of π)

σ_p Standard deviation of the sampling distribution of p

μ_p Mean of sampling distribution of a proportion

$\mu_1 - \mu_2$ Difference between two population means

$\bar{x}_1 - \bar{x}_2$ Difference between two sample means (estimator of $\mu_1 - \mu_2$)

$\mu_{\bar{x}_1 - \bar{x}_2}$ Mean of the sampling distribution of the difference between two sample means

$\mu_{\bar{p}_1 - \bar{p}_2}$ Mean of sampling distribution of the difference between two proportions

$\sigma_{\bar{x}_1 - \bar{x}_2}$ Standard error of the difference between two means

$S_{\bar{p}_1 - \bar{p}_2}$ Estimated or approximate standard error of the difference between two proportions

$p_1 - p_2$ Difference between two population proportions

$t = \dfrac{x - \mu}{s / \sqrt{n}}$ The t statistic, distributed according to the Student t distributrion with v degrees of freedom

v Number of degrees of freedom

SSB Sum of squares for blocks in a randomized block design

SSE (1) Sum of squared distances between observed and predicted values in a regression model (2) Sum of squares for error in an analysis of variance

SST Total sum of squares in an analysis of variance

$SSTr$ Sum of squares for treatments in an analysis of variance

T Grand total of all observations in a shortcut computation of an analysis of variance

T Effect of the trend factor in time-series analysis

T Wilcoxon's T statistic; the smaller of two ranked sums

T_j Total of the observations in the jth column

F F ratio: the ratio of the variance explained by treatment to the error or unexplained variance

a Y intercept calculated from a sample of observations; computed value of Y when $X = 0$ in a two-variable regression equation; computed value of Y when values of all independent variables are 0 in a multiple regression equation

b Regression coefficient calculated from a sample of observations; slope of the regression line in a simple two-variable regression equation

b_1, b_2 Sample net regression coefficients; the coefficients of independent variables X_1 and X_2

C Correction term in shortcut computation of an analysis of variance

C Effect of the cyclic factor in time series analysis

χ^2 Chi-square statistic

f_o Observed frequency in a χ^2 test

f_e Theoretical, or expected, frequency in χ^2 test

a, b Constants in a straight line trend equation

a,b,c	Constants in a parabolic trend equation	r_a^2	Corrected or adjusted sample coefficient of determination
B	Regression coefficient in a two-variable linear regression model; the value of b in a sample regression equation is an estimator of B	r_j	Number of observations in the jth column
		r_s	Rank correlation coefficient
		r_{12}	Correlation coefficient for variables X_1 and X_2
$\mu_{y.x}$	Mean of conditional probability distribution of Y given X in the linear regression model; the population value that corresponds to the Y value computed from sample observations	ρ	Population correlation coefficient
		ρ^2	Population coefficient of determination
		s_b	Standard error of regression coefficient b
$s_{y.x}$	Standard error of estimate; measures scatter of observed value of Y around the corresponding Y values on a two-variable regression line; an estimator of $\sigma_{y.x}$	ε(epsilon)	Random error component in regression models
		Y	Dependent variable in linear regression
$S_{y.12\ldots(k-1)}^2$	Sample variance around a regression equation involving $k-1$ independent variables $X_1, X_2, \ldots, X_{k-1}$ and dependent variable Y	$E(Y)$	Expected value of Y
		$\hat{Y}$	Computed value of Y in a sample regression equation
β_1, β_2	Beta coefficients in a multiple regression equation; β_1 measures the number of standard deviations of change in $\hat{Y}$ for each change of one standard of deviation in X_1, when X_2 is held constant	Y_t	Computed trend value for the time series variable Y
		S	Seasonal effect in a time series
		C	Cyclical effect in a time series
		I	Effect of irregular factor in a time series analysis
r	Number of rows in an arrangement of data to which analysis of variance is applied	i	Size of the class interval
		P_n	Price in a nonbase period in an index number formula
r	Sample correlation coefficient	P_o	Price in a base period in an index number formula
r^2	Sample coefficient of determination	Q_n	Quantity in a nonbase period in an index number formula
c	Number of columns in an arrangement of data to which analysis of variance is applied	Q_o	Quantity in a base period in an index number formula

General Mathematical Symbols

$=$	Equal to	α	Probability of a Type I error or rejecting H_o when it is true; the significance level
$\neq$	Not equal to		
$\cong$	Approximately equal to		
$<$	Less than	β	Probability of a Type II error, or accepting H_o when it is false
$\leq$	Less than or equal to		
$>$	Greater than		
$\geq$	Greater than or equal to	$n!$	n factorial, or $(n)\,(n-1)\ldots(2)(1)$
Log X	Logarithm of X to the base 10		

PART I

Descriptive Statistics in Environmental and Agricultural Sciences

chapter one

The What and Why of Statistics for Environmental and Agricultural Sciences

Students of environmental and agricultural sciences often question the value of statistics to their field of study. To address this question, it is important to understand the nature of statistics and its impact on policy decisions. Statistical analyses provide policy makers a means by which they can gauge the impact of policy choices. Environmental and agricultural policies sometimes appear to be in conflict with each other. With the advent of the Green Revolution, increased food and fiber production became highly dependent on the use of fertilizers, insecticides, herbicides, and industrialized agricultural implements. In the last few decades, nations around the world have adopted these technologies and as a consequence they have affected the environment negatively. As nations around the globe continue economic development projects, it becomes critical to assess the impact of these projects on the environment. Statistical models and techniques have an irreplaceable role in analyzing ways of reconciling long-term economic growth with the conservation of renewable natural resources, such as water, forests, soils, and living organisms, and the preservation of human health (Barnett and Turkman, 1993).

Statistics has a major role to play in quantifying effects, assessing consequences, measuring risks, and interpreting the evidence.

The objective of this text is to provide students with an understanding of the basic concepts of statistics in environmental and agricultural sciences.

1.1   What is statistics?

The word "statistics" can be singular or plural. In singular form it refers to a field of study, as one would refer to agronomy and soils, animal science, environmental sciences, etc. In the plural form it represents a group of numbers or a collection of data. The field of statistics is composed of both theory and methods that govern its application. In this text we shall look on statistics as a methodological tool in analyzing numerical data to make better decisions.

Statistics was developed to assist in those areas where laws of cause and effect are not apparent to the observer and where an objective approach is needed (Steel and Torrie, 1980). The field of statistics has contributed substantially to the evolution of agricultural research. While the history of statistics is long, we can look to the very recent past to see its contributions to agriculture and other sciences. To appreciate the important role that statistics has played in agriculture, we need look no further than the writings of Charles Darwin (1809 to 1882). Darwin used biometric or statistical theories to renew enthusiasm in his field of biology. Mendel, another noted biologist, in 1885 used a statistical approach to answer questions about plant hybridization. The writings of Darwin and Mendel provided a basis for the expansion of the field of statistics. Later in the 19th century Karl Pearson (1857 to 1936) applied mathematical concepts to Darwin's theories of evolution. Pearson's research (mostly large samples) in statistics spanned more than half a century.

Though making substantial contributions, the large-sample theory soon was found inadequate to experimentation with the small samples that necessarily occur in many fields. A most interesting factor contributed to the development of concepts dealing with small samples. In 1906 the Guinness brewery wanted to select a sample of people to do beer tasting, but the question of how large a sample to take to form a sound conclusion became financially crucial to the brewery. William S. Gossett, a mathematician, was asked to determine the appropriate sample size needed for a statistically sound beer-tasting appraisal. In 1908 Gossett, under the pen name *Student*, published the results of his work in *Biometrika*. Gossett developed the formula for the standard error of the mean, which specified how large a sample must be, for a given degree of precision, to generalize its results to the entire beer-drinking population. This formula, as well as other practical applications, has contributed to further evolution of the field of statistics. We discuss the use of Student's *t* in detail in this text when dealing with small experimental samples in agronomy, animal science, agricultural economics, agricultural engineering, and environmental sciences.

Historically, many environmental studies were more qualitative than quantitative. However, in recent years the need to develop and use quantitative mathematical and statistical analyses has become apparent to environmental researchers and policy makers (Gertz and London, 1984). Many reasons are given for the need to use more quantitative analyses in environmental studies. It suffices to mention a few here: first, the desire to have a consistent measurable base on which to establish environmental rules and regulations; second, the need to analyze and predict environmental impacts; and third, the desire to monitor the environment with minimum cost and maximum information content.

As you can see from this brief history, statistics has its roots in pragmatism and practicality.

1.2 Why study statistics for environmental and agricultural sciences

The need for better analytic tools to analyze data in agriculture led to major developments in statistics. These tools have played a significant role in the analysis and interpretation of data. In many instances we use sample data to make inferences about the entire population. For instance, a plant scientist who is interested in helping farm operators obtain higher yields from improved varieties of crops conducts field experiments, and the results of his work in the form of a recommendation are reported to the farm operators for consideration and adoption. Obviously, a sample from a field experiment done in one part of a state may be quite different if taken from another area. This limits generalizations. To improve the process of interpretation and generalization, sophisticated analytic tools have been developed. This indicates the importance of statistics to the field of agronomy. Animal scientists use statistical procedures to aid in analyzing data for decision purposes. The use of statistics is not limited to a particular area or animal under investigation. Animal scientists are constantly improving the breeding of farm animals and fowl, as well as their nutritional need. Issues of environmental and heredity effects on meat production and resistance to disease, for example, are constantly under investigation in major universities and research centers. Sound statistical procedures are needed in the design of experiments as well as in the analysis of data (Hutt, 1949; Lush, 1945). Animal nutritionists use statistical experiments to distinguish the impact of new rations on animals when the impact of such factors as age, vigor, and heredity are taken into account.

Without such statistical techniques and procedures, animal scientists would face considerable uncertainty in reporting the results of their studies and their recommendations to farmers.

Agricultural economists use forecasting procedures to determine the future demand and supply of food. Measuring demand is a difficult task in agricultural economics. Published projections of the available supply of food have an impact on the price of the commodity to consumers. As a consequence, the impact of incorrect projections is considerable for the entire economy.

Alternatively, agricultural economists use regression analysis in the empirical estimation of functional relationships between quantitative variables. In studies of this type we may be interested, for example, in the impact of farm size on net income per acre.

Agricultural engineers use statistical procedures in many areas, such as for irrigation research, modes of cultivation, and design of harvesting and cultivating machinery and equipment. Comparison of drying methods for grain and other crops uses sophisticated statistical techniques.

Chapter summary

The history of statistics reveals the substantial practicality of the subject matter. Pragmatic concerns have led to the development of techniques that are indispensable today. The importance of statistics is apparent in our daily lives. Daily news reports are filled with statistics ranging from projections by the U.S. Department of Agriculture on wheat crop, through the Dow Jones stock market reports, to unemployment figures. Such comparisons as the purchasing power of the dollar of today with the dollar of previous years are common in news reports. Reports on the farm economy and the policy decisions made by government leaders are based on statistical reports.

You discover in the following chapters a wide range of techniques that make statistics such a valuable tool in decision making.

References and suggested reading

Barnett, V. and Turkman, K. F., Eds., *Statistics for the Environment*, John Wiley & Sons, New York, 1993.

Gertz, S. M. and London, M. D., Eds., *Statistics in the Environmental Sciences*, American Society for Testing and Materials, Baltimore, 1984.

Gilbert, R. O., *Statistical Methods for Environmental Pollution Monitoring*, Van Nostrand Reinhold, New York, 1987.

Hutt, F. B., *Genetics of the Fowl*, McGraw-Hill, New York, 1949.

Lush, J. L., *Animal Breeding Plans*, Iowa State University Press, Ames, IA, 1945.

Ostle, B. and Tischer, R. G., Statistical methods in food research, *Adv. Food Res.*, 5, 161, 1954.

Pearson, K., Historical note on the origin of the normal curve of errors, *Biometrika*, 16, 402, 1924.

Steel, R. G. D. and Torrie, J. H., *Principles and Procedures of Statistics*, 2nd ed., McGraw-Hill, New York, 1980.

chapter two

Descriptive Statistical Measures

Learning objectives

In this chapter you are introduced to some of the basic techniques that environmental and agricultural scientists use to describe and summarize important characteristics of a set of data. After studying this chapter and answering the review questions, you should be able to:

- Organize and summarize data for efficient presentation.
- Compute the measures of central tendency — mean, median, and mode — and interpret their meanings.
- Compute the measures of variability (range, variance, and standard deviation) and interpret their meanings.

2.1 Introduction

Environmental and agricultural scientists and managers alike collect data for decision-making purposes. Mostly, the data are obtained from samples and are usually unorganized. It is difficult to make a decision from an unorganized set of data. It is therefore necessary to condense large sets of data into an *ordered array*. An ordered array is a listing of sampled observations from the smallest value to the largest. The data gathered on the production of any one crop by the U.S. Department of Agriculture can be used as an example. Thus, wheat production figures for all the wheat growing regions of the U.S. in a given year must be described in a condensed form before they can be published.

By looking at an ordered array, a researcher, whether he or she is an animal scientist, a plant breeder, an environmental scientist, or an agricultural economist, can get a feel for the magnitude of observations.

We can use the following example to show how an array of data may be organized in a meaningful way.

Example 2.1

A plant scientist wants to analyze the effect of thiamine hydrochloride (vitamin B_1) on vegetable transplants. A sample of 50 tomato plants treated with thiamine hydrochloride is randomly selected. Observations on the height of plants 14 d after treatment are recorded, as given next.

Effects of Thiamine Hydrochloride on Height of Tomato Plants (cm)

21.8	21.6	22.5	21.8	21.9	23.4	22.7	21.5	24.0	22.9
22.0	21.8	23.0	22.2	23.2	23.3	22.6	23.2	23.9	22.7
22.3	23.1	22.4	22.1	22.6	21.9	22.8	22.2	24.2	23.2
22.1	23.2	22.9	22.5	23.8	22.6	23.7	22.8	22.8	23.5
22.9	23.3	23.0	23.0	22.9	22.5	22.1	23.5	22.5	23.6

Solution

As can be seen, the data collected do not consist of observations that are identical. As a first step toward making sense of all the observations, we organize the numbers from the lowest to the highest in this case, as shown in Table 2.1.

This ordered array does provide some information. For instance, we can see that the growth rate varies from 21.5 to 24.2 cm, a *range* of 2.7 cm. Second, by displaying the numbers in two equal columns, we note that the lower 50% values are between 21.5 and 22.8 cm, whereas the upper 50% values are between 22.8 and 24.2 cm. Furthermore, an *array* can show the presence or absence of concentration of items around a certain value. In this case, heights of 22.5, 22.9 and 23.2 cm are more *frequent* than 21.5, 21.6, 23.1, 24.0 and 24.2 cm.

In this form, it is not easy to estimate from the array an *average* growth rate of tomato plants, especially if we are dealing with a large number of observations. We find it convenient, therefore, to use a more compact form of data organization called the *frequency distribution*. In the next section, we examine the use of frequency distribution.

Table 2.1   Ordered Array from
the Sampled Observations of
Tomato Plants

21.5	22.8
21.6	22.8
21.8	22.9
21.8	22.9
21.8	22.9
21.9	22.9
21.9	23.0
22.0	23.0
22.1	23.0
22.1	23.1
22.1	23.2
22.2	23.2
22.2	23.2
22.3	23.2
22.4	23.3
22.5	23.3
22.5	23.4
22.5	23.5
22.5	23.5
22.6	23.6
22.6	23.7
22.6	23.8
22.7	23.9
22.7	24.0
22.8	24.2

2.2   *Frequency distribution*

In Table 2.1, the data were presented in an array that made it possible for us to make some general statements about them. The data can also be presented in a *frequency distribution*, which involves grouped data that can be easily visualized. This is shown in Table 2.2. Frequency distribution gives both the values for observations (on height of the tomato plant) and their frequency of occurrence. In column one of the table, the observed heights are recorded as greater than or equal to 21.5 cm, and less than or equal to 24.2 cm. In column two, the recorded frequency of occurrence of these values tells us which heights are more often observed. This table provides us with more information than we previously had, but not precisely what the *average* biweekly growth rate of the plants has been.

To further refine the data summarization, we can use *class intervals* to condense the data. Class intervals are nonoverlapping, contiguous intervals selected arbitrarily in such a way that each and every value in the set of data can be placed in one, and only one, of the intervals. This

Table 2.2   **Frequency Distribution for Data of Table 2.1**

Tomato plant heights (cm)	Frequency
21.5	1
21.6	1
21.8	3
21.9	2
22.0	1
22.1	3
22.2	2
22.3	1
22.4	1
22.5	4
22.6	3
22.7	2
22.8	3
22.9	4
23.0	3
23.1	1
23.2	4
23.3	2
23.4	1
23.5	2
23.6	1
23.7	1
23.8	1
23.9	1
24.0	1
24.2	1

is done by dividing the *range* (lowest to highest value) into equal intervals of a given size, and then tabulating the frequencies associated with each interval. The number of class intervals depends on the number of observations (measurements) we are describing. The larger the number of observations, the more class intervals are required. It is generally advised that you use between 5 and 15 class intervals. There is a more specific guideline suggested by Sturges (1926) that can be used to determine the number of class intervals. Sturges' rule is as follows:

$$k = 1 + 3.322 \, (\log_{10} n) \qquad\qquad [2\text{--}1]$$

where k = class interval
       n = number of observations

To use Sturges' rule, the class interval for the preceding example becomes:

$$k = 1 + 3.322 \, (\log_{10} 50)$$

$$= 1 + 3.322 \, (1.6989)$$

$$\cong 7$$

For 50 observations, it is calculated that the data can be presented in seven class intervals. Table 2.3 shows the seven class intervals and the frequencies associated with each. These interval frequencies are summed from the frequency of each observation included in the interval, given previously in Table 2.2. The information presented in Table 2.3 can be interpreted as follows: five plants had a growth rate between 21.5 and 21.8 cm, eight had a growth rate between 21.9 and 22.2 cm, and so on. Note that here we do not have the same amount of information on each observation as we did in Table 2.2. However, this loss of information on each observation is compensated by the more concise presentation of data.

Table 2.3   **Frequency Distribution**
Using Class Intervals

Class	Growth (cm)	Frequency
1	21.5–21.8	5
2	21.9–22.2	8
3	22.3–22.6	9
4	22.7–23.0	12
5	23.1–23.4	8
6	23.5–23.8	5
7	23.9–24.2	3

The information given in Table 2.3 can also be presented as a *frequency histogram* or *frequency polygon*. Figure 2.1 shows the frequency histogram and polygon presentation of data for the same example. A histogram is a series of rectangles with areas proportional to the frequencies of a frequency distribution. A frequency polygon, on the other hand, is formed by joining the midpoints of the class intervals.

Presenting the data in a polygon or a histogram is a matter of personal preference. They both provide us with an estimate of the average value and also give us an idea of the amount of variability present in the data. Both the histogram and polygon are graphic ways of describing the data. Up to this point, no inferences have been made about the effects of the thiamine treatment on the tomato plants.

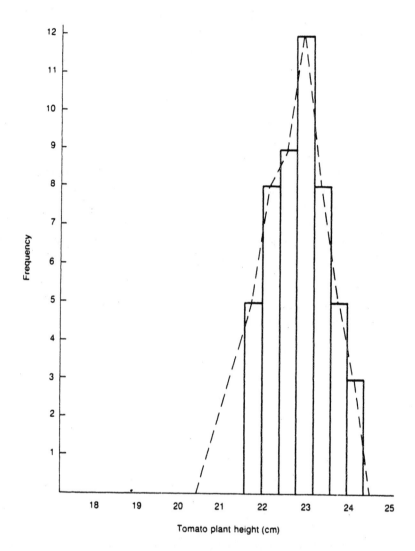

Figure 2.1   Frequency histogram and polygon plotted from Table 2.3.

Refer to Figure 2.1 and note that the frequency polygon is drawn by line segments connecting the midpoint of the class intervals that represent *discrete* groupings of data. If the class intervals were smaller and the number of data points increased, a smoother line could be drawn to approximate a *continuous* distribution of data. This is called a *frequency curve*. Because different arrays of data may be distributed differently, such frequency curves can be shaped differently.

Figure 2.2 shows several frequency curves. Most distributions are of the bell-shaped type. However, a J-shaped curve and a U-shaped curve are also observable in data from agricultural sciences. In panel A,

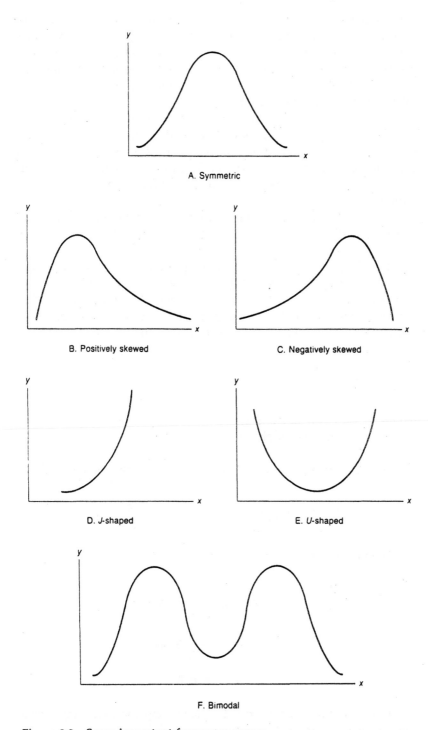

Figure 2.2   Some important frequency curves.

we observe a symmetrical curve where the left half of the curve is the mirror image of the right half. In panels B and C, positively and negatively skewed curves are illustrated, respectively. Panels D and E present J-shaped and U-shaped curves. Panel F portrays the data that have a bimodal distribution. This distribution indicates two regions of relatively high data concentration.

As an added note, the frequency histogram that provides a basis for plotting frequency curves can be used to describe a population measurement. Understanding how histogram shapes are related to the characteristics of data helps us in our knowledge of statistics and probabilities. Throughout this text, we refer to different probability distributions in decision making.

In the following pages, you find further descriptions of the particular kind of distribution associated with each type of frequency curve.

Relative frequency. Data can also be presented graphically using a *relative frequency* histogram. When frequency distributions are converted to relative frequencies or percentages, we are simply dividing the frequency of a class interval by the total number of observations (measurements) under study. A relative frequency can be stated as:

$$\text{Relative frequency} = \frac{f_i}{n} \qquad\qquad [2\text{--}2]$$

where f_i = frequency of class i
n = number of observations

The relative frequency, for example, for class two of Table 2.3 is

$$\text{Relative frequency for class two} = \frac{8}{50} = .16$$

The percentage for class two:

$$\text{Percentage} = \frac{f_i}{n} \times 100$$

$$= \frac{8}{50} \times 100$$

$$= 16\%$$

Table 2.4 shows the relative frequency and the percentage for all the classes in Example 2.1.

Table 2.4 Relative Frequencies and Percentages for
Frequency Distribution of Table 2.3

Class (growth in cm)	Frequencies (f)	Relative frequency (f/n)	Percentage (f/n × 100)
21.5–21.8	5	.10	10
21.9–22.2	8	.16	16
22.3–22.6	9	.18	18
22.7–23.0	12	.24	24
23.1–23.4	8	.16	16
23.5–23.8	5	.10	10
23.9–24.2	3	.06	6
	n = 50		

Notice that in Table 2.4 the data presented in columns three and four are similar. They both imply that 10% of the observations fall in the first class (i.e., 10% of the plants in the group have a growth rate between 21.5 and 21.8 cm), and 16% of the plants fall in the second class (with a growth rate between 21.9 and 22.2 cm), and so on. Because in day-to-day usage we talk about percentages, we should remember that in statistical language "relative frequencies" imply percentages.

2.3 Measures of central tendency

Agricultural and environmental scientists and managers alike often talk about *averages* in the context of average weight gain, average retail food price, average income of farm workers, average milk production per day, etc. These averages simply summarize, in a single value, a set of data. In other words, averages are no more than the middle, or central location, of a set of values or measurements. Averages can be presented as one of the three measures of central tendency, namely, the *mean*, the *median*, and the *mode*.

Mean. The most familiar average is the mean or the arithmetic mean symbolized as $\overline{X}$. It is found by adding all the values of a group of items and dividing the sum by the total number of items. The formula for the mean for sample and population, respectively, is as follows:
If a sample:

$$\overline{X} = \frac{\Sigma X}{N}$$ [2–3]

If a population:

$$\mu = \frac{\Sigma X}{N}$$ [2–4]

where X = measured value of item
       $\overline{X}$ = sample mean, read as X bar
       n = size of sample
       μ = population mean, read as *mu*
       N = size of finite population

The Greek capital letter Σ (sigma) is used to indicate the addition of all observed values of X. It is read as *summation of*. To distinguish between a sample and a population, we designate the population by the Greek letters, and the sample with Roman letters. The following example is used to show how we calculate the mean.

Example 2.2

An animal scientist introduces a new feed supplement to increase milk production in Holsteins. After a month of feeding the new ration, milk production per day is recorded for ten randomly selected cows. Given the information below, what is the average milk production?

Daily Milk
Production (lb.)

X	X
48	58
45	57
50	55
35	36
48	39

The sample mean is calculated as follows:

$$\text{Sample mean} = \frac{\text{sum of all values in the sample}}{\text{number of values in the sample}}$$

$$\overline{X} = \frac{\sum\limits_{i=1}^{n} X}{n}$$ [2–3]

where the symbol Σ is read as *summation from* $i = 1$ *to* $i = n$. Therefore, the sample mean is:

$$\overline{X} = \frac{48 + 45 + 50 + 35 + 48 + 58 + 57 + 55 + 36 + 39}{10}$$

$$= 47.10$$

The average milk production for the sample of ten cows is 47.10 lb.

Example 2.3

In 1996 the price of farmland in California changed erratically. In a random sample of 20 farms in the Central Valley, the following sale prices per acre were reported. Determine the average price of land per acre in the Central Valley.

Price Per Acre of Land
($)

2000	2200
3500	2250
2100	1750
3000	1800
2400	3200
3100	2900
2600	3100
2300	3600
3000	3400
2100	3800

Solution

The sample mean of these prices is

$$\overline{X} = (2,000 + 3,500 + 2,100 + 3,000 + 2,400 + 3,100 + 2,600 + 2,300 +$$

$$3,000 + 2,100 + 2,200 + 2,250 + 1,750 + 1,800 + 3,200 + 2,900 +$$

$$3,100 + 3,600 + 3,400 + 3,800) / 20$$

$$= 54,100 / 20$$

$$= \$2,705 \text{ per acre}$$

Care should be taken in interpreting the mean. In some situations, the use of the mean can lead to a distorted picture of the average value of a distribution. For instance, in the previous example of land values, suppose we had one more farm in the sample that showed a value of $10,000 for an acre of land. This would give a highly misleading picture of the average land values as the average acre price of the 21 farms would become $3,252.38.

A distribution with extreme scores or values at or near one end that are not balanced by extreme scores on the other end is referred to as a *skewed distribution*. Positively and negatively skewed distributions are depicted in Figures 2.2B and C. In a positively skewed distribution, most of the scores or values fall to the left or low end of the curve, and only a few scores or values fall on the right side or high end of the curve. The inverse is true for a negatively skewed distribution.

Whenever we have a skewed distribution, it is much more accurate to use the median instead of the mean as a measure of central tendency, especially where a small number of observations are involved. Before we turn to discussing the median and how it is calculated, let us consider a special case of calculating the mean known as the *weighted mean*.

Weighted mean.   There are times when items are grouped by a particular characteristic. It is possible to utilize the mean of such grouped values to arrive at the overall mean of all items. Because of a mathematical property of means, we are able to combine the arithmetic means of several sets of numbers into a single arithmetic mean without going back to the original data. This involves the procedure of *weighted means*. The formula for weighted mean may be written as:

$$\overline{X}_w = \frac{\Sigma wX}{\Sigma w} \qquad\qquad [2\text{--}5]$$

where w = weights applied to X values
X = values of observations to be averaged

For illustrative purposes, let us consider the following example.

Example 2.4

A plant scientist is interested in determining the yield of corn in a varietal test. He or she has computed the arithmetic mean for each of the four varieties grown in different parcels of land of varying size. The data are presented in Table 2.5. The problem is to find the arithmetic mean for all four varieties combined.

Solution

At a first glance at column one in Table 2.5, it might be assumed that the overall mean for the four varieties would be the average of the four given means, which becomes 133.75 bushels/acre. This is true only if

Table 2.5 **Corn Yield from a Varietal Test**

Variety	(1) X (bushels/acre)	(2) Number of acres	(3) (1) × (2)
A	130.7	64	8,364.8
B	144.3	63	9,090.9
C	140.0	66	9,240.0
D	120.0	61	7,320.0
Total		254	34,015.7

there are the same number of acres producing 130.7, 144.3, 140, and 120 bushels/acre. However, we have as given in column two, 64 acres that had an average yield of 130.7 bushels/acre, 63 acres of 144.3 bushels/acre, 66 acres of 140 bushels/acre, and 61 acres that had an average of 120 bushels/acre. Under these circumstances, a weighted mean should be calculated first by multiplying the values in columns one and two, and summing the products as in column three. When divided by the total number of acres, this sum yields, in this case, a weighted mean of 133.9 bushels/acre.

$$\overline{X}_w = \frac{\Sigma wX}{\Sigma w}$$

$$\overline{X}_w = \frac{34,015.9}{254} = 133.9 \text{ bushels/acre}$$

Mean of a frequency distribution. Computation of the mean of a frequency distribution (grouped data) is similar to the computation of the mean from ungrouped data. Recall that when we group the data in a frequency distribution as in Table 2.3, we lose actual values of observation in each class. Therefore, for computational purposes we make the assumption that the value of any observation in a class is equal to the midpoint of the class (some texts refer to this as classmark). Without recourse to the actual observations or measurements, the mean of a frequency distribution can be approximated as follows:

$$\mu = \Sigma f_i m_i / N \qquad \text{[2–6]}$$

where f_i = frequency of *i*th class
m_i = midpoint of *i*th class
N = total number of observations

Let us use the data from Table 2.3 on tomato plant heights to illustrate the computation of the mean of a frequency distribution. By applying the preceding equation to the distribution shown in Table 2.3, we obtain the following:

Table 2.6   **Calculation of the Mean of a Frequency Distribution**

Height (cm)	Number of plants (f)	Class midpoint (m)	(fm)
21.5–21.8	5	21.65	108.25
21.9–22.2	8	22.05	176.40
22.3–22.6	9	22.45	202.05
22.7–23.0	12	22.85	274.20
23.1–23.4	8	23.25	186.00
23.5–23.8	5	23.65	118.25
23.9–24.2	3	24.05	72.15
Total	50		$\Sigma fm = 1137.30$

To estimate the mean, the frequency (f) in each class is multiplied by the class midpoint (m). We then sum these products (fm) to arrive at Σfm. Finally, the summed value is divided by the total number of frequencies, N, to get the following arithmetic mean:

$$\mu = \Sigma f_i m_i / N = 1137.30 / 50 = 22.75$$

The arithmetic mean of biweekly growth is 22.75 cm.

Median.   The second most important measure of central tendency is the *median*, symbolized as *Md*. The median is computed the same way for the *sample* or the *population*. The median is defined as the middle point of a data set when the data set is arranged in order of size or magnitude of the measurements. It is used when distributions are markedly skewed, and when a small number of measurements are given. Usually, a median is computed when we are interested in whether cases fall within the upper or lower half of the distribution and not particularly in how far they are from the central point. When reported together with the mean, the median can indicate in which direction a distribution is skewed, as shown in Figure 2.3.

With ungrouped data, the median for an *even* number of measure-ments is the average of the values of the two middle observations, whereas the median for an *odd* number of measurements is the middle observation value.

Example 2.5

Variability of available phosphorus from soil to soil and field to field in the same locality is of concern to farmers. Various types of

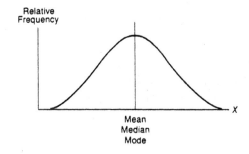

(a) Symmetrical Distribution

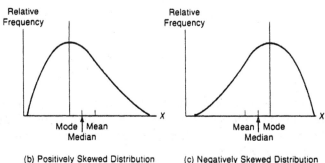

(b) Positively Skewed Distribution   (c) Negatively Skewed Distribution

Figure 2.3 **Relative position of means, medians, and modes when distributions are (a) symmetrical, (b) positively skewed, and (c) negatively skewed.**

phosphate fertilizers are used to overcome such variability. The percent of P_2O_5 in six types of fertilizers is given as follows. Determine the median of these data:

Fertilizers	Percent P_2O_5
11–46–0	46
13–39–0	39
16–20–0	20
16–48–0	48
15–62–0	62
10–34–0	34

Solution

To calculate the median, the scores (measurements) must first be arranged in either descending or ascending order. Then count down (or up) through half of the scores. The percentages of available phosphorus in the preceding data set arranged in ascending (increasing) order are 20, 34, 39, 46, 48, and 62. Because there are six observations, the median is found by determining the score that lies halfway between the two middle scores (i.e., 39 and 46). The median for this sample is

$$Md = (39 + 46)/2$$

$$= 42.5$$

Suppose, in this example, we had only data on five different types of fertilizers; then the median for the data set by definition would be

X
20
34
39 *Md*
46
48

Note: *Md* = 39.

Note that the median is 39 whether the data are arranged in ascending or descending order. Additionally, the median is unaffected by extremely low or high observations as was mentioned earlier in the discussion of the mean.

Median of a frequency distribution. In the calculation of the median of a frequency distribution (*grouped* data) the objective is the same as for calculating the median of ungrouped data, namely, to find the point that divides the distribution into equal halves. The median is estimated by locating the class interval above and below which half of the observations fall. Then, through interpolation within that class, we arrive at the median. To estimate the median, we assume that the values in each class interval are uniformly distributed throughout the class interval. The median is estimated by the following formula:

$$Md = L + \left(\frac{n/2 - CF}{f} \right)(i) \qquad\qquad [2\text{--}7]$$

where L = true lower limit of class interval that contains median
n = total number of frequencies
CF = cumulative number of frequencies in all classes immediately preceding class containing median
i = width of class interval
f = frequency in class interval containing median

For illustrative purposes, let us take the following example.

Example 2.6

In a recent survey, an agricultural economist found the following data to represent the income of beef producers in a county in California. Determine the median income of this group of cattle producers.

Table 2.7   Calculation of the Median of a Frequency
Distribution

(1) Annual income ($)	(2) Number of farmers (f)	(3) Cumulative frequencies (CF)
10,000–14,999	5	5
15,000–19,999	7	12
20,000–24,999	10	22
25,000–29,999	12	34
30,000–34,999	14	48
35,000–39,999	20	68
40,000–44,999	16	84
45,000–49,999	19	103
50,000 and over	17	120
Total	120	

Solution

To compute the median of a frequency distribution, the first step is to determine $n/2$, or the number of half of the observations. This is done by dividing the total frequency by 2; in this case, $120/2 = 60$. We expect the 60th farmer to represent the median income of the group.

The second step is to determine the *cumulative frequencies* as shown in column three of the table. This is done by adding the number of frequencies that are in each interval to the frequencies in the next class. The cumulative frequencies tell us where $n/2$ falls. We know from step one that the median is contained in the class interval in which the cumulative frequency of 60 falls. This happens to be the class interval of $35,000 to 39,999. The *true* lower limit of this class interval is $34,999.50, and the *true* upper limit is $39,999.50. Keep in mind that in our computation of the median we use the *true* lower limit of the class interval that contains the median. Because the fifth class interval corresponds to a cumulative frequency of 48, and the sixth class interval to 68, the median is located somewhere between these two classes.

The third step is to determine how many cases should be added to the lower class interval from the higher class interval to make $n/2$ or 60 cases. Because the cumulative frequency in the fifth interval is 48, we are only 12 frequencies away from 60. Based on the fact that there are 20 farmers in the sixth interval, the median becomes interpolated as 12/20 of the distance between $34,999.50 and $39,999.50. To obtain the median, we multiply 12/20 by $5,000, which is the size of class interval; and add the product to the lower true limit of the sixth class interval. To use the formula, the median is computed as follows:

$$Md = L + \left(\frac{n/2 - CF}{f}\right) (i)$$

$$= 34,999.50 + \left(\frac{120/2 - 48}{20}\right) (5,000)$$

$$= 37,999.50$$

The median income of beef producers in a California county is $37,999.50. In other words, half of the beef producers are earning incomes above $37,999.50, and the other half, below this level.

Mode. The *mode* is defined as the value or values that occur most often in a distribution. The mode of a sample is symbolized as *Mo*. In contrast to the mean that was obtained by calculation, and the median by counting, the mode is obtained by inspection. The mode of ungrouped data is estimated in the following example.

Example 2.7

An ornamental horticulturist observes the germination rate of a new variety of petunia to be 4, 5, 7, 4, 5, 5, 6, 6, 5, 4, 5, 4, 6, 5, and 5 d. Determine the mode of this set of data.

Solution

The value 5 occurs seven times. This is more often than any other value. Therefore, the mode of germination is 5 d.

It should be kept in mind that some sets of data may have no mode. For instance, if the horticulturist in the preceding example had only four observations and had recorded 4, 5, 6, and 7 d for the germination of seed, then there is no mode. Some sets of data may have two modes, in which case we refer to them as *bimodal* distributions, as shown earlier in Figure 2.2F.

Suppose in the preceding example the horticulturist had observed 4, 4, 4, 5, 6, 7, 5, 5, 6, 4, and 5 d for seeds to germinate. In this instance we have a bimodal distribution. The two modes are four and five. We may also observe a mode that is different than the central value, as in asymmetrical distributions. In a symmetrical distribution, on the other hand, the mean, median, and the mode are identical in value. Figure 2.3 shows the mode in relationship to the median and mean for symmetrical and asymmetrical distributions.

In a frequency distribution, such as the one in Table 2.7, the mode is approximated by the *modal class*. The modal class is that class interval which contains the highest frequency of observations. The mode is the midpoint of this class interval. The determination of mode in this procedure is based on the assumption that the frequencies are uniformly distributed throughout the class. This mode is also referred to as the *crude mode*. In this example, the sixth interval that has a frequency of 20 is the modal class. The mode is

$$35,000 + \frac{4,999}{2} \text{ or } \$37,499.50$$

When the frequencies are not evenly distributed throughout the class, we may use the following formula to compute the mode:

$$\text{Mode} = L + \left(\frac{d_1}{d_1 + d_2} \right)(i) \qquad [2\text{–}8]$$

where L = lower true limit of modal class
d_1 = difference between frequency of modal class and frequency of preceding class
d_2 = difference between frequency of modal class and frequency of following class
i = width of modal class

Consider the example in Table 2.7. The computed mode is

$$\text{Mode} = 34,999.50 + \left[\frac{20 - 14}{(20 - 14) + (20 - 16)} \right](5,000)$$

$$= 34,999.50 + \left[\frac{6}{4 + 6} \right](5,000)$$

$$= \$37,999.50$$

The modal income for the cattle farmers is $37,999.50.

In this section, we have discussed various measures of central tendency. The use of these measures is dependent on the nature of the problem, as well as the advantages and disadvantages associated with a particular measure. For instance, the *mean* is the most familiar of the central measures, and is considered to be the most reliable when we use a sample to make inference about the population or universe. Among the disadvantages of the mean is that it is unreliable as a representative average when the data include extreme values. Additionally, computation of the mean is difficult when we have open-ended intervals in a frequency distribution because there is no midpoint in the intervals.

The advantages of using the *median* are in the inherent characteristics of the median. As was pointed out earlier, the median takes into consideration only the number of items in a data set instead of the value of each item. Therefore, the median is not affected by extreme values in a set of data. Furthermore, we are able to compute the median of an open-ended interval that was not possible for the mean. The disadvantage of the median is that it is unreliable as a central measure when we have a sampling problem. Under these circumstances, the mean is a better measure than the median.

By definition, the *mode* is considered to be the most representative of the measures of central tendency. Because it is not affected by skewness at all, it is used as a standard against which we compare the mean and the median of a data set. The disadvantages of the mode are that it is imprecise and lacks strict mathematical properties. Additionally, under sampling conditions, the mode is least reliable.

2.4 Measures of variability

In the previous section, we discussed the measures of central tendency or location. Although these measures are important, they only tell part of the story in descriptive statistics. Equally important in our understanding of statistical methods for summarizing and describing sets of data are the measures of *variability,* or *dispersion*.

Variability, or dispersion, concerns the extent to which values of a data set differ from their computed mean. In general, the greater the spread from the mean, the greater the variability. Other terms that also convey variability are *spread, scatter,* and *variation*. In this section we discuss a number of measures of variability, from simple computation (such as of the *range*) to more sophisticated approaches such as the *average deviation,* the *variance,* and the *standard deviation*.

Range. The range is the difference between the highest and the lowest values in a data set. Therefore, the sample range is

$$R = X_{max} - X_{min} \qquad [2\text{--}9]$$

Example 2.8

Pomologists and postharvest pathologists are concerned with postharvest diseases of fruits. In a study on minimizing postharvest diseases of kiwifruit, 200 fruits from five different plots are exposed to ethylene and then stored under controlled conditions. The following table shows the *Botrytis cinerea* rot after 4 months of storage.

Determine the range of the given data set.

Botrytis Rot of Kiwifruits in
Storage for 4 Months

Plot	Pounds
1	4.2
2	6.7
3	3.4
4	2.3
5	7.9

Adapted from Summer, N. F. et al.,
Calif. Agric., 37(1–2), 16, 1983.

The range is

$$7.9 - 2.3 = 5.6$$

The range of 5.6 lb of fruit affected by the Botrytis rot, although easily computed, is not a reliable measure of variability. By definition, it only considers two values in a set of values. To overcome this limitation, other approaches to measuring the variation among the data are necessary.

Average deviation. Average deviation is defined as the average amount by which the values in a sample or population deviate from the arithmetic mean. When computed from a sample, the sum of these deviations (and hence their mean) is always equal to zero, shown as follows:

Using the data given in Example 2.8 we have:

$$\Sigma(x - \bar{x}) = (4.2 - 4.9) + (6.7 - 4.9) + (3.4 - 4.9) + (2.3 - 4.9) + (7.9 - 4.9)$$

$$= (-0.7) + (1.8) + (-1.5) + (-2.6) + (3.0)$$

$$= 0$$

Because we are interested in deviations from the mean, we must modify our procedure to give us a meaningful measure of dispersion. This is done by taking the mean of deviations while ignoring the algebraic signs. Thus, we add the absolute values of the deviations and divide by n to obtain the average deviation. Average deviation (A.D.) for a sample is, therefore, computed as follows:

$$\text{A.D.} = \frac{\Sigma|x - \bar{x}|}{n} \qquad [2\text{--}10]$$

where x = value of each observation
       $\bar{x}$ = arithmetic mean
       n = number of observations in sample
       $||$ = absolute value sign

Example 2.9

Let us use the data of Example 2.8 to show how to compute the average deviation and interpret its meaning.

$$\text{A.D.} = \frac{|4.2 - 4.9| + |6.7 - 4.9| + \ldots + |7.9 - 4.9|}{5}$$

$$= \frac{.7 + 1.8 + 1.5 + 2.6 + 3.0}{5}$$

$$= 1.92 \text{ lb}$$

To interpret the meaning of average deviation, we would say that on the average, the amount of rot in the sample differs from their mean by 1.9 lb.

Variance.   Like the average deviation, the variance uses all the deviations of values from their mean. Recall that in computing the average deviation, we used absolute values. In computing the variance, however, we avoid negative differences by squaring the deviations from the arithmetic mean. The formula for the variance of the population and sample are given:

$$\text{For population} \ldots \sigma^2 = \frac{\Sigma(X - \mu)^2}{N} \qquad [2\text{-}11]$$

where σ^2 = population variance (σ is the lower case Greek letter sigma)
X = value of observations in population
μ = mean of population
N = total number of observations in population

$$\text{For sample} \ldots s^2 = \frac{\Sigma(x - \bar{x})^2}{n - 1} \qquad [2\text{-}12]$$

where s^2 = sample variance
x = value of observations in sample
$\bar{x}$ = mean of sample
n = total number of observations in sample

Example 2.10

Environmental scientists are concerned with the increasing rate of ocean pollution from oil spills. Studies have shown that petroleum pollution contributes to the growth of some types of bacteria. Analysis of seawater off the coast of Alaska after a major oil spill shows the following petroleumytic microorganisms (bacteria per 100 ml) in 20 samples of seawater:

Number of bacteria per 100 ml of seawater

32	65	72	45	52	74	53	42	58	61
59	40	36	76	55	75	62	48	35	39

What is the population variance?

Solution

To determine the population variance, the deviations from the population mean (μ) is first determined as follows:

$$\mu = \frac{\Sigma X}{N} = \frac{1079}{20} = 53.95$$

X	$X - \mu$	$(X - \mu)^2$
32	−21.95	481.8
65	11.05	122.1
72	18.05	325.8
45	−8.95	80.1
52	−1.95	3.8
74	20.05	402.0
53	−0.95	0.9
42	−11.95	142.8
58	4.05	16.4
61	7.05	49.7
59	5.05	25.5
40	−13.95	194.6
36	−17.95	322.2
76	22.05	486.2
55	1.05	1.1
75	21.05	443.1
62	8.05	64.8
48	−5.95	35.4
35	−18.95	359.1
39	−14.95	223.5
1079	0.0	3780.9

The population variance becomes:

$$\sigma^2 = \frac{\Sigma(X - \mu)^2}{N}$$

$$= \frac{3,780.9}{20}$$

$$= 189.0$$

Interpretation of the variance is difficult for a single set of observations. However, the variance can be used to compare dispersion between two or more sets of observations (the variance of two or more sets is discussed in later sections of this book). The reason for the difficulty of interpretation is that 189.0 in the preceding example is not in terms of number of bacteria, but instead in *number of bacteria squared*. To make a better sense of such a number, we transform the variance into the standard deviation. This gives us the same unit of measurement as the original data. For instance, by taking the square root of 189.0 we obtain:

$$\sigma = \sqrt{189.0} = 13.7 \text{ bacteria}$$

The variance can be computed using a shortcut method. The shortcut approach is not based on deviations from the mean, but instead on the actual values of the observations. The method is less cumbersome because it does not require a large number of subtractions. The formula for the shortcut method is

$$\sigma^2 = \frac{\Sigma X^2}{N} - \left(\frac{\Sigma X}{N}\right)^2 \qquad [2\text{--}13]$$

Applying the preceding formula to Example 2.10, we obtain:

$$\sigma^2 = \frac{(32)^2 + (65)^2 + \ldots + (39)^2}{20} - \left(\frac{32 + 65 + \ldots + 39}{20}\right)^2$$

$$= 189.0$$

Standard deviation. The standard deviation is an important measure of dispersion. As was indicated, the variance is a useful measure when we compare variations of samples from their mean. The standard deviation is also used to compare the dispersion in two or more sets of observations. The standard deviation is defined as a *measure of variability that indicates by how much all the values in a distribution typically deviate from the mean*, or it may be defined as *the positive square root of the population variance*. The larger the value of the standard deviation, the more the individual observations are spread out around the mean; the smaller the standard deviation, the less the individual observations or values are spread out around the mean. Two methods for calculating the standard deviation are presented here. The first is called the *deviation* method, and the second is the *computational* or shortcut method. Conceptually, the deviation method provides for a better understanding of standard deviation. However, the computational method is easier to use.

Symbolically, standard deviation for population as well as sample may be represented as follows:

$$\text{For population}: \ \sigma = \sqrt{\frac{\Sigma(X - \mu)^2}{N}} \qquad [2\text{--}14]$$

$$\text{or the shortcut method}: \ \sigma = \sqrt{\frac{\Sigma X^2}{N} - \left(\frac{\Sigma X}{N}\right)^2} \qquad [2\text{--}15]$$

$$\text{For sample}: \ s = \sqrt{\frac{\Sigma(X - \overline{X})^2}{n-1}} \qquad\qquad [2\text{--}16]$$

$$\text{or the shortcut method}: \ s = \sqrt{\frac{\Sigma X^2 - \dfrac{(\Sigma X)^2}{n}}{n-1}} \qquad\qquad [2\text{--}17]$$

Note that in the formula for the sample standard deviation, the denominator is divided by $n - 1$ instead of n. It can be proved (see Mason, 1982) that if we were to use n instead of $n - 1$ in the denominator, the sample standard deviation would be a *biased* estimator of the population standard deviation. This is especially true if the sample size is small. Thus, to have an *unbiased* estimator of the population standard deviation, we will use $n - 1$ in the denominator.

Example 2.11

The monthly gross sales of a group of agricultural machinery firms are selected at random. The sample gross sales in thousands of dollars are 24, 32, 28, 22, 20, 26, 28, and 20. What is the standard deviation of the sample?

Solution

Deviation method:

$$\overline{X} = \frac{\Sigma X}{n}$$

$$= \frac{200}{8}$$

$$= 25$$

X	$X - \overline{X}$	$(X - \overline{X})^2$
24	−1	1
32	7	49
28	3	9
22	−3	9
20	−5	25
26	1	1
28	3	9
20	−5	25
Total	0	128

$$s = \sqrt{\frac{\Sigma(X - \overline{X})^2}{n-1}}$$ [2–16]

$$= \sqrt{\frac{128}{8-1}}$$

$$= \sqrt{18.28}$$

$$= 4.28$$

Because the sales were measured in thousands of dollars, the computed standard deviation implies a $4280 deviation from the mean.

Turning now to the shortcut or computational method:

X	X^2
24	576
32	1024
28	784
22	484
20	400
26	676
28	784
20	400
200	5128

$$s = \sqrt{\frac{\Sigma X^2 - \frac{(\Sigma X)^2}{n}}{n-1}}$$ [2–17]

$$= \sqrt{\frac{5128 - \frac{(200)^2}{8}}{8-1}}$$

$$= \sqrt{\frac{5128 - 5000}{7}}$$

$$= \sqrt{\frac{128}{7}}$$

$$= 4.28$$

For a better understanding of the interpretation and use of the standard deviation, let us suppose that the sample for the preceding example was from a group of farm machinery dealers in Illinois. Furthermore, assume that we had taken a sample of monthly gross sales of farm machinery dealers in Nebraska and found that the means of sales from Illinois and Nebraska were about the same, whereas the standard deviation of sales in Nebraska was $5200. The standard deviation of $4280 indicates that the sales in Illinois are not dispersed as much as those in Nebraska. It follows that the mean in Illinois is more reliable and useful than the average for gross monthly sales in Nebraska.

Another way of illustrating the relationship between the mean and standard deviation is given by Chebyshev's theorem. This theorem states that:

> *Given any set of observations, at least [1 − (1/k²)] of the observations will fall within k standard deviations of the mean of the set of observations.*

By working out the theorem we learn that at least 75% of the values in a set of observations can be expected to fall within two standard deviations above and below the mean, at least 88.9% within three standard deviations of the mean, and at least 96% within five standard deviations of the mean. Using the formula given by Chebyshev, we can see how the two, three, and five standard deviations translate to the percentage values.

For about 75%, we have:

$$1 - \frac{1}{k^2} = 1 - \frac{1}{2^2} = 1 - \frac{1}{4} = .75$$

For about 88.9%, we have:

$$1 - \frac{1}{k^2} = 1 - \frac{1}{3^2} = 1 - \frac{1}{9} = .88$$

For about 96%, we have:

$$1 - \frac{1}{k^2} = 1 - \frac{1}{5^2} = 1 - \frac{1}{25} = .96$$

This theorem is applicable to any set of observations; that is, we can use it either for sample or population. To see how we can apply the theorem, let us use Example 2.11

The arithmetic mean of the monthly gross sales in Example 2.11 was computed as 25, and the standard deviation was 4.28. Given this information and Chebyshev's theorem, we can expect at least 75% of the values to be between $(25 - 4.28 \times 2)$ or 16.44 and $(25 + 4.28 \times 2)$ or 33.56. At least 88.9% of the values are between $(25 - 4.28 \times 3)$ or 12.16 and $(25 + 4.28 \times 3)$ or 37.84. At least 96% are between $(25 - 4.28 \times 5)$ or 3.60 and $(25 + 4.28 \times 5)$ or 46.40. It should be stressed that Chebyshev's theorem applies to any sample of measurements *regardless* of the shape of the frequency distribution. There is another convention for interpreting the standard deviation called the *empirical rule*, which is discussed in connection with the normal curve in Chapter 3, Section 3.5.

Standard deviation of grouped data. The calculation of the sample standard deviation for grouped data or a frequency distribution is similar to the ungrouped data. We use the following formula to compute the standard deviation:

$$s = \sqrt{\frac{\Sigma f x^2 - \frac{(\Sigma f x)^2}{n}}{n-1}} \qquad [2\text{--}18]$$

where x = midpoint of class
f = class frequency
n = number of observations in sample

Example 2.12

California's controversial Peripheral canal that was planned to bring more water for the agricultural communities of the southern San Joaquin Valley from northern California has been the center of debate by agriculturists and environmentalists. In a recent referendum, the pro- and antialliances sought the support of the public by asking for contributions for their cause. The following data show the donors by size of contribution for building the Peripheral canal. What is the standard deviation for these grouped data?

Donors by Size of Contribution for Building the
Peripheral Canal ($1000)

Size of contribution	Number of donors
0.00 < 5.00	30
5.00 < 10.00	35
10.00 < 15.00	42
15.00 < 20.00	10
20.00 < 25.00	3
25.00 < 30.00	5
30.00 < 35.00	7

Solution

To solve the problem, we first find the midpoints of each class interval as shown in column three of Table 2.8, and then multiply them by their frequencies as given in column four. Finally, the products of columns three and four are summarized as shown in column five. Substituting these sums into the formula gives us the standard deviation of the sample shown as follows:

Table 2.8 Calculation of Sample Standard Deviation with Grouped Data

(1) Donations	(2) Number (f)	(3) Midpoint (x)	(4) (fx)	(5) $fx \cdot x$ (fx^2)
0.00 < 5.00	30	2.5	75.00	187.50
5.00 < 10.00	35	7.5	262.50	1,968.75
10.00 < 15.00	42	12.5	525.00	6,562.50
15.00 < 20.00	10	17.5	175.00	3,062.50
20.00 < 25.00	3	22.5	67.50	1,518.75
25.00 < 30.00	5	27.5	137.50	3,781.25
30.00 < 35.00	7	32.5	227.50	7,393.75
Total	132		1,470.00	24,475.00

$$s = \sqrt{\frac{\Sigma f x^2 - \frac{(\Sigma f x)^2}{n}}{n-1}} \qquad [2\text{--}18]$$

$$= \sqrt{\frac{24,475 - \frac{(1,470)^2}{132}}{132-1}}$$

$$= \sqrt{\frac{24,475 - \frac{2,160,900}{132}}{132-1}}$$

$$= \sqrt{\frac{24,475 - 16,370.45}{131}}$$

$$= \sqrt{61.87}$$

$$= \$7.87$$

In calculating the standard deviation by this method, the computation becomes extremely tedious. This is especially true when the midpoints are large and there are a large number of frequencies. To remedy this problem, you may use the *coded* method that was discussed in the computation of the mean of grouped data. Recall that we used class deviations (d) from the mean instead of class midpoints (X). The formula for the coded method is

$$s = (i)\sqrt{\frac{\Sigma fd^2 - \frac{(\Sigma fd)^2}{n}}{n-1}}$$

[2–19]

where i = class interval
f = frequencies
d = class deviations

To use Example 2.12, we have:

f	d	fd	fd^2
30	-2	-60	120
35	-1	-35	35
42	0	0	0
10	+1	10	10
3	+2	6	12
5	+3	15	45
7	+4	28	112
132		-36	334

By substituting these sums into the formula, we get the standard deviation of the sample by the coded method as follows:

$$s = (i)\sqrt{\frac{\Sigma fd^2 - \frac{(\Sigma fd)^2}{n}}{n-1}}$$

$$= (5)\sqrt{\frac{334 - \frac{(-36)^2}{132}}{132-1}}$$

$$= (5)\sqrt{\frac{334 - 9.8}{131}}$$

$$= (5)\sqrt{\frac{334 - 9.8}{131}}$$

$$= (5)\sqrt{247}$$

$$= 5(1.57)$$

$$= 7.87$$

As can be seen, we can arrive at the same answer using either method.

Chapter summary

In this chapter, we have discussed the use of descriptive statistics in summarizing data. The use of frequency distribution, frequency polygon, and histograms for efficient presentation of data was explained. This chapter was mainly concerned with the descriptive measures of *central tendency* and *dispersion*. Whereas measures of central tendency define the concept of averages, the measures of dispersion describe the variability of the data.

The specific properties and uses of the four measures of central tendency, namely, the *mean, weighted mean, mode,* and *median* were discussed.

The four major measures of variability or dispersion discussed in this chapter are the *range,* the *average deviation,* the *variance,* and the *standard deviation.* Methods of determining or computing each of these measures were illustrated.

Case Study

Describing the Use of Agrochemicals in Indian Agriculture

Agrochemicals have had a great impact on human health and the production and preservation of food, fiber, and other cash crops around the world. With their introduction, farm practices have undergone revolutionary changes leading to the possibility that hunger may vanish from the earth (Dhaliwal and Pathak, 1993). There are at least 1250 agrochemicals registered throughout the world, of which 25% have been phased out or banned (Tomar and Parmar, 1993). The rate of increase in the use of agrochemicals in developing countries is considerably higher than that in developed countries. The agrochemical industry in India has seen much progress. By starting with a meager beginning, the country now produces sophisticated pesticides, fertilizers, plant hormones, and other allied chemicals to meet the food and fiber requirements of its expanding population (Dhaliwal and Singh, 1994).

The population of India in 1995 was estimated to be 931 million (World Resources Institute, 1995). It is estimated that by the turn of the century the population will cross the billion mark. Food requirements by 2001 will be on the order of 240 million tons (Dahliwal and Singh, 1994). The 1990 to 1992 average cereal production in the country was around 196 million tons. The additional food grain production will have to be realized by increasing the cropping intensity and productivity of the land. Agrochemicals play a major role in increasing food production. One aspect of agrochemical use is fertilizer consumption. It is of interest to policymakers who are concerned with production to know about the extent of fertilizer use in different parts of the country. The table following shows the consumption of plant nutrients per unit of gross cropped area for the various states in India. These types of data can be analyzed with descriptive statistics.

Consumption of Plant Nutrients ($N + P_2O_5 + K_2O$) per Unit of Gross Cropped Area in 1992 to 1993

State	Consumption (kg/ha)
Andhra Pradesh	114.6
Assam	7.5
Bihar	57.2
Gujarat	72.7
Haryana	107.7
Himachal Pradesh	31.5
Jammu and Kashmir	40.0
Karnataka	66.4
Kerala	66.5
Madhya Pradesh	35.3
Maharashtra	57.3
Orissa	21.9
Rajasthan	27.4
Tamil Nadu	113.6
Uttar Pradesh	86.0
West Bengal	87.6

From Fertilizer Association of India, *Fertilizer Statistics for 1992 to 1993*, New Delhi, India, 1993.

We can use the data to:

1. Determine the mean and the standard deviation of the data.
2. Apply Chebyshev's theorem to describe the distribution of measurements.

References

Dahliwal, G. S. and Singh, B., Agrochemicals in Indian agriculture: retrospect and prospect, in *Management of Agricultural Pollution in India*, Dahliwal, G. S. and Kansal, B. D., Eds., Commonwealth, New Delhi, India, 1994.

Dahliwal, G. S. and Pathak, M. D., Pesticides in the developing world: A boon in bane. In Dahliwal, G. S. and Singh, B., Eds., *Pesticides: Their Ecological Impact in Developing Countries*, Commonwealth, New Delhi, India, 1993, 1.

Tomar, S. S. and Parmar, B. S., *Evolution of Agrochemicals in India — An Overview*, Indian Agricultural Research Institute, New Delhi, India, 1993.

World Resources Institute, *World Resources: 1994–1995*, Oxford University Press, New York, 1995.

Review questions

1. A(an) _____ is an arrangement of observations from largest to smallest or from smallest to largest.

2. The term skewness refers to the _____ in a frequency curve.

3. The following observations on the daily growth (in inches) of seedlings in a greenhouse are recorded by an agronomist.

.38	.42	.50	.36	.62	.46	.41
.34	.75	.55	.42	.78	.79	.69
.65	.72	.49	.56	.61	.66	.73
.66	.55	.88	.82	.66	.76	.48

   a. Arrange the data in an ascending array.
   b. What is the range of values?
   c. Compute the arithmetic mean, median, and mode for the given data.

4. The following distribution gives the annual income and the number of employees who earn such incomes with an agricultural chemical firm.

Income ($)	Number of employees
15,000 < 17,000	32
17,000 < 19,000	22
19,000 < 21,000	15
21,000 < 23,000	16
23,000 < 25,000	11
25,000 < 27,000	5
27,000 < 29,000	10
29,000 < 31,000	9

a. Construct a histogram and a frequency polygon of this income distribution.
b. Compute the mean and the median, and interpret their meanings.
c. Compute the mode and interpret its meaning.
d. Compute the standard deviation and interpret its meaning.

5. Per capita pork consumption fluctuates throughout the world. Given in the following table is per capita pork consumption of selected countries around the world in 1983. Find the mean and median pork consumption.

Country	Carcass weight (kg)
U.S.	30.1
Canada	27.7
Mexico	14.2
Germany (FRG)	50.0
France	34.6
Netherlands	39.5
Germany (GDR)	58.3
Poland	36.6
U.S.S.R.	20.6
Taiwan	29.4
Japan	13.9

From U.S. Department of Agriculture, *World Agriculture: Outlook and Situation Report*, USDA/ERS/WAS 35, March 1984, 13.

6. The U.S. Department of Agriculture reports the following agricultural production indices for selected countries.

Country	Agricultural production index
U.S.	106
Canada	136
Japan	99
South Africa	115
Oceania	120
Western Europe	123
South and Central America	148
North Africa	124
Sub-Sahara[a]	119
Middle East	147
Southern Asia[b]	160
U.S.S.R.	123
Eastern Europe	125
China	179

a Excludes South Africa.

b Excludes China and Japan.

From U.S. Department of Agriculture, *World Agriculture: Outlook and Situation Report*, USDA/ERS/WAS 35, March 1984.

    a. Set up an array.
    b. Use the Sturges formula to determine a class interval.
    c. Construct a frequency distribution.
    d. Calculate the mean.
7. A food broker determines the mean daily sales of oranges to be 200 boxes with a standard deviation of 25.
    a. What minimum percentages of the time can the broker expect to sell between 180 and 220 boxes per day?
    b. Between what two bounds can the broker expect daily sales to lie at least 95% of the time?
8. A plant scientist collects the following set of data on digestible protein (%) from an improved variety of alfalfa hay (early bloom).

11.4	11.2	11.0	11.8	10.9
10.8	11.2	11.4	11.5	11.8
11.0	11.9	10.8	11.3	11.6

    a. Determine the mean, mode, and median of the data.
    b. Compute the standard deviation and the variance.
9. Water is an essential nutrient for all animals. The actual water requirements for various animals are highly variable. An animal scientist observes the following water requirement for swine (market hogs). Find the mean, mode, and median water requirement.

Water Requirement for Swine (gal/d/head)

1.0	1.2	1.4	1.5	1.8	2.0	2.2	2.5
2.1	2.2	1.9	1.8	1.9	2.1	2.3	2.4
2.1	2.3	2.4	2.5	2.1	2.8	2.7	1.9
1.9	1.8	2.0	2.0	2.2	2.2	2.3	2.3
2.3	2.0	1.8	2.0	2.1	1.9	1.7	1.8

10. Animal scientists are constantly working on new research in genetic improvements to produce more efficient animals. A recent study shows the following gestation period for a group of genetically improved cows.

276	276	281	280	281	283	279	276
284	288	280	281	280	281	283	282
285	283	284	279	276	281	275	276
277	273	272	275	280	276	279	278

   a. Determine the mean, median, and mode for the given data.
   b. Compute the standard deviation.
11. The accompanying data are the weights of marketable hens.

3.6	3.3	3.5	3.6	4.1	3.2	3.8	3.6
3.8	3.6	3.6	3.7	3.2	3.4	3.4	3.5
3.7	3.8	3.9	3.2	3.3	3.3	3.6	3.6
3.2	3.3	3.3	3.4	3.6	3.3	3.6	3.8
3.6	3.4	3.4	3.5	3.7	3.7	3.6	3.8

   a. Determine the mean and the median.
   b. Calculate the standard deviation and the variance of the data.
12. With increasing pressure from the environmentalist, timber-producing firms in the northwest U.S. are facing challenges in harvesting timber. The firm's environmental engineer suggests that no trees less than 20 inches in diameter be harvested. To determine how much lumber can be harvested from a tract of 100-by-100 ft squares from trees with 20-in diameters, the environmental engineer gathers the following data from 40 tracts.

12	8	6	8	10	11	7	5
10	8	7	4	6	8	9	11
9	9	7	8	6	5	10	9
7	10	12	10	9	7	8	12
8	6	9	9	7	12	11	8

    a. Construct a relative frequency histogram for the data.

    b. Compute the sample mean $\overline{X}$ as an estimate of the population mean μ.

    c. Compute the standard deviation of the data.

13. The 24-h milk yield from a population of crossbred cows is given as follows. What is the population variance?

24-h Milk Yield from
Crossbred Cow Group

Breed	kg
HA	7.0
SA	7.5
SH	6.5
BA	8.0
BH	6.5
JA	6.5
JH	7.0

Note: A = Angus, H = Hereford,
S = Simmental, B = Brown
Swiss, and J = Jersey.

14. To determine parasitic contamination in Lake Michigan, a random sample of 100 water specimens examined by environmental researchers produces the following frequency tabulation:

Number of parasites, x	0	1	2	3	4	5	6	7
Number of samples, f	54	15	10	8	6	5	2	0

    a. Construct a relative frequency histogram for the number of parasites in the sample.

    b. Calculate the mean $\overline{x}$ and s for the sample.

    c. What percent of the parasite counts fall within one standard deviation of the mean? Do the results agree with Chebyshev's theorem?

References and suggested reading

Anderson, D. R., Sweeney, D. J., and Williams, T. A., *Statistics for Business and Economics*, 2nd ed., West Publishing, St. Paul, MN, 1984.

Dixon, W. J. and Massey, F. J., Jr., *Introduction to Statistical Analysis*, 3rd ed., McGraw-Hill, New York, 1969.

Hamburg, M., *Statistical Analysis for Decision Making*, 4th ed., Harcourt Brace Jovanovich, New York, 1987.

Lapin, L. L., *Statistics for Modern Business Decisions,* 4th ed., Harcourt Brace Jovanovich, New York, 1987.

Mason, R. D., *Statistical Techniques in Business and Economics,* 5th ed., Richard D. Irwin, Homewood, IL, 1982.

Ott, L., *An Introduction to Statistical Methods and Data Analysis,* 3rd ed., PWS-KENT, Boston, 1988.

Plane, D. R. and Opperman, E. B., *Business and Economic Statistics,* Business Publications, Plano, TX, 1981.

Rosner, B., *Fundamentals of Biostatistics,* 3rd ed., PWS-KENT, Boston, 1990.

Snedecor, G. W. and Cochran, W. B., *Statistical Methods,* 7th ed., The Iowa State University Press, Ames, IA, 1980.

Sturges, H. A., The choice of class interval, *J. Am. Stat. Assoc.,* 21, 65, 1926.

Summer, N. F., et al., Minimizing postharvest disease of kiwifruit, *Calif. Agric.,* 37, 16, 1983.

U.S. Department of Agriculture, *World Agriculture: Outlook and Situation Reports,* USDA/ERS/WAS 35, March 1984.

PART II

Inferential Statistics in Environmental and Agricultural Sciences

chapter three

Probability Theory

Learning objectives

After completing this chapter and working through the review questions, you should be able to:

- Define probability and explain what objective and subjective probabilities are.
- Measure probability, using the counting techniques.
- Explain the rules for combining probabilities.
- Explain Bayes' theorem and how it is used.
- Explain what a probability distribution is and how to compute the binomial distribution probabilities.
- Use the table of areas for a standard normal curve to determine probabilities of a normal distribution.

3.1   Introduction

In the preceding chapter, you were introduced to descriptive statistics. Descriptive statistics, however, do not permit us to infer population values from sample characteristics. To make such generalizations, we must use *inferential* statistics. In this chapter, the concept of probability that has served as the cornerstone of inferential statistics is explained.

Whether in applied or basic research, the researcher or manager is often faced with limited information to make a decision. Because complete information is usually difficult or costly to obtain, decision makers faced with such uncertainty base their judgments on the likely outcomes of an event. The theory of probability is concerned with the concept and measurement of uncertainty, and has its inception in the games of chance developed by European mathematicians of the 17th century.

Decisions made by researchers and managers with limited information carry a certain amount of risk. The risk is usually associated with the *uncertainty* that accompanies such a decision. To analyze risks, managers and researchers use probability theory.

Defining probability has been a center of controversy among mathematicians. However, no matter how we define probability, methods of measuring probability are basically the same. For the purposes of this book, we define probability as *a measure of the strength or degrees of belief one has in a particular outcome of an event or experiment.*

One's belief in an event taking place may be subjective or objective probabilities. The preceding definition implies subjective probabilities. This means that the outcome of an event is based on an individual's personal belief that an outcome or event will occur. For example, the chief executive officer of an agribusiness firm may predict the likelihood of a price increase in his or her commodities based on personal judgment and past experiences.

Objective probabilities, on the other hand, are based on either advance knowledge of the process in question (*a priori* probabilities), or experimental evidence (*empirical probabilities*).

An *a priori* probability can be determined without experimentation. Coin tossing and other games of chance where all possible outcomes and occurrence of certain events are known are examples of *a priori* probabilities. An empirical probability is determined with experimentation.

By way of empirical probability, assume 50 animals that have been vaccinated with a certain vaccine are subsequently examined to determine if they have developed immunity to the disease against which they were theoretically protected. This sample of observations serves as the basis for determining the probability of successful immunization programs in the future. Thus probability can be defined in the empirical mode as *the relative frequency of occurrence of an event after repetitive trials or experiments.*

Such a definition implies that an event will occur a proportion of the time. In this context, if the event does not occur, the probability of the event is zero; and if the event does occur, it has a probability of one. Thus, probability falls between values of zero and one.

Probability concepts play an important role in many statistical analyses. The subject of probability is large, and consequently can only be covered in a general way in a single chapter. However, a rudimentary understanding of the basic concepts is absolutely essential before we can go on to other topics such as sampling, hypotheses testing, and decision making.

3.2 *Measurement of probability*

Regardless of how we define probabilities, the mathematical rules in performing probability calculations are the same. Before describing various computational approaches, however, it is important to become familiar with some of the terms and basic rules that are used in probability analyses.

Event.   An event may be defined as an uncertain outcome of an experiment, or one or more possible outcomes of a given observation or experiment. For instance, if five seeds are planted for a germination experiment, the outcome of *five germinated seedlings* is one of the possible events that can occur.

Sample space.   This refers to a complete listing of the events that can occur in an experiment or observation. For instance, in a 5-seed germination experiment, the sample space would contain the possibility of 0, 1, 2, 3, 4, and 5 germinated seeds.

Union of events.   The *union* ($\cup$) of events is defined as *the outcome of an event that contains all sample points belonging to* A *or* B *or both.* The union of events is expressed as $A \cup B$.
   Figure 3.1 shows the union of events A and B. The diagram is referred to as the Venn diagram, named after the English logician J. Venn (1834 to 1888) who portrayed graphically the outcome of such experiments. The interior of the rectangle represents the sample space that contains the sample points. The total of all sample points account for the sample space. The two circles represent events A and B. The fact that the circles overlap (as shown in panel b) indicates areas that are contained in both A and B. To obtain the desired probability of events A or B we must make sure that we subtract the probability of the overlapped region when we add the probability of event A to the probability of event B.

Intersection of events.   The *intersection* of two events is defined as *the event that contains the sample points that belong to both A and B.* The intersection is expressed as $A \cap B$.

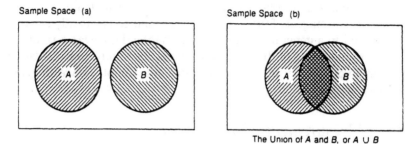

Figure 3.1   Union of events *A* and *B* when the events are (a) **mutually exclusive** and (b) **when they are not mutually exclusive.**

Figure 3.1(b) is the Venn diagram depicting the intersection of two events. Intersection refers to the area where the two circles overlap.

Basic rules of probability.   Probability computations basically follow either *addition* or *multiplication* rules. When we have two events and are interested in knowing the probability that at least one event in which we are interested occurs, we use the addition rule. Another way of stating the addition rule is if we are interested in knowing the probability that event *A* or event *B* or both occur, we simply add the elementary probabilities.

On the other hand, if we are interested in finding the probability of two (or more) events occurring simultaneously, we apply the multiplication rule.

Let us now illustrate the use of these rules. When applying the *addition rule*, it is important to know whether we have *mutually exclusive* or not mutually exclusive events. If the occurrence of one event precludes the occurrence of the other event, we say that the two events are mutually exclusive. For example, if a plant breeder who in a single experiment on chlorophyll transformation gets a yellow colored seedling instead of green one, he is faced with a mutually exclusive event. For two events that are mutually exclusive as graphically portrayed in Figure 3.1a, the equation for the addition rule may be stated as:

$$P(A \text{ or } B) = P(A) + P(B) \qquad [3\text{--}1]$$

or

$$P(A \cup B) = P(A) + P(B) \qquad [3\text{--}2]$$

The preceding probability statement is read, the *probability that either A or B will occur is equal to the sum of the probability of A and B.* **For** example, if we roll a single die, the probability of getting a two or a three would be

$$P(2 \text{ or } 3) = P(2) + P(3)$$
$$= 1/6 + 1/6$$
$$= 2/6$$
$$= .33$$

Situations in which two events are not mutually exclusive are portrayed as the shaded area in Figure 3.1b minus the intersected shaded area in Figure 3.2. As can be seen, it is possible for both events to occur; thus the addition rule applicable to such an occurrence is

$$P(A \text{ or } B) = P(A) + P(B) - P(A \text{ and } B) \qquad [3\text{–}3]$$

Sample space

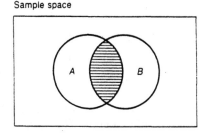

Figure 3.2   Intersection of events *A* and *B*.

For example, if we draw a card from a deck, the probability of getting either a king or a heart will be

$$P(\text{king or heart}) = P(\text{king}) + P(\text{heart}) - P(\text{king and heart})$$
$$= 4/52 + 13/52 - 1/52$$
$$= 16/52$$
$$= .31$$

If we are interested in determining the probability of two (or several) events occurring at the same time (joint probability), we use the *multiplication rule*. The multiplication rule is applied differently depending on whether we are dealing with events that are *dependent* on or *independent* of each other. Two events are statistically independent of each other if the occurrence of one in no way affects the probability of the second event occurring.

To illustrate this concept, assume you have a pair of dice. One die is white and the other is black. The appearance of a three on the white die is completely independent of the appearance of a three on the black die. Symbolically, the probability of independent events is stated as:

$$P(A \text{ and } B) = P(A) \times P(B) \qquad\qquad [3\text{--}4]$$

or

$$P(A \cap B) = P(A) \times P(B) \qquad\qquad [3\text{--}5]$$

We can use Equation 3–4 to determine the probability of throwing a three with the dice as

$$
\begin{aligned}
P(3 \text{ on white and } 3 \text{ on black}) &= P(3 \text{ on white}) \times P(3 \text{ on black}) \\
&= 1/6 \times 1/6 \\
&= 1/36 \\
&= .028
\end{aligned}
$$

When two events (A and B) are dependent on each, the probability of occurrence of one event is dependent or *conditional* on the occurrence or nonoccurrence of the other event. The notation |, which means *given*, is used to indicate the conditional nature of the probability statement. For instance, $P(A \mid B)$ is read, *probability of event A given the condition that event B has occurred*. As you notice, the multiplication rule is based on conditional probability. Symbolically, the multiplication rule for dependent events is

$$P(A \cap B) = P(A) \times P(A \mid B) \qquad\qquad [3\text{--}6]$$

or

$$P(A \cap B) = P(B) \times P(B \mid A) \qquad\qquad [3\text{--}7]$$

Equation 3–6 is read, *the probability of simultaneous or successive occurrence of events A and B is equal to the product of probability of event A and the probability of event A given that event B has occurred*. We may use either Equation 3–6 or 3–7 that is its inverted form to compute $P(A \mid B)$.

To illustrate this rule, consider the following example.

Example 3.1

In her presentation to a group of sales agents, the vice president of the sales division of an agricultural chemical company points out that past experience has shown that the probability of making a sale to a farm on the second visit is .30. Her sales records show that 52% of such sales are for amounts in excess of $200. What is the probability that a sales agent will make a sale in excess of $200?

Solution

Before using Equation 3–6 or 3–7 to solve for the probability, we set up the problem with the following notation:

$P(A)$ = probability that a sale is made

$P(A \mid B)$ = probability that the sale is in excess of $200 given that a sale is made

The probability that a sales agent will make a sale in excess of $200 is

$$P(A \text{ and } B) = P(A) \times P(A \mid B)$$
$$= (.30)(.52)$$
$$= .16$$

The multiplication rule can be applied to any number of events. The generalization of the multiplication rule to account for more than two events is as follows:

$P(A, \text{ and } B, \ldots, \text{ and } N) = P(A)P(B \mid A)P(C \mid B \text{ and } A), \ldots,$

$P(N \mid N - 1 \text{ and}, \ldots, A)$ [3–8]

To apply the multiplication rule to more than two events, let us consider another example.

Example 3.2

An animal scientist is working with three different cattle breeds to improve the efficiency of beef production. Suppose he has ten steers, five of which are Angus, three are Brahman, and two are Hereford. Suppose he selects three at random from a list numbered one to ten.

He selects each number one at a time, so that at each drawing all remaining numbers have an equal chance of being selected. What is the probability that all three steers selected are Angus?

Solution

To keep our notation straightforward, let us designate the events of the first-drawn, second-drawn, and third-drawn Angus steers as A, B, and C, respectively. The joint probability is

$$P(A, B, \text{ and } C) = P(A)P(B \mid A)P(C \mid B \text{ and } A)$$
$$= (5/10)(4/9)(3/8)$$
$$= .08$$

To make sure that we understand the basis for the ratios on the right-hand side of the equation, consider what the probability for event A and the conditional probabilities are. Because there are 10 steers and 5 of them are Angus, the probability of the first Angus drawn will be $P(A) = 5/10$. Now we are left with 9 steers, so the conditional probability $P(B \mid A) = 4/9$. Similarly, after the second Angus is drawn, there are eight remaining steers, three of which are Angus. Now we have accounted for each of the ratios.

In answer to our question, the probability for all steers selected being Angus is .08.

This example sheds light on two important points. First, it should be noted that the *joint* probability of the simultaneous occurrence of events A, B, and C is the same as their successive occurrence. This implies that the multiplication rule may be used to obtain the joint probability of the simultaneous or successive occurrence of two or more events.

Second, if we have drawn a random sample *with replacement*, the probability for each event would remain the same. For instance, in the preceding example, if sampling had been done with the replacement of another Angus steer, the joint probability for events A, B, and C is

$$P(A, B, \text{ and } C) = (5/10)(5/10)(5/10)$$
$$= 0.13$$

Notice that the ratios for each event remain the same because after each draw we replace the number for each steer to be drawn back into the same pool.

Another way of looking at the relationship between sampling and the nature of events is that sampling *without replacement* is similar to *dependent* events in probabilities, and sampling with replacement is similar to *independent* events.

Conditional probability. The probability of occurrence of one event given that another related event has occurred is called a *conditional probability*. Equation 3–9 is the way in which the conditional probability of two events A and B is noted mathematically:

$$P(A|B) = \frac{P(A \cap B)}{P(B)} \qquad [3\text{–}9]$$

or

$$P(B|A) = \frac{P(A \cap B)}{P(A)} \qquad [3\text{–}10]$$

provided that $P(B)$ and $P(A)$ are not equal to zero. Let us use an example to illustrate conditional probabilities.

Example 3.3

Avian coccidiosis, a poultry disease, is estimated to cost the poultry industry $300 million yearly (Danforth and Augustine, 1985). A poultry scientist is testing a new vaccine to remedy the problem. The following data show the result of the exposure of birds to the vaccine. The question to be solved is, what is the probability that a bird selected randomly will be infected, given that it is vaccinated?

| Result | Treatment | | |
of exposure	Vaccinated	Control	Total
Infected	22	44	66
Noninfected	78	56	134
Total	100	100	200

Solution

Before we compute the conditional probabilities, let

I = event that bird is infected

H = event that bird is not infected

V = event that bird is vaccinated

C = event that bird is not vaccinated

Because each cell of the table indicates the intersection of two events, we are able to compute the *joint probabilities* of these events. The joint probabilities are computed by dividing the observed values of each cell in the preceding table by the total number of birds, as follows:

$$P(I \cap V) \quad = 22/200 \quad = .11$$

$$P(I \cap C) \quad = 44/200 \quad = .22$$

$$P(H \cap V) = 78/200 \quad = .39$$

$$P(H \cap C) = 56/200 \quad = .28$$

Table 3.1 displays the *joint* and *marginal probabilities* for this example.

Table 3.1   Joint and Marginal Probabilities of Controlled and Vaccinated Birds

	Vaccinated	Control	Marginal probabilities
Infected	.11	.22	.33
Noninfected	.39	.28	.67
Marginal probabilities	.50	.50	1.00

Marginal probabilities refer to the sum of joint probabilities of each separate event. The marginal probability of .33 and .67 indicate that 33% of the birds are infected, and 67% are not infected. The marginal probabilities of .50 and .50 indicate that 50% are vaccinated and 50% are not vaccinated.

To compute the conditional probability that a bird is infected given that the bird is vaccinated, we have:

$$P(I|V) = \frac{P(I \cap V)}{P(V)} \qquad [3\text{--}11]$$

The joint probability $P(I \cap V)$ from Table 3.1 is .11. Also note that .50 is the marginal probability that a bird is infected given that it is vaccinated, $P(I) = .50$. Given this information, the conditional probability is

$$P(I|V) \quad = \frac{.11}{.50}$$

$$= .22$$

The conditional probability solution states that given a bird is vaccinated there is a 22% chance of infection.

Tree diagram. Graphically we can present a sample space either as a rectangular diagram as in the previous examples, or as a *tree diagram*. You have already been exposed to the use of the Venn rectangular diagram. Similarly, tree diagrams are equally useful. However, they are much more manageable than the rectangular diagrams.

We construct a tree diagram by first drawing a heavy dot or a square to represent the trunk of the tree as shown in Figure 3.3. The outcome of each successive sample observation is represented by a branch. Let us illustrate the use of a tree diagram with an example.

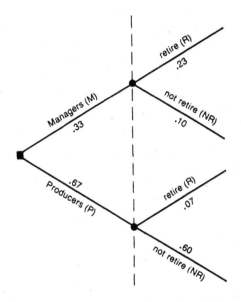

Figure 3.3 Tree diagram for 300 agriculturally related job holders who prefer to retire after age 60.

Example 3.4

Recent studies indicate that people involved in farming do not retire until after age 60. An agricultural economist is surveying either managers of agricultural businesses or producers to determine their plans after age 60. The results of the survey are

| | Plans after age 60 | | |
| | Retire | Not retire | |
Nature of job	(R)	(NR)	Total
M: managers	70	30	100
P: producers	20	180	200
			300

The problem is to draw a tree diagram and determine the joint probabilities.

Solution

For this problem, two main branches extend from the tree trunk. One represents the managers and the other the producers. The probability for each tree branch is given as .33 and .67, respectively, and is shown in Figure 3.3.

To simplify the discussion, we designate M, P, R, and NR, respectively, as managers, producers, retire, and not retire. Thus, the joint outcome that an individual who is a manager and one who retires after age 60 is denoted as M and R, so the joint probability of this outcome is $P(M$ and $R)$. Similarly, the joint probability of the outcome that the individual is a producer and retires after age 60 is $P(P$ and $R)$.

Thus the joint probability that a randomly selected individual is a manager and retires after age 60 is

$$P(M \text{ and } R) = \frac{70}{300} = .23$$

Similarly, the joint probability that an individual is a producer and retires after age 60 is

$$P(P \text{ and } R) = \frac{20}{300} = .07$$

The joint probabilities for other outcomes are shown in Table 3.2. The resultant joint probabilities in Table 3.2 are probabilities calculated as relative frequencies.

Table 3.2 Joint Probability for 300 Individuals in Agriculturally Related Jobs with Preference for Retirement

Nature of job	Retire	Not retire	Marginal probabilities
M: managers	.23	.10	.33
P: producers	.07	.60	.67
Marginal probabilities	.30	.70	1.00

3.3 Optional topic: principles of counting

In some experiments the sample space is so large that the total number of outcomes in an experiment is complex and difficult to determine. In such a situation it is helpful to understand some counting techniques. The counting principle is used only in cases of more than two events.

In its simplest form, the counting principle states that if a first event can be performed in i different ways and a second event can be performed in j different ways, the total number of possible outcomes is simply the product of i and j. That is

$$\text{Total number of outcomes} = i \times j$$

For instance, if there are 2 ways of shipping a wheat crop from a farm to a grain elevator, and 3 ways of shipping it from the grain elevator to the wholesaler, and 4 ways of shipping it from the wholesaler to the retailer, in total there are $2 \times 3 \times 4 = 24$ ways to move the wheat crop from the farm to the retailer.

The principles of *multiplication, permutation*, and *combination* provide us with sophisticated counting techniques that help us determine the probabilities of the various ways in which the grain is shipped.

Multiplication principle. When order is important in the outcome of an experiment (i.e., in tossing a coin, for example, HT = TH, where H = heads and T = tails), the multiplication principle is used to count the total number of possible outcomes. The formula for multiple choices from an experiment is given in Equation 3–12:

$$_n M_r = n \qquad\qquad [3\text{–}12]$$

where $_n M_r$ is read, *the number of multiple choices of n events taken r times.* The multiple choices that we may have from combining different characteristics are best illustrated with an example.

Example 3.5

An ornamental horticulturist is conducting a breeding program where she is working to develop a cultivar having yellow colored flowers, large flowers, and long stem. She has assigned (Y) to yellow flowers, (L) to large flowers, and (S) to flowers with long stems. How many multiple choices can we obtain from these letters in sets of two?

Solution

The number of multiple choices we have is

$$_3 M_2 = 3^2 = 9$$

or

$$YY \quad LY \quad SY$$
$$YL \quad LL \quad SL$$
$$YS \quad LS \quad SS$$

Permutation. If an outcome excludes repetition but *order* is still important in such outcome, we are dealing with permutation. We may define permutation as *an arrangement of a set of objects in which there is a first, a second, a third, and an n order.*

We may apply the counting principle to count the arrangements; however, an easier way is to use the following equation:

$$_nP_r = \frac{n!}{(n-r)!} \qquad [3\text{--}13]$$

where $_nP_r$ is read, *the number of permutation of n things taken r at a time.*

Before illustrating how we may use Equation 3–13 to solve a permutation problem, we should explain the notation (!) introduced in equation 3–13. This notation, called a *factorial*, is used both in permutation and combination principles of counting. An n factorial ($n!$) simply means the product of n $(n-1)(n-2)(n-3), ..., [n-(n-1)]$. For instance, 4! is found as:

$$4! = 4(4-1)(4-2)[4-(4-1)]$$
$$= 4 \times 3 \times 2 \times 1$$
$$= 24$$

You should note that when we have the same numbers in the numerator and the denominator, they cancel each other out as follows:

$$\frac{5!3!}{4!} = \frac{5.4.3.2.1.(3.2.1)}{4.3.2.1}$$
$$= 30$$

We use Example 3.5 to illustrate the concept of permutations. For instance, we may ask how many permutations of letters $Y, L,$ and S can be obtained, taking them two at a time?

Using Equation 3–13, we have:

$$_3P_2 = \frac{3!}{(3-2)!}$$

$$= \frac{3!}{1!}$$

$$= \frac{(3.2.1)}{1}$$

$$= 6$$

Combination. When order is not important, we use the combination principle in grouping objects. The combination formula is

$$_nC_r = \frac{n!}{r!(n-r)!} \qquad [3\text{--}14]$$

Considering Example 3.5 once more, how many combinations of letters Y, S, and L can we have, taking two letters at a time?

$$_nC_r = \frac{n!}{r!(n-r)!}$$

$$_3C_2 = \frac{3!}{2!(3-2)!}$$

$$= \frac{3.2!}{2!}$$

$$= 3$$

3.4 Bayes' theorem

This theorem is named after an English clergyman, the Reverend Thomas Bayes (1702 to 1761), who proposed a means of revising probabilities based on gathered additional information in the course of a study. In many research situations, we face uncertainty with respect to the outcome of certain events, or we have made prior probability estimates that are used as the basis for the analysis of outcomes. However, in the course of research or a study project, we may come across additional information about the occurrence or nonoccurrence of some events. Given this additional information, Bayes' theorem suggests a means for calculating the revised or posterior probabilities.

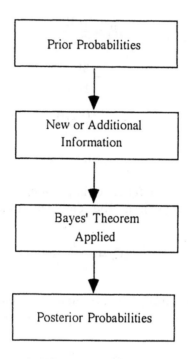

Figure 3.4   Revising probabilities with Bayes' theorem

Figure 3.4 shows the steps in revising the probabilities.

Bayes' work has received considerable attention since its first publication in 1763. *Bayesian decision theory* has been expanded in the last 40 years thanks to its great utility in solving problems of decision making under uncertainty. The theorem, a statement of conditional probabilities, may be stated as

$$P(A_i|B) = \frac{P(B|A_1)P(A_i)}{P(B|A_1)P(A_1) + P(B|A_2 + \ldots + P(B|A_n)P(A_n)} \qquad [3\text{--}15]$$

where $i = 1, 2, \ldots n$.

Generally, in most decision problems, A_i represents events that have occurred before event B is observed. In its simplest form Bayes' theorem can be stated as:

$$P(A_1|B) = \frac{P(A_1)P(B|A_1)}{P(B)} \qquad [3\text{--}16]$$

The following example illustrates the nature of the theorem.

Example 3.6

A food and nutrition specialist on arrival in a developing country observes malnutrition among some of the people. She assumes that 3% of the population of the country suffers from malnutrition. In checking the government's past health records, she finds that through testing (event B) the government has determined the conditional probability (that malnutrition exists, given that a person tested is malnourished) to be

$$P(B \mid A_1) = .95$$

and the corresponding probability, given that the person is not malnourished, is

$$P(B \mid A_2) = .04$$

Simple malnutrition is often difficult to identify because it commonly occurs with, and may be aggravated by, other diseases. Therefore, what is the probability that a person selected at random is suffering from malnutrition only?

Solution

Let A_1 represent those individuals who are suffering from malnutrition and A_2 represent those who are not suffering from malnutrition but some other disorder. The probability for each of the two events are

$$P(A_1) = .03$$
$$P(A_2) = .97$$

In Bayesian terms these two probabilities are referred to as the *prior probabilities* because they were assigned by our nutritionist's preliminary observations. That is, in the absence of empirical evidence she assigned these probability values to the population of interest. To determine the revised probability, we must compute the joint probabilities in the numerator as well as the denominator of Equation 3–16. This is accomplished using the multiplication rule as follows:

$$P(A_1 \cap B) = P(A_1)P(B \mid A_1) \tag{3–17}$$

and

$$P(A_2 \cap B) = P(A_2)P(B \mid A_2) \tag{3–18}$$

Therefore

$$P(B) = P(A_1)P(B \mid A_1) + P(A_2)P(B \mid A_2) \qquad [3\text{–}19]$$

We now substitute Equations 3–17 and 3–19 into the numerator and denominator, respectively, of Equation 3–16 to obtain the revised probabilities as follows:

$$P(A_1|B) = \frac{P(A_1)P(B|A_1)}{P(A_1)P(B|A_1) + P(A_2)(P(B|A_2)} \qquad [3\text{–}20]$$

For the present problem, the revised probabilities are

$$P(A_1|B) = \frac{(.03)(.95)}{(.03)(.95) + (.97)(.04)}$$

$$= \frac{.0285}{.0673}$$

$$= .42$$

As can be seen, the posterior probability that an individual is malnourished given that the test indicated existence of malnourishment is .42.

The use of Bayes' theorem calculations are presented in a tabular form as shown in Table 3.3.

Table 3.3   **Calculation of Bayes' Theorem for Example 3.6**

(1) Events A_i	(2) Prior probabilities $P(A_i)$	(3) Likelihoods $P(B \mid A_i)$	(4) Joint probabilities $P(A_i)P(B \mid A_i)$	(5) Revised probabilities $P(A_i \mid B)$
A_1: malnourished	.03	.95	.0285	.0285/.0673=.42
A_2: nourished	.97	.04	.0388	.0388/.0673=.58
	1.00		$P(B)$=.0673	1.00

The steps in calculating the revised or posterior probabilities in tabular form are

Step one — Prepare a five-column table where in the first column all mutually exclusive events are listed. In the second and third columns, list the prior and conditional probabilities, respectively. Columns four and five contain the joint and posterior probabilities, respectively, which are explained in steps two and three.

Step two — Given the new information available to the researcher, the joint probabilities for each event are computed and placed in column four. This is simply the product of the prior and conditional probabilities. That is, probabilities in column two are multiplied by the conditional probabilities in column three.

Step three — Sum the joint probability column to determine the probability associated with the new information, $P(B)$. The joint probabilities for the preceding example, given additional information such as the clinical test (designated as event B), is .0673.

Step four — Compute the posterior (revised) probabilities using the basic relationship of conditional probability as given by Equation 3–15. That is, the joint probability for each event is divided by the $P(B)$, which is the sum of the joint probabilities.

Modern decision theory uses Bayesian analysis to revise probabilities as additional information becomes available to researchers. These decision-making techniques are beyond the scope of this book. For those interested, however, see the references at the end of this chapter for elaborations on Bayesian analysis.

3.5   *Probability distributions*

The previous sections dealt with concepts of probabilities and how to compute the probability of an event. When we wish to know the probability of each and every outcome in a set of events, we are dealing with *probability distribution*. Probability distributions are a special form of frequency distribution where frequencies can be looked on as probabilities. We may define a probability distribution as *a complete listing of all possible outcomes of an experiment together with their probabilities where outcomes are mutually exclusive.*

There are several probability distributions that are important in environmental and agricultural research.

We shall now examine the binomial distribution for discrete variables and the normal (Gaussian) distribution for continuous variables.

Binomial distribution.   This is the most widely used probability distribution of a discrete variable. It describes the distribution of probabilities when there are only two possible outcomes for an event or experiment.

For example, the distribution of number of boars in litters of size n is binomial because each piglet must be either a male or a female. On a field plot test, an agronomist finds the use of an herbicide either effective or ineffective. Alternatively, an animal scientist typically finds an experimental vaccine either effective or ineffective. All these problems have two possible outcomes. The two possible outcomes in any one trial are conventionally known as *success* and *failure*.

If a random event may take two forms, and the probability of an event occurring (success) is identified as p and q, which is $1 - p$ or the probability of the event not occurring (failure), then the probability of r successes in n trials is noted mathematically as:

$$P(r) = (_nC_r)(p)^r(q)^{n-r} \qquad\qquad [3\text{–}21]$$

for $r = 0, 1, 2, ..., n$.

The symbol $(_nC_r)$ stands for the number of possible combinations of n things taken r at a time. Let us illustrate the computation of binomial probabilities with an example.

Example 3.7

An environmentalist knows from past experience that 80% of conifer seedlings survive being transplanted. If we take a random sample of six seedlings from the current stock, what is the probability that exactly two seedlings will survive?

Solution

For this problem $p = .80$, and $q = 1 - p = .20$; thus the probability of observing two seedlings which survive transplanting is

$$P(2) = {_6C_2} (.80)^2 (.20)^4$$

$$= \frac{6!}{(2!)(6-2)!}(.80)^2(.20)^4$$

$$= \frac{6.5.4!}{2.1.4!}(.80)^2(.20)^4$$

$$= 15(.64)(.0016)$$

$$= 15(.001024)$$

$$= .01536$$

Therefore the probability of survival of exactly two seedlings after transplanting, given a random sample of six, is .01536. The probabilities for other possible outcomes are given in Table 3.4.

The probability distribution for all possible outcomes in this sample of six are given in Figure 3.5. The probability values may also be obtained from the general binomial tables found in Appendix A.

Table 3.4   Probability Distribution for the
Number of Seedlings Surviving Transplanting

Number of surviving seedlings	Probability
0	0.00006
1	0.00156
2	0.01536
3	0.08192
4	0.24576
5	0.39322
6	0.26214
Total	1.00000

The binomial distribution is used in situations where we wish to determine the probabilities of outcomes characterized by independence (that is, the outcomes of any given trial or trials do not affect the outcomes on subsequent trials), and for each trial there are two mutually exclusive outcomes. To meet the condition specified as *independence*, we either sample from an infinite universe or sample with replacement.

Normal distribution.   *Normal distribution,* also referred to as *Gaussian distribution,* is one of several major continuous probability distributions. It is called a normal distribution because it approximates many kinds of natural phenomena. In other words, the normal distribution approximates the behavior of a large number of random variables that are continuous in a large number of cases. This distribution is applicable in many situations, and is central to many statistical problems.

As shown in Figure 3.6, its probability density function when plotted is bell shaped and symmetrical, with tails approaching, but not touching, the horizontal axis. (The latter feature is referred to as the asymptotic nature of the normal curve.)

The normal curve is defined by its mean, μ, and its standard deviation, σ. These are important characteristics of the normal curve because we are able to compute the entire distribution based on the knowledge of the mean and its standard deviation. Figure 3.7 shows three normal curves with the same mean, but different standard deviations. Note, in particular, the effect the standard deviation has on the general shape of the normal curve.

The three curves have different sizes, but certain relative relationships are common in all three. First, the total area under the curve of a continuous probability is equal to 1. Second, each curve has one half of its area above the mean and the other half below the mean (described by the vertical midline from the base to the apex of the curve). Third, the values of X range from minus infinity to plus infinity. However,

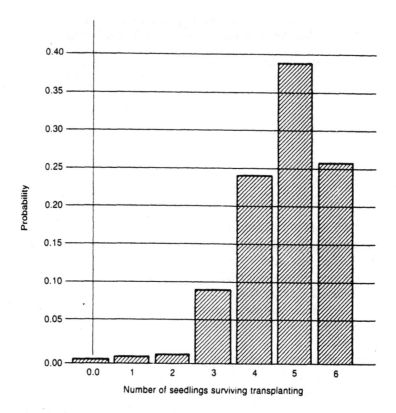

Figure 3.5   Probability distribution of the number of seedlings surviving transplanting.

for practical purposes we need not be concerned about X values lying beyond three or four standard deviations from the mean. The convention that is followed is called the *empirical rule*, which applies only to samples with frequency distributions that are bell shaped. By this rule, approximately 68, 95, and 99% of the values lie in the respective ranges of $\mu \pm 1$, $\mu \pm 2$, and $\mu \pm 3$, as shown in Figure 3.6.

Standard normal curve.   We now turn our attention to the use of standard normal curves upon which tables of area are based, and the way in which any normal curve can be converted into the standard normal curve. The properties of standard normal curve are that it has a mean equal to zero and a standard deviation equal to one. Because normally distributed variables may come in a variety of units of measurement such as centimeters (cm), kilograms (kg), pounds (lb), hours (h), days (d), inches (in), and other units, it is best to convert such values into a standard unit known as the *standard normal deviate*.

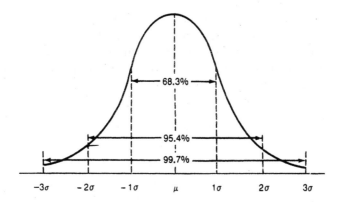

Figure 3.6 Bell-shaped curve of the normal distribution.

The normal deviate z that measures the distance separating a possible normal random variable value x from its mean and then states this deviation in multiples of standard deviation is found by the following expression

$$z = \frac{x - \mu}{\sigma}$$

[3–22]

where x = value of observation
μ = mean of distribution
σ = standard deviation of distribution:

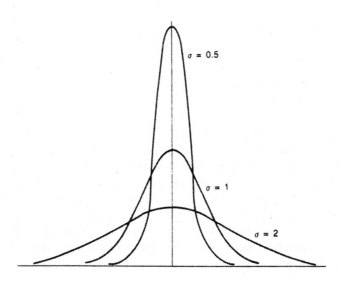

Figure 3.7 Normal distributions with equal means but different standard deviations.

To illustrate how we may use the table of areas, let us assume that the average daily milk production from a sample of normally distributed cows is 65 lb. with a standard deviation of 15. Let us also assume that one of the values of the variable from which the mean of 65 was computed is 87. How many z values from the mean will 87 fall?

To obtain the z value we have

$$z = \frac{x - \mu}{\sigma}$$

$$z = \frac{87 - 65}{15}$$

$$= 1.47$$

Figure 3.8 shows the conversion of pounds of milk into standard normal deviate.

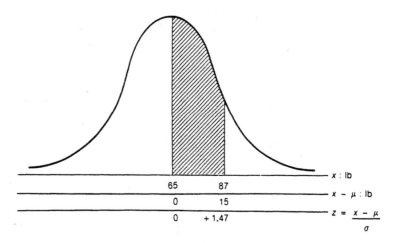

Figure 3.8   Normal curve conversion.

It is apparent from this figure that a value of 87 will lie 1.47 standard deviations to the right of the mean 65. To find the area for z = 1.47, turn to Appendix B, which is a replication of a table of areas developed by statisticians to give portions of the total area lying between or beyond ordinates erected at various standard scale values. Now go down the first column to 1.4 and across to the column labeled .07. The area under the curve is found at the cross section of these two columns, and is .4292. This latter number represents the percentage (42.92%) of the area contained within the curve defined by the distribution of milk output by the dairy herd.

Because the normal curve is symmetrical, what is true for one half of the curve is also true for the other. Therefore, the table values define only one half of the curve. Let us now use several examples to illustrate the use of table of areas under the normal curve.

Example 3.8

An agronomist measures the dry weight of a crop species as 10 g, with standard deviation of 2 g. What is the proportion of dry weights between 10 and 14 g?

Solution

We can obtain the value for the proportion of dry weight between 10 and 14 g by consulting the table in Appendix B, which gives areas under the normal curve lying between the mean and at the specified point above the mean as shown in Figure 3.9. The z value is calculated as:

$$z = \frac{14-10}{2}$$

$$= 2$$

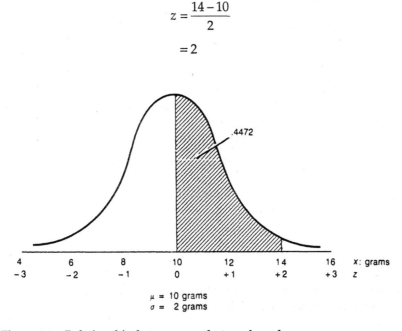

Figure 3.9 Relationship between *x* values and *z* values.

Now, we go to Appendix B and find that the area under the curve for a z of +2 is 0.4772. Therefore, we can state that 47.72% of the area in a normal distribution lies between the mean and a value two standard deviations above it or to the right of the mean. We can conclude that .4772 is the proportion of dry weights between 10 and 14 g.

Example 3.9

An ornamental horticulturist improves the *freshness quotient* of a particular flower by genetic breeding. This means that the improved variety stays fresh longer after it is cut. From a normally distributed sample, it is determined that the flowers on the average remain fresh for 168 h (μ = 168), with a standard deviation of 30 h. What is the probability that a cut flower will last between 192 and 216 h?

Solution

For 192 h:

$$z = \frac{192 - 168}{30}$$

$$= .80$$

For 216 h:

$$z = \frac{216 - 168}{30}$$

$$= 1.60$$

The areas for each of the above z values are

$$z = 1.60, \text{ the area} = .4452$$

$$z = .80, \text{ the area} = .2881$$

The probability is simply the difference between the two areas, that is, .1571. This difference is shown in the shaded area in Figure 3.10.

Example 3.10

Given the data from Example 3.9, what is the probability that a cut flower will last between 144 and 204 h?

Solution

Figure 3.11 shows the shaded area in which we are interested.
    For 144 h:

$$z = \frac{144 - 168}{30}$$

$$= -.8$$

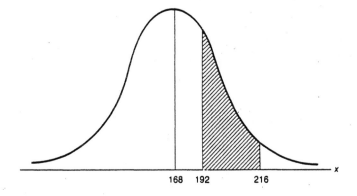

Figure 3.10   Probability of a cut flower lasting between 192 and 216 h.

For 204 h:

$$z = \frac{204 - 168}{30}$$

$$= 1.20$$

The areas for the respective z values are

$$z = -.8, \text{ the area} = .2881$$

$$z = 1.20, \text{ the area} = .3849$$

The probability that the cut flower will last between 144 and 204 h is equal to the sum of the two areas, that is, .6730.

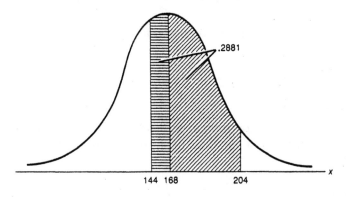

Figure 3.11   Probability of a cut flower lasting between 144 and 204 h.

Example 3.11

Again, using the information provided in Example 3.9, what is the probability that a cut flower will last more than 240 h?

Solution

The z value for 240 h will be

$$z = \frac{240 - 168}{30}$$

$$= 2.4$$

For z = 2.4, the area under the normal curve is .4918. Thus, if .4918 is the area between 168 and 240 h, the area beyond 240 h is the difference between .5000 (the total area to the right of the mean) and .4918. The probability that a cut flower will last more than 240 h is .0082, and is shown in Figure 3.12.

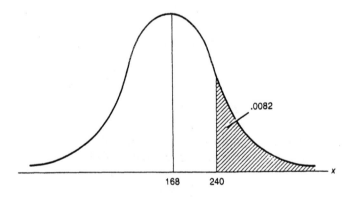

.0082

168   240

Figure 3.12   Probability of a cut flower lasting more than 240 h.

Example 3.12

Given the information from example 3.9, what is the probability that a cut flower will last less than 192 h (see Figure 3.13)?

Solution

The calculated z for 192 h is

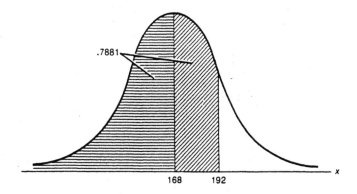

Figure 3.13 Probability of a cut flower lasting less than 192 h.

$$z = \frac{192 - 168}{30}$$

$$= .80$$

The area less than the mean of 168 is equal to .5000, and the area for $z = .80$ is equal to .2881. Therefore, the probabilities that are the sum of the two areas equal .7881.

Example 3.13

We may also ask, what is the probability for a cut flower lasting more than 120 h as illustrated in Figure 3.14?

Solution

For 120 h the z value is

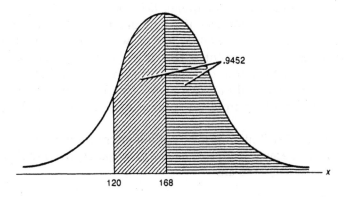

Figure 3.14 Probability of a cut flower lasting more than 120 h.

$$z = \frac{120 - 168}{30}$$

$$= -1.6$$

The area greater than mean of 168 is equal to .5000, and the area for $z = -1.6$ is equal to .4452. Therefore, the probability that a cut flower will last more than 120 h is equal to the sum of the two areas. That is, the probability is equal to .9452.

Chapter summary

In the chapter, you have been introduced to the *concept of probabilities*, and how they are computed. Probabilities are defined in terms of the chance that an event will occur. In numerical terms, probabilities take on values ranging from zero to one. The probability of an event that is certain to occur is one, and the probability of an event that will not occur is zero. The probability of an event can be computed either by summing the probabilities of experimental outcomes or using the relationships established by laws of probability. You were also introduced to the *addition* and the *multiplication rules of probability*.

Some *counting rules* (*permutations* and *combinations*) were also introduced to ease your task of counting certain events as you compute probabilities associated with these events.

To calculate probabilities, it is important to distinguish between the concepts of *mutually exclusive events, independent events, conditional probability, marginal probability,* and *joint probability. Bayes' theorem* can be used to revise probabilities when given additional information.

Finally, the concept of *discrete* and *continuous probability distributions* was discussed. You were introduced to the concept of *binomial distribution* and the *normal distribution* as examples of the discrete and the continuous probability distributions, respectively.

Binomial distribution requires *dichotomous classification* and *independence*. Dichotomous classification describes the outcome of an experiment that can take either one of two forms, but not both. The independence requirement means that the probabilities associated with any particular trial or experiment are not affected by the outcome of another trial or experiment. This requirement can be met either by sampling from an infinite universe or sampling with replacement.

Normal distribution is the most important continuous probability distribution because it approximates a great variety of natural phenomena. Its characteristics are

1. The curve is bell shaped, which means that half of the values fall to the right side (or above) of the mean, while the other half falls to the left of (or below) the mean.
2. The mean, median and mode are equal to each other.
3. The curve is symmetrical, that is, the two halves on either side of the mean are identical.
4. The normal curve is asymptotic, meaning that the curve approaches more and more closely to the X-axis, but the tails never touch it.

The normal distribution is a family of distributions in which each distribution is identified by the values of μ and σ. The *standard normal distribution*, which has a mean equal to zero and a standard deviation equal to one, is the most important member of the normal distribution family.

You learned how to transform any normal distribution to the standard normal distribution by using the following, as given in Equation 3–16.

$$z = \frac{x - \mu}{\sigma}$$

Equation 3–16 transforms any value of z in the original distribution to the corresponding z in the standard normal distribution. After we have calculated the z value, which is basically the number of standard deviations from the mean, we can then use the table of normal deviates given in Appendix B (but available in any number of books of statistical tables) to determine the area under the curve for each z value or the associated probabilities.

Case Study

Making Probability Statements about Soil Erosion

Soil erosion has become a major environmental problem for many countries around the world. Without effective monitoring, the problems will continue to grow (Brown et al., 1995). The U.S. is one of the few countries that closely monitors loss of top soil. The responsibility for this task is given to the Natural Resources Service (NRS) or what was formerly known as the Soil Conservation Service. The NRS monitors the rate of wind and water erosion at more than 800,000 sites.

Erosion in the U.S. is mainly concentrated in two regions: the humid midwestern grain belt, where soil swept from cropland by water accounts for most of the problem; and the Great Plains and western states, where the dry climate, cropping, and excessive livestock grazing contribute to high rates of wind erosion (U.S. Department of Agriculture, 1994).

Since 1982, the registered amount of soil lost to water erosion dropped from 1700 million to 1150 million tons. The amount of land vulnerable to wind erosion increased slightly, but the average rate has fallen from 12 to 11 ton/ha/year (U.S. Department of Agriculture, 1994).

As in the U.S., similar monitoring programs could guide conservation policy in other agricultural regions of the world. However, many countries around the world lack financial and technical resources to monitor soil erosion. Techniques such as satellite mapping and monitoring of the sediment loads of major rivers can provide data on soil erosion around the world. Unfortunately, these monitoring tools are expensive for most countries.

International organizations and industrialized nations need to work cooperatively to provide the leadership in preserving the natural resource base around the world by enabling countries to monitor soil erosion. Measures of soil erosion can be used to gauge the extent of an environmental problem as well as the degree to which it has been checked. The implications of the data are usually presented in the form of a probability statement.

A table follows that summarizes annual sediment load for various rivers around the world in million tons per year.

River	Sediment load[a] (million tons/year)
Huang He (Yellow)	1866
Ganges/Brahmaputra	1669
Amazon	928
Indus (Pakistan)	750
Yangtze	506
Orinoco (Venezuela)	389
Irrawaddy (Myanmar)	331
Magdalena (Colombia)	220
Mississippi	210
MacKenzie (Canada)	187

[a] Annual average weight of solid and dissolved materials.

Adapted from Milliman, J. D. and Meade, R. H., *J.Geol.*, 91, 1, 1983.

The data given in this table are from 1983. Suppose a normally distributed sample taken from the Amazon River today shows the mean sediment load to be 928 million tons/year with a standard deviation of 20 million tons/year.

1. What is the probability that the sediment load on the Amazon is between 950 and 1000?
2. What is the probability that the sediment load on the Amazon is between 800 and 850?

References

Brown, L., Lenssen, N., and Kane, H., *Vital Signs 1995*, W. W. Norton, New York, 1995, 15.

U.S. Department of Agriculture, *Soil Conservation Service, Summary Report*, 1992 National Resources Inventory, Washington, D.C., July 1994.

Review questions

1. If we assume male and female births are equally likely, what is the probability that exactly three of ten calves born will be male?
2. An animal scientist has eight sheep, three of which are Suffolk, and the other five are Rambouillet. He wishes to choose different groupings for his research.
   a. A three-member group is to be chosen from the eight sheep at random. How many possible groups are there?
   b. What is the probability that a randomly selected group will have two Rambouillet and one Suffolk?
3. The Green Thumb Seed Company is faced with the problem of pricing a package of grass seed mix. The appropriate pricing is dependent on the amount of rye grass found in the mix. Because we know that the package in question has come from one of two bins that contain one fourth and three fourths rye grass mixture, respectively, what is the probability that the package came from each of the two bins?
4. By assuming male and female litters are equally likely, graph the probability distribution of the random variable x that is the number of female litters that will occur if there are three pregnancies. (Hint: This problem is mathematically equivalent to tossing three fair coins.)
5. In a statistically independent genetic experiment (i.e., the results of one experiment do not affect the outcome of another experiment), an ornamental horticulturist found two possible colors

in the flowers of the offsprings. The probabilities associated with the colors were red 50%, and pink 50%. Among six offspring, what is the probability that:

a. At least one half will be red?

b. None will be pink?

6. A produce distributor in New York receiving shipments of lettuce from Mexico has found that 3% are usually rotten.

a. What is the probability of finding exactly two rotten heads in a sample of size four chosen from the carton?

b. What is the probability that three heads of lettuce will be rotten?

7. The average growing season for a new variety of corn is normally distributed with a mean of 95 d and a standard deviation of 10 d. What is the probability that the growing season from planting to maturity will be

a. Between 95 and 104 d

b. Between 85 and 100 d

c. Between 85 and 105 d

d. More than 110 d

8. In a study (Cochran et al., 1984) it was found that the birth weights of Dorset crossbred ewes are normally distributed with a mean of 4.29 kg and a standard deviation of .13. What is the probability that a ewe will have a weight:

a. Between 4.5 and 4.8 kg

b. Between 3.8 and 4.4 kg

c. More than 4.6 kg

9. An environmental scientist is concerned about the daily discharge of suspended solids from a phosphate mine into a river in Nevada. It is assumed that the discharge is normally distributed with a mean daily discharge of 38 mg/l and a standard deviation of 22 mg/l. What proportion of days will the daily discharge be below 30 mg/l?

10. Bacterial pollution of the drinking water has become a major concern for many cities in the U.S. Suppose that the number of a particular type of bacteria in 1 ml of drinking water is normally distributed with a mean of 94 and a standard deviation of 14. What is the probability that a given 1 ml sample will contain more than 110 bacteria?

References and suggested reading

Cochran, K. P., Notter, D. R., and McClaugherty, F. S., A comparison of Dorset and Finish Landrace crossbred ewes, *J. Anim. Sci.,* 59, 329, 1984.

Danforth, H. D. and Augustine, P., Avian coccidiosis vaccine: a first step, *Anim. Nutr. Health,* 40, 18, 1985.

Feller, W., *An Introduction to Probability Theory and Its Applications*, Vol. 1, 3rd ed., John Wiley & Sons, New York, 1968, chaps. 1, 4, and 5.

Hoel, P. G., Port, S. C., and Stone, C. J., *Introduction to Probability Theory*, Houghton Mifflin, Boston, 1971.Mendenhall, W. and Beaver, R. J., *Introduction to Probability and Statistics*, 8th ed., PWS-KENT, Boston, 1991, chaps. 3, 4, and 5.

Mendenhall, W. and Beaver, R. J., *Introduction to Probability and Statistics*, 8th ed., PWS-KENT, Boston, 1991, chaps. 3, 4, and 5.

Mendenhall, W., Wackerly, D., and Scheaffer, R. L., *Mathematical Statistics with Applications*, 4th ed., PWS-KENT, Boston, 1990.

Mostteller, F., Rourke, R. E. K., and Thomas, G. B., Jr., *Probability with Statistical Applications*, 2nd ed., Addison-Wesley, Reading, MA, 1970.

Winkler, R. L., *An Introduction to Bayesian Inference and Decision*, Rinehart & Winston, New York, 1972, chap. 2.

chapter four

An Introduction to Sampling Concepts

Learning objectives

After mastering this chapter and working through the review questions, you should be able to:

- Explain the importance of sampling in statistical analyses.
- Define the key sampling concepts: population, parameter, sample, and statistic.
- Distinguish between different sampling procedures, and assess when and how to use them.
- Explain the nature of sampling distributions, compute the mean and standard deviation, and determine the pattern of distribution of a random sampling distribution of means.
- Explain the nature of a sampling distribution of percentages, and compute their mean and standard deviation.
- Define the Central Limit Theorem and explain the relationship between the standard error of the mean and the size of the sample.

4.1   Introduction

Environmental and agricultural scientists in collecting data to analyze a problem have a choice between making a complete enumeration (census) or taking a sample. In most experiments, whether in agriculture, other sciences, or management studies, the objective is to determine a model that best describes the experimental results. Although one may have taken all variables into account in a model, obtaining information on the variables for the entire population may not be possible. For instance, what qualities do farmers of a particular region of the country value most in a new insecticide? Is it the long-lasting effect of the chemical or the environmental impact of the insecticide, or can it be the price?

We can, of course, find an answer to this question by asking all the farmers in that region. Most often, however, it is impractical and uneconomical to obtain answers to such questions by asking each farmer for his or her opinion. Therefore, we must obtain answers to our questions by sampling only a small percentage of farmers, and using these findings to generalize about the entire population. Because we are using a small fraction of the farm population to generalize about the entire population, great care must be taken in determining the type of sample we draw from the population, that is, the characteristic of small sample population we question. In this chapter, you are introduced to those sampling techniques that have desirable properties for making appropriate generalizations.

There are a number of advantages to sampling:

1. It reduces the amount of time, money, and other resources required to gather information that would lead to a useful and sturdy conclusion. For example, to gather information on how the farming community feels about the new farm bill would require enormous amounts of money, time, and people if all the farmers in the U.S. had to be polled.
2. It provides limits on the destructive nature of certain tests that are inherently destructive. For example, only a few seeds are tested for germination prior to planting season.
3. It provides a universe that is conceptually infinite. For instance, the universe to a manufacturer of a swine vaccine not only is the present generation but also conceivably includes uncountable future generations not yet born that will receive the vaccine.
4. Interestingly a sample provides greater overall accuracy than can an observation of the entire population. For example, the U.S. Department of Agriculture uses samples of prices of selected food items gathered nationwide in calculating the Food Price

Index. The price index does not differ from the published figures if the price of items other than those such as radishes and eggplants is added to those items included in the computation of the index.

4.2 Key sampling concepts

There are several important sampling concepts that must be clarified before sampling techniques can be understood.

Population. A *population* or *universe* is defined as *an entire group of persons, things, or events having at least one trait in common.* It may also be defined as a *totality consisting of all items that might be surveyed in a particular problem if a complete enumeration were made.*

The population concept is arbitrary in the sense that it is based on whatever the researcher chooses to call the common trait. An animal scientist may be interested in a population of Herefords; an environmental scientist may talk about a population of striped dolphins (*Stenella coeruleoalba*); an agronomist may be interested in a population of IR8 variety rice; an agricultural economist may talk about a population of wheat farmers, etc.

A population may be either *finite* or *infinite*. A finite population contains an exact upper limit, whereas in an infinite population there is no limit to its size.

Parameter. A *parameter* may be defined as an *estimate of one or more population characteristics.* If we gather information on the yearly income of farmers and divide the value by the total number of farmers, the resulting mean income is a parameter for that *specific* population, that is, the population of farmers.

Keep in mind that even if this were possible, the time and expense involved would be substantial. For this reason, most parameters are estimated, or inferred.

Sample. A *sample* is defined as *a smaller part of a population selected by some rule or plan.* If an ornamental horticulturist has a greenhouse with a total orchid population of 500, and he selects 50, 25, or 10 from that population, only the orchids selected form the sample. Thus, a sample is a part of a population. That is, if the horticulturist selects 499 of the orchids, he still has a sample. Not until he has included the last orchid could we say he has observed an entire population. Because samples are more accessible than populations for reasons discussed earlier, we select and measure the sample, and from these we can make sound inferences about the population.

Statistic.   A *statistic* is defined as *a measured characteristic of a sample.* If the ornamental horticulturist selects 50 orchids, measures their bud size, and calculates the mean bud size, the resulting value is a statistic. Inferential statistics is the methodology of predicting unknown parameters from known statistics.

4.3   Sampling techniques in environmental and agricultural sciences

In any sampling, an important element is the *representativeness* of the sample from a population. The sample must include characteristics as close as possible to the value that a researcher might have obtained had he or she been able to observe the universe or population.

The difference between an estimate and the true value of the parameter being estimated constitutes the *sampling error*. Hence, a good sampling technique is one that gives a small sampling error. In this chapter, we discuss various sampling techniques in environmental and agricultural sciences that provide a representative sample.

Basically, there are two methods of selecting a sample — *probability* and *nonprobability* methods. In the probability method, all items in the population of interest have an equal chance of being selected. The probability samplings of interest discussed in this chapter are *simple random sampling, systematic sampling,* and *stratified random sampling.*

In the nonprobability method, the researcher uses judgment in selecting items to be included in the sample. The nonprobability samplings of interest are *quota sampling* and *convenience sampling.*

4.3.1   Probability sampling

Simple random sampling.   The most common type of probability sample is the simple random sample. A random sample is one chosen from the universe in such a way that all items have an equal chance of appearing in the sample.

For instance, the population (N) may consist of 50 corn plants in a plot from which a researcher wishes to take a sample of 20 for tissue analysis. One way of ensuring that every corn plant has a chance of being selected is to write a number for each plant and deposit it in a box. After mixing thoroughly, the first selection is made. This process is repeated until the sample size of 20 is reached. This method of sampling is also referred to as the *goldfish bowl* technique.

A more convenient method is to use the *random digits technique*. To illustrate the use of a table of random numbers, such as the one given in Appendix D, let us consider the previous example that asked for a sample of 20 from a population of 50, and proceed through the following steps:

Step one — Each plant is numbered from 001 to 050.

Step two — Select a random starting point in the table of random numbers in any way you wish. For purposes of example, start at the left-hand corner of the table.

Step three — The first random number is 85967. A partial table of random numbers is shown in Table 4.1. Note that this table shows the first two digits of the first column of numbers given in Appendix D. Because we need only two digits (we have only 50 plants in our population), our first sample would be plant number 85, a number we cannot use because it exceeds 50. We proceed down the column to the second number, 07. This is a number we can use, so plant number 07 is included in the sample. Proceed in this fashion until all 20 plants have been selected as shown in Table 4.1.

The underlined values are the selected 20 for the sample in this example. Notice that 28 appeared twice in the table of random numbers. Because this is sampling without replacement, we do not want to use the same number twice. Replacement of 28 with another plant raising our population to 51 is one of many alternative ways of selecting a random sample.

Systematic sampling: Selection of a random sample as discussed previously can become tedious, especially when the sample to be selected is large, and the population from which it is selected is also very large. To maintain randomness and minimize tedious selection, *systematic sampling* is used.

Table 4.1 **A Partial Table of Random Numbers**

85	74	58
07	03	61
96	75	18
49	09	11
97	75	85
90	21	66
28	65	53
25	84	14
28	46	37
84	59	85
41	31	52
67	82	42
72	01	26
92	32	95
29	59	39

In this sampling technique, also known as *multistage random sampling* (Gomez and Gomez, 1984), we select our sample by drawing every 10th or 20th item from the population. Suppose that an animal scientist wants a random sample of $n = 400$ beef cows out of a population of $N = 1600$ for a study of a feed ration. Assume further that the population is in random order, so that we can select a systematic sample by drawing every fourth cow to be subjected to the feed ration. Because $1600/400 = 4$, every fourth cow is selected for the sample. That is, cows number 4, 8, 12, 16, etc., are selected until we reach our desired 400 for the sample.

Stratified sampling. In this type of sampling, the sampling unit or population is *stratified* into k strata or subpopulations before we make a selection of a random sample from each subpopulation or stratum. Stratification is especially useful where there is expected variation between subpopulations. By stratifying, a researcher is trying to group sampling units into different homogeneous strata in such a way that the variability within a subpopulation (stratum) is smaller than that between sampling units in other subpopulations.

For example, an animal scientist who is interested in drawing a random sample may use the age of the animals as a variable of interest in stratification. That is, he or she has different subpopulations representing different age groups as strata. The random sample drawn from each age group is a stratified sample.

An agronomist, on the other hand, may stratify a plot of land based on its known fertility level, and then take a sample of plants from each different stratum to measure the yield. In agricultural research, stratification is equivalent to blocking in experimental design, which is discussed in later chapters.

There are a number of advantages to stratified sampling. An effective stratification strategy provides more accurate estimators that have smaller variances than estimators from simple random sampling. This results from the fact that effective stratification produces more homogeneous subpopulations as compared with the original population. Additionally, stratified sampling is cost efficient and administratively convenient. Cost efficiency is realized by easily locating observational units in a stratified sample as compared with a simple random sample. It is administratively convenient because the strata may be smaller subdivisions of a large geographic area. For instance, in large experimental plots devoted to varietal tests for soil salinity tolerance, plots may be stratified according to salinity levels before plant samples are collected from each stratum.

Agricultural and environmental scientists use variations of the stratified sampling. An example of such a stratified sampling is referred to as the *stratified multistage random sampling*. A corn researcher, for instance, may use this type of sampling to measure the average number of corn husks per plant. In this instance, the individual rows in the plot may be used as the primary sampling unit, and the individual corn plant as the secondary sampling unit. This means that we divide the plants in each selected row (the primary sampling unit) into k strata, based on their relative position in the row, and then take a simple random sample of m plants from each stratum separately.

Suppose that we divide the plants in each selected row into two strata (namely, the short and the tall strata) and select a random sample of five plants from each stratum; then the total number of plants selected for the sample using three selected rows in a plot is

$$(A) \times (B) \times (C) = s$$

where A = stratum
B = desired sample size from each stratum
C = total number of randomly selected rows per plot
s = total number of plants selected per plot for sample

Using the preceding formula, we have a sample that includes 30 plants, shown as follows:

$$(2)(5)(3) = 30 \text{ sample plants}$$

4.3.2 Nonprobability sampling

As the name implies, samples selected this way do not permit all the items in the population to have an equal chance of being selected. The arguments often heard in favor of using nonprobability sampling are based on the cost and efficiency with which such samples are selected.

The most common of the nonprobability sampling techniques are *judgment sampling, quota sampling,* and *convenience sampling.*

Judgment sampling.   In this technique the subjective judgment of the researcher is the basis for selecting items to be included in a sample. For instance, a large tractor firm uses a judgment sample as the basis for making inferences about the buying habits of farmers who purchase tractors in a certain part of the country. The researcher selects a sample of farmers that in his or her judgment represents all farmers in that part of the country, and then collects the necessary data for analysis.

Quota sampling.   In this sampling technique, major population characteristics play an important role in the selection of a sample. For example, an animal scientist, recognizing that the variability in daily milk production may result from age differences, a characteristic of the cows, selects his or her sample from different age groups. For instance, if 30% of the cows are between the ages of 4 and 6 years old and the remaining 70% are between 6 and 8 years old, a quota sample must reflect those same percentages.

Convenience sampling: As the name implies, this technique is simply convenient to the researcher in terms of time, money, and administration. Though convenient and occasionally used in special circumstances, this sampling method should not be used to make inferences about populations.

As has been indicated, nonprobability sampling does not provide the objectivity offered by probability sampling. Sample estimates are subject to greater variability than the probability sampling. For this reason, we emphasize probability sampling in this text.

In the next section we examine a basic tool known as the sampling distribution that helps us in our understanding of problems related to sampling.

4.4 Sampling distributions

We can define a *sampling distribution* as *a frequency distribution showing the probability of all possible sample results computed from samples of the same size drawn from the same population.*

As indicated before, we use sample statistics to make decisions about population parameters. To accomplish this, it is important that we understand the sampling distribution of the population of interest. Because the sampling distribution is a probability distribution, it helps us judge the quality of our decision.

Sampling distributions can be constructed from discrete, finite populations. Three properties of a given sampling distribution are important to us:

1. The mean of the distribution
2. The standard deviation
3. The functional form or the pattern of the distribution

A separate sampling distribution can be constructed for arithmetic means, for percentages, and for standard deviations, as explained in the following sections.

Sampling distribution of the mean. In our earlier discussions it was mentioned that we use sample means and other measured characteristics of a sample (statistics) to make inferences about the characteristics of a population (parameters). Given that the mean of a sample drawn from a population is rarely equal to the population mean, and that it only approximates the mean of the population, the researcher has at his or her disposal other approaches that improve on his or her ability to make inferences about a population.

One approach is to consider all possible sample combinations and compute their mean and variance. In reality this is a difficult and time-consuming task.

Another approach is to use the sampling distribution to make inferences about the population of interest. Because the sampling distribution of means is a *normal distribution*, we can use the percentages on the normal curve from the z score table. This means we are able to make probability statements concerning the distribution of means.

The following example illustrates the relationship that exists between the mean and variance of a sampling distribution and the population mean and variance. You observe that the mean of the sampling distribution is equal to the mean of the population of interest. Additionally, it is shown that the histogram of the sampling distribution of means does tend to approximate the *normal curve*.

Example 4.1

Suppose we have a population of five sows, and we wish to take a sample of two sows for an experiment. The random variable of interest, X, is the weight in kilograms. The values of the variable are as follows:

$$X_1 = 50, X_2 = 45, X_3 = 50, X_4 = 45, \text{ and } X_5 = 40$$

Solution

Let us first compute some descriptive measures for this population. The arithmetic mean (μ) of this population is

$$\mu = \frac{\Sigma X}{N}$$

$$= \frac{50 + 45 + 50 + 45 + 40}{5}$$

$$= \frac{230}{5}$$

$$= 46 \text{ kg}$$

The average weight of the population of five sows is 46 kg. We ask, if we were to take any sample combination of two sows, how close would this sample average weight be to the population average weight?

To find the answer to this question, we first construct the sampling distribution of the mean and compute its mean. Then we compare this computed mean with the population mean.

To construct a sampling distribution of means, we determine all possible samples of size two without replacement by using the equation given in Chapter 3 in the following manner:

$$_nC_r = \frac{n!}{r!(n-r)!}$$

$$_5C_2 = \frac{5!}{2!\,3!}$$

$$= 10$$

There are ten possible sample combinations that can be drawn from this particular population. Table 4.2 shows these possible sample combinations and their means.

From Table 4.2 you can see that the possible values of the sample means *tend toward* the population mean. This implies that the sample means of a sampling distribution (column four of the table) are closer to the population mean (of 46) calculated earlier. Since the sample means have frequencies of occurrence, the sampling distribution is nothing but a probability distribution as shown in Table 4.3.

Table 4.2   **Sample Means for All Possible Samples of Size Two**

Sample number	Sample combination value	Total (Σx)	Sample mean $\bar{x}$
X_1, X_2	50, 45	95	47.5
X_1, X_3	50, 50	100	50.0
X_1, X_4	50, 45	95	47.5
X_1, X_5	50, 40	90	45.0
X_2, X_3	45, 50	95	47.5
X_2, X_4	45, 45	90	45.0
X_2, X_5	45, 40	85	42.5
X_3, X_4	50, 45	95	47.5
X_3, X_5	50, 40	90	45.0
X_4, X_5	45, 40	85	42.5

The mean of the sampling distribution is found by adding all the sample means and dividing them by the number of possible samples given as follows:

Table 4.3 **Probability Distribution of Sample Means**

Sample mean $(\bar{x})$	Frequency (f)	$(f\bar{x})$	Probability
42.5	2	85	2/10 = .20
45.0	3	135	3/10 = .30
47.5	4	190	4/10 = .40
50.0	1	50	1/10 = .10
Total	10	460	10/10 = 1.00

$$\mu_{\bar{x}} = \frac{(\bar{x}_1 + \bar{x}_2 + \bar{x}_3 + \ldots + \bar{x}_n C_r)}{{}_nC_r}$$

$$= \frac{(47.5 + 50 + 47.5 + \ldots + 42.5)}{10} \qquad [4\text{-}1]$$

$$= 46 \text{ kg}$$

The average weight of swine in the sampling distribution of means is 46 kg as expected. Therefore we may say that $\mu_{\bar{x}} = \mu$. The mean of a sampling distribution of means is equal to the population mean.

The question that may arise in your mind is, does $\mu_{\bar{x}}$ always equal μ? In computing the mean of the sampling distribution, we take all possible sample combinations into account, which is no different than a complete enumeration of the population of interest. Because in reality researchers do not take all possible sample combinations into consideration, is the mean of a sampling distribution in close proximity to the population mean? The answer is yes, because as the result of repeatedly testing this relationship statisticians have proved that the mean of the sampling distribution always tends toward the population mean.

Standard deviation of the sampling distribution of means. As was pointed out in Chapter 2, to determine the extent to which a sample mean varies from the population mean we have to use a measure of dispersion such as the standard deviation. In the case of the sampling distribution of the means, we are also interested in the likely deviation of a sample mean from the mean of the sampling distribution. The standard deviation of the sampling distribution is referred to as the *standard error of the mean*. It tells us about the accuracy of our estimate. The larger the standard error, the poorer the estimate.

To illustrate how the standard error of the mean is calculated, let us use Example 4.1. It should be noted that the computation of the standard error of the mean is similar to the calculation of any other standard deviation.

$$\sigma_{\bar{x}} = \sqrt{\frac{\Sigma(\bar{x} - \mu_{\bar{x}})^2}{N}} \qquad [4\text{--}2]$$

$$\sigma_{\bar{x}} = \sqrt{\frac{(47.5 - 46)^2 + (50 - 46)^2 + \dots + (42.5 - 46)^2}{10}}$$

$$= \sqrt{\frac{52.50}{10}}$$

$$= 2.29$$

In the preceding example, we have taken all possible sample combinations into account. In reality, seldom do we take all possible sample combinations to make inferences about the population. Therefore, to overcome such a cumbersome procedure we use the following equation to compute the standard error of the mean:

$$\sigma_{\bar{x}} = \frac{\sigma}{\sqrt{n}} \qquad [4\text{--}3]$$

where $\sigma_{\bar{x}}$ = standard error of the mean
σ = population standard deviation
n = sample size

The preceding formula is used to compute the standard error of the mean when we have an *infinite population*. However, for a *finite population* we have to consider a correction factor in our computation as follows:

$$\sigma_{\bar{x}} = \frac{\sigma}{\sqrt{n}} \sqrt{\frac{N - n}{N - 1}} \qquad [4\text{--}4]$$

where σ = population standard deviation
N = population size
n = sample size

$\sqrt{\dfrac{N - n}{N - 1}}$ = finite population correction factor

The finite population correction factor $\sqrt{(N - n / N - 1)}$ is approximately equal to 1 when the population size N is large relative to the sample size n. This implies that when our sample size n is from a much larger (but finite) population, the standard error of the mean $\sigma_{\bar{x}}$ is equal to $\sigma/\sqrt{n}$. In practice, the finite population correction factor is usually ignored whenever n is less than 10% of N. The implication of Equation 4–4 is that as long as we have a large population relative to the sample, the sampling precision becomes a function of sample size alone, and it does not matter what proportion of the population was sampled.

Notice that the preceding equation requires a knowledge of the standard deviation of the population. From the data given in Example 4.1, we can compute the standard deviation of the population as was given in Chapter 2 (Section 2.4).

$$\sigma = \sqrt{\frac{\Sigma(X - \mu)^2}{N}}$$

$$= \sqrt{\frac{(50 - 46)^2 + (45 - 46)^2 + \ldots + (40 - 46)^2}{5}}$$

$$= \sqrt{\frac{70}{5}}$$

$$= 3.74$$

Because this example involves a finite population, we have to employ the correction factor. Therefore, the standard error for this set of data becomes:

$$\sigma_{\bar{x}} = \frac{3.74}{\sqrt{2}} \sqrt{\frac{5-2}{5-1}}$$

$$= 2.64 \, (.8660)$$

$$= 2.29$$

This exercise shows that computing the standard error of the mean by either method yields the same answer. It should be noted that the standard deviation of the population is not equal to the standard deviation of the sampling distribution of means. However, the standard deviation of the sampling distribution is equal to the standard deviation of the population divided by the square root of the sample used to obtain the sampling distribution. That is

$$\sigma_{\bar{x}} = \frac{\sigma}{\sqrt{n}} \qquad [4\text{--}3]$$

This equation reveals two interesting relationships that exist between the standard error and the sample size. First, as the sample size increases, the standard error decreases. This implies that with an increase in sample size we have more information on which to estimate the population mean, and therefore the difference between the true value and the sample value decreases.

Second, we are able to determine the standard error of the sampling distribution of means with a knowledge of the *population standard deviation* (σ), the *sample size* (n), and the *population size* (N). At times, researchers do not know what the standard deviation of the population is. In situations like this, the standard deviation of the sample or samples is used to approximate the population standard deviation. The formula for the standard error then becomes:

$$\sigma_{\bar{x}} = \frac{s}{\sqrt{n}} \qquad\qquad [4\text{--}5]$$

A number of observations can be made about the sampling distribution of the mean and the population:

1. The population mean and the mean of the sampling distribution of the means are equal. This is always true if all possible samples are drawn from the population of interest.
2. The dispersion in the distribution of sample means is *less than* the dispersion in the population or universe. As can be seen in Table 4.2, the population values (as exhausted by the sample combinations) vary from 40 to 50 kg, while the sample means vary from 42.5 to 50 kg.
3. The histogram of the sampling distribution of means is tending to approximate a normal curve even though the population is not normally distributed as shown in Figure 4.1.

Note that although the relative frequency of individual values of X are equal, and hence the relative frequency distribution is flat (Figure 4.1a), the distribution of the sample means is somewhat bell shaped (Figure 4.1b). An important point to keep in mind is that whether the population is normally distributed or not, the sampling distribution of the mean tends to approximate the normal distribution. This point is directly related to the *Central Limit Theorem*, which states:

> *As the sample size n becomes sufficiently large, the sampling distribution of the mean tends toward a normal distribution.*

What constitutes a sufficiently large sample is debatable. However, in this text a sample size of 30 items is considered to be sufficiently large to assume that the sampling distribution of means is normal.

(a) Relative frequency distribution for X

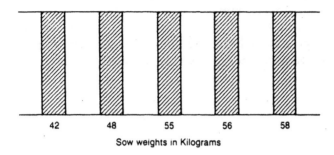

(b) Relative frequency distribution of the
sample means

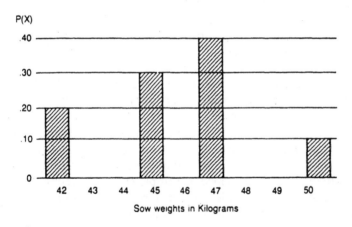

Figure 4.1   Histogram of the sample means and the population values.

Conceptually, the Central Limit Theorem provides us with a framework for *statistical estimation* and *hypotheses testing*. Chapter 5 is devoted to estimation, and later chapters deal with tests of hypotheses.

Sampling distribution of the percentages. Environmental and agricultural scientists are often interested in estimating population percentages (proportions). For example, an agronomist may be interested in the percentage increase in yield of a new hybrid; an agricultural engineer may want to estimate the percentage of defective items produced by a machine; an agricultural economist may be interested in the percentage of farmers receiving a farm subsidy; an environmental scientist may want to estimate the percentage of logged timber from a forest.

In this section, you are introduced to the concept of sampling distributions of the percentage, and how we estimate a population percentage from a sample.

We use the Greek letter π (pronounced pi) to denote the population percentage, and the lowercase letter p to denote the sample percentage. In equation form, they are expressed respectively as:

$$\pi = \frac{\Sigma X}{N} \qquad [4\text{--}6]$$

$$p = \frac{\Sigma x}{n} \qquad [4\text{--}7]$$

where X = number of occurrences of interest
  N = population size
  x = number of occurrences of interest in a sample
  n = sample size

The sampling distribution of the percentage may be defined as *a distribution of percentages of all possible samples where each sample drawn randomly has a fixed size n.*

The pattern of distribution of the sampling distribution of the percentage is a binomial distribution that can be approximated by the normal curve. Recall from Chapter 3 that the curve of the binomial distribution tends to be bell shaped as sample sizes (n) become larger. This implies that for large sample sizes, the binomial distribution approaches the shape of the normal curve. Based on this fact, we can make probability statements about the possible value of a sample statistic if we have a knowledge of the population percentage. For instance, we are able to make a probability statement such as, there is a 68.3% chance that a sample percentage will fall within one standard deviation (σ_p) of the population percentage (π), there is approximately a 95.4% chance that the sample percentage will lie within two standard deviations of π, or there is an approximately 99.7% chance that a sample percentage will fall within three standard deviations of the population percentage.

The mean of the sampling distribution of the percentage is expressed mathematically as:

$$\mu_p = \frac{\Sigma p}{{}_nC_r} \qquad [4\text{--}8]$$

where Σp = sum of sample percentages

$_nC_r$ = number of sample combinations

The sampling distribution of the percentages has a mean (μ_p) that equals the mean of the population (μ). The following example illustrates this equality.

Example 4.2

An animal scientist has a population of five cows and wishes to take a simple random sample of three to estimate the true percentage of cows that produce over 40 lb of milk daily. Given the following data on cow population and the daily milk production, what would the sampling distribution look like?

Cow Population and Daily Milk Production

Cow	Production (lb)	Production >40 lb
A	52	Yes
B	48	Yes
C	35	No
D	45	Yes
E	38	No

Note: X = 3 (the number of cows producing over 40 lb of milk).

Solution

It appears from the data that three cows produce more than 40 lb of milk daily, while two do not. Therefore the population percentage is

$$\pi = \frac{\Sigma X}{N}$$

$$= \frac{3}{5} \text{ or } 60\%$$

To construct the sampling distribution of the percentage, we follow the same procedures as we did in the construction of the sampling distribution of the means; that is, we have to first determine the sample combinations, and then compute the percentages. The sample combinations are ten in this example, and are given in Table 4.4.

The mean of the sampling distribution of the percentage is calculated as follows:

Table 4.4 **Sampling Distribution of Percentages**

Sample combination	Sample data	Sample percentages
A, B, C	Yes, Yes, No	0.667
A, B, D	Yes, Yes, Yes	1.000
A, B, E	Yes, Yes, No	0.667
A, C, D	Yes, No, Yes	0.667
A, C, E	Yes, No, No	0.333
A, D, E	Yes, Yes, No	0.667
B, C, D	Yes, No, Yes	0.667
B, C, E	Yes, No, No	0.333
B, D, E	Yes, Yes, No	0.667
C, D, E	No, Yes, No	0.333

$$\Sigma p = 6.000$$

$$\mu_p = \frac{\Sigma p}{{}_nC_r}$$

$$= \frac{6.00}{10} \text{ or } 60\%$$

Thus, the mean of the sampling distribution of the percentage is equal to the mean of the population. That is, $\mu_p = \pi$.

Standard deviation of the sampling distribution of percentages. Similar to the standard deviation of the sampling distribution of the means, the standard deviation of the random sampling distribution of percentages is called the *standard error of the percentage*. Computation of this standard deviation depends on our knowledge of the population percentage (π), population size (N), and the sample size (n). For a finite population, σ_p is expressed mathematically as:

$$\sigma_p = \sqrt{\frac{\pi(100 - \pi)}{n}} \sqrt{\frac{N - n}{n - 1}} \qquad [4\text{--}9]$$

where π = population percentage having desired characteristic
100 – π = population *not having* desired characteristic
n = sample size
N = population size

$\sqrt{\dfrac{N - n}{n - 1}}$ = finite population correction factor

The σ_p for our Example 4.2 is

$$\sigma_p = \sqrt{\frac{60(100-60)}{3}} \sqrt{\frac{5-3}{5-1}}$$

$$= \sqrt{\frac{60(40)}{3}} \sqrt{\frac{2}{4}}$$

$$= 28.28 \ (0.71)$$

$$= 20.08\%$$

Let us use the data in Example 4.2, and ask another question. What are the chances that the value of p will be between $\pi \pm 5\%$?

The format to determine the width of an interval is $\pi \pm z\sigma_p$. In this case, $z\sigma_p$ equals 5%. Because $\sigma_p = 20.08\%$, we can solve for z as follows:

$$z\sigma_p = 5\%$$

$$z(20.08\%) = 5\%$$

$$z = \frac{5}{20.08}$$

$$= 0.24$$

With a z value of 0.24, the area under the normal curve from Appendix B corresponds to .0948. This represents .0948 standard error on each side of the population mean. Therefore, the likelihood of a sample percentage within 5% of the population percentage is 18.9%.

Chapter summary

Environmental and agricultural sciences use inferential statistics to generalize from a small sample to the entire population or universe. A sample is perceived to have the basic characteristics of the population. The measured characteristics of the sample are called *statistics*. From these statistics, scientists make inferences about the characteristics of the population. The measured characteristics of the population are referred to as *parameters*.

Sampling is used extensively to make managerial as well as research decisions. The two sampling techniques discussed are probability and nonprobability sampling. In probability sampling, every item in the population has an equal chance of being selected. In nonprobability sampling, the selection of the sample is based on either the judgment of the researcher or some population characteristic that enforced its inclusion in the sample. Researchers using nonprobability methods must be careful that the sample be a representative sample.

The three types of probability sampling discussed are *simple random sampling, systematic sampling,* and *stratified sampling.* The concept of sampling distribution is introduced to help you understand the relationship that exists between the sampling distribution and the population. When we take successive random samples from a single population and calculate the means of each sample, the result is a frequency distribution called the *sampling distribution of means.* When a simple random sample is used, the mean of the sampling distribution is the parameter value to be estimated.

Furthermore, the mean of the sampling distribution of means tends toward the population mean. This property of the sampling distribution is attributable to the Central Limit Theorem. This theorem states that as the sample size is increased the sampling distribution more nearly approximates the normal distribution, whether the samples taken are from a normal population or not. With the use of examples, you are shown how you may construct sampling distributions for the mean and the percentage, and compute the mean and the standard deviation of such distributions.

Case Study

Applying Sampling Concepts to Water Use Efficiencies

Life on earth depends on water. Population clusters around the world are often found along ocean, lake, and river shores. Water is essential not only for farming and industry but also for many other human activities. Its consumption is rising and water tables are falling in many parts of the world.

Fresh water is unequally distributed among countries and people of the world. Although global renewable water resources average 7420 m^3 per capita per year, many countries have far more limited resources (World Resources Institute, 1995). At the current rate of consumption and with increasing population growth in many parts of the world, it is expected that water use will soon surpass the renewable supply. It is estimated that in Egypt this could happen in the next year or two; in Uzbekistan, in some 10 years; in Afghanistan, 24 years; in Azerbaijan, 30 years; and in Tunisia, renewable water resources could run out in 34 years (Brown et al., 1995).

To overcome the problems associated with shortages and overdrafting, environmental and agricultural scientists have to work together for solutions. Agricultural scientists have been working on how to improve the yield potential of many crops. Now they face the added problem of reducing the water intake by plants. Efficiency in water use is an important goal.

Plants differ in their water requirements. Different crops require different amounts of water, and the amount required often determines where these crops can be grown. The water requirements of any plant depends on its environment and nutrition. Soils that are rich in nutrients produce healthy crops with extensive root systems. In such a case, the total amount of water lost by transpiration increases, but water is used more efficiently. In a corn experiment, it was found that corn grown in a poor quality soil had a water use efficiency of 2000 as compared with 350 when the corn crop was grown in a well-fertilized soil (Chrispeels and Sadava, 1994). Additionally, placing plants closer together also contributes to water use efficiency. Agronomic practices have to be aimed at reducing water use and minimizing the loss from evapotranspiration.

The following table shows some examples of water use efficiency, expressed as the number of kilograms of water used per kilograms of dry matter produced.

Plant	Water used/kg per dry matter produced
Alfalfa	850
Soybeans	650
Wheat	550
Corn	350
Sorghum	300

Adapted from Chrispeels, M. and Sadava, D., *Plants, Genes, and Agriculture*, Jones & Bartlett, Boston, 1994, 190.

Data such as these drawn from samples can be used to determine the percentage of crops in the populations that are considered efficient users of water.

The sampling distribution can be used to make inferences about the corresponding parameters in the population. Suppose we are interested in determining the percentage of crops that use less than 500 kg of water to produce 1 kg of dry matter. If we were to take a sample from the group of crops shown in the table,

1. How would the sampling distribution appear?
2. What are the chances that the value of p will be between $p \pm 5\%$?

References

Brown, L., Lenssen, N., and Kane, H., *Vital Signs 1995,* W. W. Norton, New York, 1995.

Chrispeels, M. and Sadava, D., *Plants, Genes, and Agriculture,* Jones & Bartlett, Boston, 1994.

World Resources Institute, *World Resources: 1994–1995,* Oxford University Press, New York, 1995.

Review questions

1. Parameter is to population as_____ is to sample.
2. A probability distribution for all possible values of a sample statistic is a _____ .
3. _____ sampling is one in which every item in the population has an equal chance of being selected.
4. An agronomist has a population of five experimental rice plots. The yield per plot is as follows:

Plot	Rice yield (kg)
A	1400
B	1650
C	1800
D	1700
E	1500

A simple random sample of three plots is to be taken and the average yield per plot is to be estimated.
   a. Obtain the sampling distribution of x.
   b. Calculate the mean and the standard deviation of the sampling distribution of means.
5. An agricultural machinery firm manufactures a special cable for tractors. The company takes a sample of 50 and wants to see if the thickness of the cable meets the minimum specification set by the agricultural engineer of the company. Assume that $\mu = 0.25$ in, with a standard deviation of 0.02 in.
   a. Calculate the mean and the standard deviation of the sampling distribution.

    b. Within what range of values does the sample mean have a 95.4% chance of falling?

6. Farmland prices have been fluctuating in recent years. A population of five farms with class A soils has the following price per acre:

Farm	Price/acre
A	3000
B	2800
C	1000
D	2500
E	3200

A simple random sample of two is to be taken and an average price per acre is to be estimated.
    a. What is the mean of the sampling distribution of means?
    b. Calculate the standard deviation of the sampling distribution.

7. HI-PRO Feed Company wants to estimate the average tonnage of protein supplement sold monthly. The company takes a sample of 30 months. Assume that the true average tonnage per month is 180 tons with a standard deviation of 25 tons. What are the chances that the sample mean will have a value within 5 tons of the true mean?

8. Suppose that a random sample of ten observations is selected from a population that is normally distributed with mean equal to 2 and standard deviation equal to .48.
    a. Give the mean and the standard deviation of the sampling distribution of $\bar{x}$.
    b. Within what range of values does the sample mean have a 99.7% chance of falling?

9. Food and nutrition scientists are interested in the daily intake of potassium in adults. The normal daily intake of potassium varies between 2500 and 5000 mg depending on the type of food consumed. Some foods and drinks vary in the level of potassium they contain. For example, there are approximately 630 mg in a banana, 300 mg in a carrot, and 440 mg in a glass of orange juice. Suppose that the distribution of potassium in a banana is normally distributed with a mean of 610 mg and a standard deviation of 30 mg per banana. Suppose that you eat two bananas per day, and T is the total number of milligrams of potassium you receive from consuming them.
    a. What is the mean and standard deviation of T?
    b. What is the probability that your total daily intake of potassium from two bananas will exceed 2500 mg?

10. Recent studies show that the tropical forests are lost at a rate of 15.4 million hectares a year (UNEP, 1994). Environmental scientists are concerned that with the increasing use of fossil fuels greater amounts of carbon dioxide are emitted into the atmosphere. Because green plants absorb carbon dioxide, it is important to know the total amount of vegetative biomass on the earth's surface. Vegetative biomass in woodland forests are estimated to be 40 kg/m^2. Suppose that you measure the vegetative biomass in 100 randomly selected square meter plots, and your sample average is 35.5 kg/m^2.

   a. Approximate σ, the standard deviation of the biomass measurements.

   b. What is the probability that your sample average is within two units of the true average of vegetative biomass?

References and suggested reading

Cochran, W. G., *Sampling Techniques,* 3rd ed., John Wiley and Sons, New York, 1977.

Gomez, K. and Gomez, A. A., *Statistical Procedures for Agricultural Research,* 2nd ed., John Wiley & Sons, New York, 1984.

Keith, L. H., Ed., *Principles of Environmental Sampling,* American Chemical Society, 1988.

Raj, D., *Sampling Theory,* McGraw-Hill, New York, 1968.

Scheaffer, R. L., Mendenhall, W. L., and Ott, L., *Elementary Survey Sampling,* 2nd ed., Duxbury Press, North Situate, MA, 1979.

Sukhatme, P. V., Sukhatme, B. V., Sukhatme, S., and Asok, C., *Sampling Theory of Surveys with Applications,* 3rd ed., Iowa State University Press, Ames, IA, 1984.

United Nations Environmental Program (UNEP), *World Resources: 1994–1995,* Oxford University Press, New York, 1994.

Weiss, N. A., *Elementary Statistics,* Addison-Wesley, Reading, MA, 1989.

chapter five

Estimation of Parameters: Means and Percentages

Learning objectives

The material dealt with in this chapter revolves around estimation. The basic learning objective is to be able to use sample data to estimate population means and percentages. After mastering this chapter and working through the review questions, you should be able to:

- Explain the underlying concepts in point and interval estimation.
- Construct confidence intervals for a population mean and a population proportion.
- Estimate a population mean when σ is both known and unknown.
- Describe the t distribution and determine when its use is appropriate.
- Determine sample size for estimating means.

5.1   Introduction

In the previous chapter, we discussed how difficult and time consuming it is to enumerate an entire population to make generalizations about it. We learned that instead we can use a sampling distribution to generalize about the population mean and standard deviation. By employing our knowledge from a single sample, inferences can be made about the population of interest.

The process of making inferences generally takes one of two forms. The first is making *estimates*. For instance, an agricultural chemical firm wishes to forecast its sales for the following year. A numerical answer of so many dollars is needed.

The second form of inference can be drawn from *testing hypotheses*. In this type of sampling problem, the answer required is a yes or no. For instance, an environmental scientist may want to determine if the average weekly growth in inches of a new hybrid pine exceeds a certain figure. In this chapter, you are introduced to estimation; and in Chapter 6, to hypotheses testing.

To estimate the population mean, we may follow either one of two approaches: *point estimation* or *interval estimation*.

5.2   Point estimation

In this approach a single value is used to estimate. A researcher often selects a random sample from the population, determines the sample mean, and uses the resulting value to predict the population mean.

For example, in a study (Marshall et al., 1984) of 105 two-breed cows (Simmental and Angus) the average total intake of total digestible nutrient (TDN) by cow and calf for a 365-d drylot period was 2309 kg. This is a point estimate because the value is only a point along a scale of possible values. This estimate is useful whether we compare it against other estimates of the same crossbreed or other crossbreeds. However, we can be sure that if we were to take another sample of 105 cows and determine the average intake of TDN, the average total intake of TDN will probably be different than 2309 kg. Our reluctance to say that another sample estimate may yield an average intake of 2309 kg stems from the fact that there is a margin of error in making estimates. To allow ourselves some margin of error, we qualify our point estimate with appropriate standard error or make estimates in the form of interval instead of point.

5.3   Interval estimation

The *interval estimate* considers a range of values that is assumed to contain the true mean. In this estimation procedure we do not base our judgment on only a single value (point estimate) but instead present an interval that contains the true mean. Because a statement of confidence is attached to the estimate, we refer to the interval estimate as the *confidence interval*. The boundaries of the intervals are called *confidence limits*.

As was pointed out earlier, it is highly unlikely that any particular sample mean is exactly the same as a population mean. The precision of an estimate is determined by the degree of sampling error. The interval estimate generally differs from the parameter being estimated.

For example, the average TDN intake in the previous example may be stated as falling in the interval of 2300 and 2320 kg. The amount of confidence one has in the interval estimate is referred to as *degree of confidence*.

Interval estimate is based on the Central Limit Theorem, which states that as the sample size increases, the sampling distribution of means approximates a normal distribution. Given a normal distribution, we can make probability statements based on the *empirical rule*, which states:

> For a symmetrical, bell-shaped frequency distribution (normal distribution), approximately 68% of the observations will fall within one standard deviation of the mean; about 95% of the observations will fall within two standard deviations of the mean; and practically all (99.7%) will fall within three standard deviations of the mean.

The relationship between the standard deviations of the mean and the area under the normal curve are graphically portrayed in Figure 5.1.

We can construct confidence intervals for any level of confidence. The equation used to estimate the universe mean is

$$\mu = \bar{x} \pm z\sigma_{\bar{x}} \qquad\qquad [5\text{--}1]$$

where μ = population mean
  $\bar{x}$ = sample mean
  z = value determined by probability associated with interval estimate
  σ_x = standard error of mean

In general, the larger the confidence level, the more certain we can be that the true mean is included in the estimate. However, as we increase our confidence level, we also increase the width of the interval, and therefore the estimate loses some precision. Table 5.1 displays the relationship that exists between confidence level and the interval width.

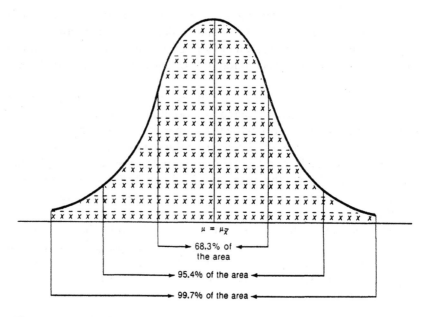

Figure 5.1　Sampling distribution of means.

Table 5.1　Commonly Used Levels of Confidence and
Confidence Intervals for Large Sample

Confidence level	z value	Interval width
90	1.64	$\mu = \bar{x} \pm 1.64\,\sigma_{\bar{x}}$
95	1.96	$\mu = \bar{x} \pm 1.96\,\sigma_{\bar{x}}$
99	2.57	$\mu = \bar{x} \pm 2.57\,\sigma_{\bar{x}}$

5.4　Estimating the population mean: σ known

In the previous section we discussed the theoretical basis for interval estimation. The starting point for interval estimation is the point estimator around which an interval is formed with a probability statement. Equation 5–1, which constructs an interval estimate, requires computation of the standard error. Recall that to compute $\sigma_{\bar{x}}$ we must have knowledge of the population standard deviation or be able to approximate it accurately. If σ is known, we simply substitute the population standard deviation in the formula and construct the interval estimate. In the example to follow, we can see how an interval estimate is constructed when we have a knowledge of the population standard deviation (σ).

Example 5.1

The U.S. Department of Agriculture (USDA) routinely projects future grain production, based on estimates from a sample. A sample of 100 plots (50 acres each) produced a mean of 35 bushels/acre. USDA assumes a population standard deviation of 4.5 bushels.

    a. Calculate a 95% confidence interval.
    b. Interpret the findings.

Solution

1. $\bar{x} \pm 1.96 \dfrac{\sigma}{\sqrt{n}} = 35 \pm 1.96 \dfrac{4.5}{\sqrt{100}}$

$$= 35 \pm 1.96 \dfrac{4.5}{10}$$

$$= 35 \pm 0.882$$

$$= 34.118 \text{ and } 35.882 \text{ bushels/acre}$$

2. The probability is .95 that the mean bushel per acre of wheat is in the interval of 34.118 and 35.882. In this instance, about 5 out of 100 confidence intervals would not contain the population mean (μ) bushels per acre.

The preceding statement is summarized graphically in Figure 5.2.

5.5   *Estimating the population mean: σ unknown*

In many practical situations it is difficult to determine the population mean and the standard deviation. Under these circumstances, a sample standard deviation that is an estimator of the population standard deviation is used to construct an interval estimate with the following formula:

$$\mu = \bar{x} \pm z\sigma_{\bar{x}} \qquad\qquad [5\text{--}1]$$

The standard error in Equation 5–1 is calculated by substituting the population standard deviation with the sample standard deviation:

$$\mu = \bar{x} \pm z\dfrac{s}{\sqrt{n}} \qquad\qquad [5\text{--}2]$$

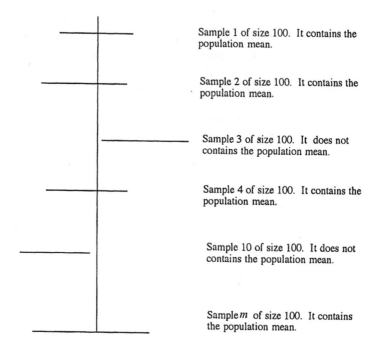

Sample 1 of size 100. It contains the population mean.

Sample 2 of size 100. It contains the population mean.

Sample 3 of size 100. It does not contains the population mean.

Sample 4 of size 100. It contains the population mean.

Sample 10 of size 100. It does not contains the population mean.

Sample m of size 100. It contains the population mean.

Figure 5.2   Confidence intervals for a population mean μ resulting from different random samples.

Let us use another example to illustrate how we construct a confidence interval when σ is unknown.

Example 5.2

Environmental scientists are concerned with the increasing use of fossil fuels and the amount of carbon dioxide that is emitted into the atmosphere from such use. Estimates of the earth's biomass, the total amount of vegetation held by the earth's forests, are important in determining the amount of unabsorbed carbon dioxide in the earth's atmosphere (World Resources Institute, 1994). To determine the mean biomass in a Brazilian rainforest, an environmental scientist takes a random sample of 50 plots (1 m^2 each). The mean biomass from this sample was 5.6 kg/m^2 with a standard deviation of 1.2 kg/m^2.

1. Construct a 95% confidence interval.
2. What are the 95% confidence limits?
3. Interpret the findings.

Solution

1. $\bar{x} \pm z\dfrac{s}{\sqrt{n}} = 5.6 \pm 1.96\dfrac{1.2}{\sqrt{50}}$

$= 5.6 \pm 1.96\dfrac{1.2}{7.07}$

$= 5.6 \pm 1.96\,(0.17)$

$= 5.6 \pm .33$

$= 5.27 \text{ and } 5.93 \text{ kg/m}^2$

2. The confidence limits are 5.27 and 5.93.
3. The probability is .95 that the biomass mean is in the interval 5.27 and 5.93 kg/m².

5.6 Use of Student t distribution

In constructing the interval estimates for the mean and the percentage, we assume a large sample. Recall that the Central Limit Theorem states that as the sample size increases, the sampling distribution approaches a normal distribution. However, samples of small size do not follow the normal distribution curve, but instead another distribution called the *t distribution*. The *t* distribution, like the normal distribution, is symmetrical in shape, and has a mean equal to zero. However, the shape of the *t* distribution is flatter than the z distribution as shown in Figure 5.3.

It is apparent from Figure 5.3 that the *t* distribution has a greater spread than the z distribution. However, as the sample size approaches 30, the flatness associated with the *t* distribution disappears and the curve approximates the shape of the z distribution. Another way of looking at the *t* distribution is to see it as a family of distributions, because there is a different distribution for each *degree of freedom (d.f.)* value.

The expression *degree of freedom* refers to the amount of data available for estimating a parameter. The degrees of freedom associated with a sample standard deviation is equal to $n - 1$. For example, when we calculate the standard deviation of a sample (as in Chapter 2), we subtract the sample mean from each of the n data observations to determine the deviations from the mean. As we complete the second last subtraction, the final deviation is automatically determined, because the deviations prior to squaring must sum to zero. Therefore, the last deviation is not free to vary; only $n - 1$ is free to vary. In other

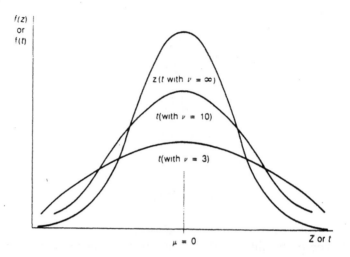

Figure 5.3   **The relationship between the sample size and the shape of the *t* distribution.**

instances, the number of degrees of freedom is different than $n - 1$, as discussed in Chapter 12.

To construct an interval estimate when σ is unknown and the sample size is small, the following equation is used:

$$\mu = \bar{x} \pm t_{\alpha/2}\, \sigma_{\bar{x}} \qquad\qquad [5\text{--}3]$$

To be able to determine an appropriate *t* value, we must know the confidence level and the degrees of freedom. Statisticians have determined the area under the curve associated with different degrees of freedom and the level of confidence that is presented in tabular form, called the *t* table. Now we turn to a *t* table that is found in Appendix E.

t Tables are based on the *t* distribution percentiles for selected tail areas. Note that in the table the chances that μ is not included in the estimate are emphasized; the chance of an error arising by not including μ in an estimate is given by the shaded area under the curve and is labeled α (alpha), and the confidence coefficient is $(1 - \alpha)$. For example, if we are interested in a confidence level of 95% in a given problem, the corresponding value to this confidence level as given in the *t* table is found by looking at the α value that is $1.00 - .95$ or .05. Because .05 represents the total chance of error and the *t* table only deals with areas on one side of the distribution, the chance of error in each tail is .025. Thus, the column $t_{0.025}$ should be used for a confidence level of .95.

The second important piece of information needed to determine an appropriate *t* value in a *t* table is the degree of freedom. Column one of the *t* table provides the degree of freedom associated with different sized samples. Let us consider the following example by way of illustration.

Example 5.3

An agricultural chemical retail firm wants to estimate the average number of gallons of a weed killer sold per day for purposes of accurately forecasting and controlling inventory. Twelve business days are monitored, and average daily sales of 10 gal are recorded. The sample yields a standard deviation of 2 gal. The problem is to calculate the confidence limits at the 95% level.

Solution

The sample provides us with the following information:

$$\bar{x} = 10 \text{ gal}$$
$$s = 2 \text{ gal}$$
$$n = 12 \text{ d}$$
$$\text{confidence level} = .95$$

Given a confidence level of 95% and a sample size of 12, the $t_{.025}$ value is 2.201.

The interval estimate is

$$\mu = \bar{x} \pm t_{\alpha/2}\sigma_{\bar{x}}$$

$$= 10 \pm 2.201 \frac{2}{\sqrt{12}}$$

$$= 10 \pm 2.201 \frac{2}{3.46}$$

$$= 10 \pm 1.27$$

$$= 8.72 \text{ and } 11.27 \text{ gal}$$

The confidence limits are 8.72 and 11.27 gal of the weed killer.

You should note that the choice of the confidence level in a given circumstance is made by the researcher or the manager of the experiment. Most confidence intervals are constructed by using one of the confidence levels given in Table 5.1.

5.7 Interval estimation of population percentage

Theoretically and procedurally, interval estimation of a population percentage is similar to the way we constructed the interval estimate for the mean when sample sizes are sufficiently large to allow the use of the normal curve. As was indicated in the previous chapter, the mean of the sampling distribution of the percentage is equal to the population percentage. Similarly, the sample percentage (p) is an unbiased estimator of the population percentage (π).

To construct the interval estimate of the population percentage, we use the sample percentage as the basis for such an interval. To determine the sample percentage, we draw a random sample of size n and observe the number of samples having the characteristic (r) in question. The sample percentage is described mathematically as:

$$p = \frac{r}{n} \times 100 \qquad [5\text{–}4]$$

Once we have calculated the sample percentage, we then estimate the standard error of the percentage as follows:

$$\sigma_p = \sqrt{\frac{p(100-p)}{n}} \qquad [5\text{–}5]$$

The confidence interval for the population proportion is estimated by:

$$p \pm z \sqrt{\frac{p(100-p)}{n}} \qquad [5\text{–}6]$$

where p = sample proportion
$\quad\quad\ \ z$ = z score for selected degree of confidence
$\quad\quad\ \ n$ = sample size

Example 5.4

An Agricultural Credit Association wants to estimate the percentage of farmers who are late in their monthly loan payments. A sample of 80 accounts showed that 18 were late in their monthly payments.

1. Estimate at the 95% confidence level the true percentage of farmers who are late in their monthly loan payments.
2. Interpret the findings.

Solution

1. In this problem the characteristic of concern (r) is the delayed monthly loan payments, and the percentage is computed as:

$$p = \frac{18}{80} \times 100 = 22.5\%$$

The estimate of the standard error is

$$\sigma_p = \sqrt{\frac{p(100-p)}{n}} \qquad\qquad [5\text{-}5]$$

$$= \sqrt{\frac{22.5(100-22.5)}{80}}$$

$$= 4.67$$

The 95% confidence interval is

$$\pi = p \pm z\sigma_p$$

$$= 22.5 \pm 1.96(4.67)$$

$$= 22.5 \pm 9.15$$

$$= 13.35 \text{ and } 31.65\%$$

2. We are 95% sure that the population percentage (the true number of farmers who are late in their monthly payments) is in the interval between 13.35 and 31.65%.

5.8 Determining sample size

The advantages of using a sample in statistical studies are given in Chapter 4. It is indicated that because of the cost and administrative difficulties of enumerating the entire population, we must depend on samples for estimating a population characteristic. How large a sample should be given to provide a statistically sound estimate of the population parameter is the subject of this section.

Researchers are often concerned with sampling error that results from not taking every item in the population into consideration. Sampling error, however, can be controlled by selecting a sample of adequate size.

In determining the sample size, three factors play an important role. These are the *degree of confidence*, the *maximum allowable error*, and the *standard deviation* or *variance* of the population. The degree of confidence associated with an estimate is usually set at 90, 95, or 99%, but any level may be used. The maximum allowable error sets the limits on total error in predicting a population characteristic. Finally, the variation in the population characteristic as measured by the standard deviation or variance plays a significant role in determining the sample size.

The size of a sample is determined by solving for n in the formula of the standard error given as follows:

$$\sigma_{\bar{x}} = \frac{\sigma}{\sqrt{n}} \text{ or } \sigma_{\bar{x}} = \frac{s}{\sqrt{n}}$$

To solve for n we have:

$$\sigma_{\bar{x}}\sqrt{n} = \sigma$$

$$\sqrt{n} = \frac{\sigma}{\sigma_{\bar{x}}}$$

$$n = \frac{\sigma^2}{\sigma_{\bar{x}}^2} \text{ or } n = \frac{s^2}{\sigma_{\bar{x}}^2} \qquad [5\text{--}7]$$

Example 5.5

A plant scientist wants to know the average nitrogen uptake by a vegetable crop. A pilot study of midrib nitrate nitrogen uptake provides a standard deviation (s) estimate of 120 parts per million (ppm). Use a confidence level of 95%, and determine for the scientist the sample size required to estimate μ within 80 ppm.

Solution

First, let us look at what the plant scientist actually wants. For sample mean, she wants the interval estimate to have limits that are no more or less than 80 ppm. She also would like to have a 95% confidence level that the interval estimate contains the true average. Thus, the confidence limit desired is

$$\bar{x} \pm 80 \text{ ppm}$$

The general form of the confidence limit was given earlier as:

$$\mu = \bar{x} \pm z\sigma_{\bar{x}} \qquad\qquad [5\text{--}1]$$

Therefore, in this example, $z\sigma_{\bar{x}} = 80$ ppm. Because the scientist wants a 95% confidence interval, the z value from Appendix C is 1.96.

With the preceding information and Equation 5–1, we can determine the sample size by first calculating the standard error as follows:

$$z\sigma_{\bar{x}} = 80$$

$$1.96\,\sigma_{\bar{x}} = 80$$

$$\sigma_{\bar{x}} = \frac{80}{1.96}$$

$$= 40.8$$

Because $\sigma_{\bar{x}} = 40.8$, we compute the sample size by using formula 5–7:

$$n = \frac{s^2}{\sigma_{\bar{x}}^2}$$

$$= \frac{(120)^2}{(40.8)^2}$$

$$= \frac{14,400}{1,664.6}$$

$$= 8.6 \text{ or } 9$$

The necessary sample for the plant scientist's desired level of precision is 9. Notice that we have rounded the sample size from 8.6 to 9. When rounding is necessary, we always round the required sample size up.

The sample size can also be determined using a more convenient formula given as follows:

$$n = \left(\frac{z \cdot s}{E}\right)^2 \qquad\qquad [5\text{--}8]$$

where n = sample size
z = z score associated with selected degree of confidence
s = sample standard deviation of pilot study
E = allowable error

For the preceding example, we now have:

$$n = \left(\frac{1.96 \times 120}{80}\right)^2$$

$$= \left(\frac{235.2}{80}\right)^2$$

$$= (2.94)^2$$

$$= 8.6 \text{ or } 9$$

Note that the sample size obtained is the same as in the previous procedure.

Determining the sample size for the estimation of population percentage. Procedurally, determining the sample size for estimation of population percentage is similar to the determination of sample size for population mean.

The formula for determining the sample size of a percentage is

$$n = p(1-p)\left[\frac{z}{E}\right]^2 \qquad [5\text{--}9]$$

where n = sample size
 p = estimated percentage based on past records or from pilot study
 E = allowable error
 z = z score associated with selected degree of confidence

Example 5.6

The market research department of an agricultural farm machinery firm wants to know what percentage of farmers in a given region is planning to buy new tractors in the coming year to guide the manufacturing operation in planning production. The researcher needs his estimate to be within 2% of the true percentage at a 95% confidence level. A pilot study in the region shows that 8% of the farmers are planning to buy new tractors in the coming year. How large a sample should the researcher consider?

Solution

$$n = p(1-p)\left[\frac{z}{E}\right]^2$$

$$n = .08(1-.08)\left[\frac{1.96}{.02}\right]^2$$

$$= .08\,(.92)[98]^2$$

$$= .07(9604)$$

$$= 672.3$$

$$= 672$$

In this problem, the researcher should take a sample of 672 farms to be 95% confident that the sample will produce a percentage that will be within 2% of the true percentage.

Chapter summary

The discussion in this chapter has dealt with the estimation of parameters. As the preceding chapter points out, a study of an entire population requires enormous resources in terms of time and money. For this and other reasons, we must take samples from the population of interest to make inferences about some characteristic of that population. Inferences can be made through either *point estimates* or *interval estimates*.

A point estimate provides a single number estimation of a population parameter such as the mean or the percentage. An interval estimate, on the other hand, provides an interval within which the population parameter probably falls. The Central Limit Theorem serves as the foundation for interval estimates. The *degree of confidence* associated with each interval estimate indicates the probability that the estimate may include the population parameter. For instance, when a probability value of 95% is used, it is assumed that an accurate prediction can be made 95% of the time.

In constructing an interval estimate, we recognize the need for substituting the standard deviation of a pilot study when the population standard deviation is not known. If, on the other hand, we do have the population standard deviation available to us, it is used in the formula to construct the interval estimate. Procedurally, the construction of an interval estimate for the mean is similar to the interval estimate for the percentage.

The use of Student t distribution was introduced in the context of sample size. It is important to remember that sample size affects the width of the confidence interval. For large samples ($n > 30$), the sampling distribution approximates the normal distribution, in which case we can use the distribution. However, when $\sigma_{\bar{x}}$ must be estimated and the sample size is 30 or less, the sampling distribution follows the t distribution.

Determining the size of a sample is based on knowledge of three factors:

1. The degree of confidence a researcher must place on the results
2. The maximum allowable error that the researcher can tolerate
3. An estimate of the variation (the standard deviation) within the population, such variation estimated with a pilot study or from past experience

Case Study

Estimating the Impact of the Green Revolution on Cereal Production in Central America

Agricultural production has increased significantly in output as a result of the Green Revolution. The Green Revolution uses the principle of improved cultivar selection along with a variety of other inputs to improve crop productivity. Scientists estimate that roughly 40% of all increases in crop productivity are a result of plant breeding; and the other 60% are a result of the use of fertilizer, pesticide, energy, irrigation, and management practices.

When first introduced, the Green Revolution was hailed as a solution to many of the world food and fiber production problems. Forty years after the first adoption cycles of the Green Revolution in Mexico, the world is not much better off. A similar percentage (10 to 15%) of the world population is undernourished in the 1990s (Chrispeels and Sadava, 1994). The global trends for each of the three major cereals — wheat, rice, and corn — show that since 1990, neither wheat nor rice yields have shown a strong upward trend (U.S. Department of Agriculture, 1994). World grain yield in 1994 averaged 2.6 ton/ha (U.S. Department of Agriculture, 1994). With each passing year, accumulating evidence indicates that rapid sustained rise in land productivity that characterized the period from the mid-1900s onward may be coming to an end (Brown, 1995). The question becomes, can the Green Revolution offer new solutions to areas that are in need?

In the case of Central America, rice, which is a major staple in the diet of many areas, has to be produced in sufficient quantities to meet the needs of a rising population. Without increases in the production of rice there is a potential for food deficit. A high-yield variety may be part of the answer. A new type of rice is under experimentation, for example, at the International Rice Research Institute in the Philippines, which should be ready for widespread commercial use around the end of this decade. This new variety has the potential to raise production by an estimated 20 to 25% (Rensberger, 1994).

To determine the impact of the Green Revolution on cereal production in Central America, policy makers can use estimates to infer if higher yields have been achieved. To accomplish this, policy makers can make an estimate or predict a range of values for the estimate.

The following table shows the average yields of cereals in Central American countries when high yielding crops are planted:

Country	Average yields of cereals(kg/ha) in 1990 to 1992
Belize	1490
Costa Rica	2531
El Salvador	1876
Guatemala	1891
Honduras	1377
Nicaragua	1498
Panama	1796

From World Resources Institute, *World Resources: 1994–1995*, Oxford University Press, New York, 1995, Table 18.1.

Using the data given:

1. Compute the mean and the standard deviation of cereal production in Central America.
2. Identify the range of values within the 95% confidence limits.

References

Brown, L., Grain yields remain steady, in Brown, L., Lenssen, N., and Kane, H., *Vital Signs 1995*, W. W. Norton, New York, 1995.

Chrispeels, M. and Sadava, D., *Plants, Genes, and Agriculture*, Jones & Bartlett, Boston, 1994.

Rensberger, B., New 'super rice' nearing fruition, *Washington Post*, October 24, 1994.

U.S. Department of Agriculture, *Production, Supply and Demand View* (electronic database), Washington, D.C., November 1994.

World Resources Institute, *World Resources: 1994–1995,* Oxford University Press, New York, 1995.

Review questions

1. A random sample of 45 corn plants reveals that on the average 12 husks are formed on each plant. The standard deviation of the sample is two husks. Using the .99 confidence level, construct the confidence interval within which the population mean falls.

2. The average number of eggs produced daily is to be estimated. A .95 degree of confidence is required. A pilot study from this egg farm produces a mean of 450 eggs per day, with a standard deviation of 50 eggs. If we wish to be within 24 eggs from the population mean, how many laying hens should be sampled?

3. An agricultural ecologist selects 16 (1 m³ volume) samples from a river into which drainage canals empty water. The researcher determines the average salt content in parts per million (ppm). This sample yields a mean of 800 ppm with a standard deviation of 50 ppm. Construct an interval estimate with .99 degree of confidence.

4. A poultry processing company receives a shipment of 2000 Cornish hens, and the firm's quality manager wants to estimate the true average weight of the hens to determine if they meet quality standards set by the firm. A sample of 35 Cornish hens produces an average weight of 2.2 lb with a standard deviation of 0.4 lb.
   a. Construct an interval estimate with 95% confidence level.
   b. Interpret the meaning of this estimate.

5. A recent survey of 600 farm interest groups regarding sale of grain to Russia reveals that 320 approved of the grain sale. Using a 99% confidence level, determine the interval within which the true percentage may fall.

6. An agricultural economist is interested in determining the number of farm operators owning their farms. She randomly samples 250 farms and finds that 110 farms are owned by the farmers. Determine at 95% confidence limits the true percentage of farm owners.

7. An ornamental horticulturist develops a new variety of grass seed for dry climates (less than 10 in. of rain per year). Testing on 150 plots reveals that 70% of the seeds germinated in dry conditions. Determine at the 90% confidence interval the true percentage of the seeds that germinate under equivalent conditions.

8. A farmer is interested in a survey of wheat prices around the country. He requests his county extension agent to do the job for him. The farmer asks the agent to be 95% certain that the true mean price per bushel be within an error of no more than $0.10. A quick check of wheat prices shows that they range from $4.50 to $4.95 per bushel. What size sample should the county agent use for the farmer?

9. A Washington apple producer claims that the average weight of his apples is 7 oz. A sample of 60 from his orchard produces a mean of 6 oz, with a standard deviation of 1.5 oz. Construct a 95% confidence interval for the population mean.

10. An agronomist develops a new variety of barley (Trebi-CP) that is considered superior in yield. A sample of 10 plots produces an average yield of 45.5 bushels/acre. The standard deviation of this sample is 2.5 bushels/acre. Construct a 95% confidence interval for μ on the assumption that you have a random sample from a normal population. Interpret this confidence interval.

11. A poultry scientist is interested in estimating the average weight gain over a month for 100 chickens introduced to a new diet. To estimate this average weight gain, she takes a random sample of 40 and measures the weight gain. The mean and the standard deviation of this sample are $\bar{x}$ = 60 g and s = 5 g.
    a. Construct a 99% confidence interval.
    b. What are the 99% confidence limits?
    c. Interpret the findings.

12. Canadian environmentalists are concerned with the increasing amount of acid rain falling into the Great Lakes. Acid rain affects marine life and has a negative ecological impact. Pure rain falling through clean air registers a pH value (pH is a measure of acidity) of 5.7. A scientist takes 30 random samples of rainfalls that result in a mean pH value of 3.5 and a standard deviation of 0.8. By using the 95% confidence level, construct the confidence interval within which the population mean falls. Interpret the meaning of this estimate.

References and suggested reading

Fisher, R. A., Theory of statistical estimation, *Proc. Cambridge Philos. Soc.*, 22, 1925.

Freund, J. E. and Walpole, R. E., *Mathematical Statistics*, 4th ed., Prentice-Hall, Englewood Cliffs, NJ, 1987.

Good, J. J., What are the degrees of freedom, *Am. Stat.*, 27, 227, 1973.

Gossett, W. S. [Student], The probable error of a mean, *Biometrika*, 6, 1, 1908.

Gurland, J. and Tripathi, R. C., A simple approximation for unbiased estimation of the standard deviation, *Am. Stat.*, 25, 30, 1971.

Marshall, D. M., Frahm, R. R., and Horn, G. W., Nutrient intake and efficiency of calf production by two-breed cross cows, *J. Anim. Sci.*, 59, 317, 1984.

Mendenhall, W. and Beaver, R. J., *Introduction to Probability and Statistics*, 8th ed., PWS-KENT, Boston, 1991.

Mendenhall, W. and Reinmuth, J. E., *Statistics for Management and Economics*, 4th ed., Dubury Press, Belmont, CA, 1983.

Mendenhall, W., Wackerly, D., and Scheaffer, R. L., *Mathematical Statistics with Applications*, 4th ed., PWS-KENT, Boston, 1990.

Scheaffer, R. L., Mendenhall, W., and Ott, L., *Elementary Survey Sampling*, 4th ed., PWS-KENT, Boston, 1990.

Snedecor, G. W. and Cochran, W. G., *Statistical Methods*, 7th ed., Iowa State University Press, Ames, IA, 1980.

World Resources Institute, *World Resources: 1994–1995*. Oxford University Press, New York, 1994.

PART III

Tests of Comparison: Parametric Methods

chapter six

Hypothesis Testing: One-Sample

Learning objectives

After mastering this chapter and working through the review questions, you should be able to:

- Explain how the null and alternative hypotheses are formulated.
- Explain the necessary steps in hypothesis testing.
- Perform large and small sample tests using the mean and the proportion.
- Explain the types of errors in hypothesis testing.
- Explain the relationship between the α risk and the β risk.

6.1 Introduction

In the previous chapter it is pointed out that statistical inference basically deals with two types of problems: (1) making estimates and (2) testing hypotheses. Because we discussed estimation in the previous chapter, we turn next to testing hypotheses.

A hypothesis may be looked on as a speculation concerning an observed phenomenon in the world around us. If the speculation is translated into a statement about some condition of a population or populations, that statement is a *statistical hypothesis*. Examples of hypotheses are

1. The reduction of sulfur emission reduces acid rain.
2. A particular feed ration increases the weight gain in steers.
3. The use of a new soil fumigant drastically reduces nematode problems.
4. A new advertising program raises the average sales volume of cotton.

A *hypothesis test* is a procedure we use to conclude that the characteristic measured in a sample either agrees reasonably well with the hypothesis or does not. Our decision rule tells us to accept the hypothesis or reject it. The decision rule is based on the researcher's (in sciences) or the manager's (in business) willingness to incur errors of judgment in accepting or rejecting the hypothesis.

With this understanding of the general logic of testing hypotheses, we can move to the basic steps in testing a hypothesis.

6.2 Steps in hypothesis testing

Step one — stating the null and alternative hypothesis. To test a hypothesis, we must first state the *null hypothesis*, commonly denoted as H_o, and the *alternative hypothesis*, denoted as H_1.

The null hypothesis refers to a beginning hypothesis that the researcher wishes to disprove, because disproving it provides a stronger conclusion for the researcher or manager.

The alternative hypothesis, on the other hand, refers to the complement of the null hypothesis, or the research hypothesis that is used to verify the null hypothesis. There are three basic forms of the hypothesis statement about the population μ:

$$\text{Form I} \quad \begin{aligned} H_0&: \quad \mu = \mu_0 \\ H_1&: \quad \mu \neq \mu_0 \end{aligned}$$

$$\text{Form II} \quad \begin{aligned} H_0&: \quad \mu \geq \mu_0 \\ H_1&: \quad \mu < \mu_0 \end{aligned}$$

$$\text{Form III} \quad \begin{aligned} H_0&: \quad \mu \leq \mu_0 \\ H_1&: \quad \mu > \mu_0 \end{aligned}$$

Step two — selecting the level of significance. The next step in hypothesis testing is the selection of α, the level of significance. The level of significance establishes a criterion for rejection or acceptance of the null hypothesis. We discussed the significance level in the previous chapter on estimation. Because this represents the probability of incurring a *Type I error* (rejecting the null hypothesis when it is true), it would seem reasonable to select a small value of α. However, it should be mentioned that a smaller α increases the β risk, or the probability of incurring a *Type II error* (accepting a false null hypothesis). Later in this chapter we discuss the relationship between the α risk and the β risk. Table 6.1 summarizes the nature of Type I and Type II errors in hypothesis testing.

Table 6.1 **Table of Decisions in Hypothesis Testing**

Decision regarding alternative course of action	Status of null hypothesis	
	True	False
Accept	Correct Decision	Type II error (Accepting a false null hypothesis)
Reject	Type I error (rejecting a true null hypothesis)	Correct Decision

Step three — determining the test distribution to use. The test statistic is the value used to determine whether the null hypothesis should be rejected or accepted. The choice of the appropriate statistical test is based on the use of the appropriate sampling distribution (the normal or any other distribution such as the t distribution). If our sample statistic can be assumed to be normally distributed, we can convert the values of a sample mean into a z value; we then use this z value as the test statistic. As you recall from previous chapters, a sample is considered to be normal when the mean ($\mu_{\bar{x}}$) of the distribution of $\bar{x}$ is equal to the mean of the population from which the sample was drawn, and the variance $\sigma_{\bar{x}}^2$ of the distribution of $\bar{x}$ is equal to the variance of the population divided by the sample size. It should be clear that even if the universe is not normal, the Central Limit Theorem states that the sampling distribution will approach normality as the size of the sample is increased. If our sample size is less than 30 ($n < 30$), then we use the *t distribution*. The method of converting the sample mean into a z value is presented in Section 6.3 where a formal illustration of hypothesis testing is presented.

Step four — defining the rejection or critical regions. Because we have stated the null and alternative hypotheses and selected the level of significance and the type of test statistic to be used, it is now possible to define the rejection or critical regions of the sampling distribution. The critical value, which is the point of demarcation between the acceptance and rejection regions, can be presented in standard units of measurement (z *scale*) or in actual units of measurement (x *scale*).

Figure 6.1 shows the critical regions of a test statistic when the sampling distribution is assumed to be normally distributed, and the risk of erroneous rejection in each tail is .025. In this situation the sum of the two shaded areas equals .05.

If the test statistic value falls in the acceptance region, the null hypothesis is accepted; if its value falls in the rejection region, the null hypothesis is rejected.

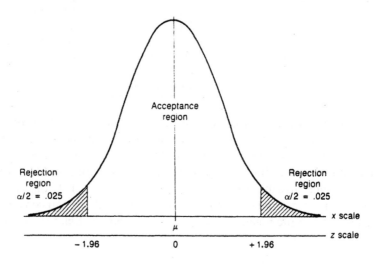

Figure 6.1   Acceptance and rejection regions for a two-tailed test at the 5% level.

Step five — performing the statistical test. We are now in a position to use our sample data and calculate the parameter estimates. If we are testing a hypothesis about the population mean (μ), we must first calculate the sample mean ($\bar{x}$). From the sample, we can estimate the standard error ($s_{\bar{x}}$) and establish the test ratio in standard units as:

$$z = \frac{\bar{x} - \mu_0}{\sigma_{\bar{x}}} \text{ if } \sigma \text{ known} \qquad [6\text{--}1]$$

$$z = \frac{\bar{x} - \mu_0}{s_{\bar{x}}} \text{ if } \sigma \text{ unknown} \qquad [6\text{--}2]$$

Alternatively, the acceptance or rejection boundaries (critical values) can be represented in actual units of measurement (x scale) as:

$$\bar{x}_U^* = \mu_o + z_{\alpha/2}\sigma_{\bar{x}} \qquad [6\text{--}3]$$

$$\bar{x}_L^* = \mu_o + z_{\alpha/2}\sigma_{\bar{x}} \qquad [6\text{--}4]$$

where $\bar{x}_L^*$ = lower critical rejection boundary

$\bar{x}_U^*$ = upper critical rejection boundary

Step six — drawing the statistical conclusion. Because we have performed the statistical test, it is now possible to determine whether the sample information agrees reasonably well with the hypothesis and to make an inference about the population. The acceptance or rejection of the null hypothesis is the conclusion of the test. How confident we are in our conclusion is gauged by the *observed significance level*, or the *p* value.

6.2.1 *p Values*

Observed significance level or *p* value means the probability of observing a value of the test statistic that is at least as contradictory to the null hypothesis when the null hypothesis is true. In other words, the smallest value of α for which test results are statistically significant is called the *p* value. Most statistical computer programs compute the *p* values for different levels of significance. However, if you are using statistical tables to determine the *p* value, you will only be able to approximate its value. This is because most statistical tables give the critical value of α at .01, .025, .05, .10, and so on. The use of *p* values reflects the desire to report the results and not to rely too heavily on arbitrarily significance levels such as .05 and .01, so that users of these results can draw their own conclusions.

The following sections illustrate a number of tests of hypotheses.

6.3   Tests of a mean: large sample

6.3.1   Two-tailed test of a mean

Example 6.1

Major Mills, a manufacturer of cereals, wants to test the hypothesis that a filling machine is functioning properly. This machine is set to load boxes with a mean weight of 15 oz. A weight exceeding 15 oz is costly to Major Mills; and if below 15 oz, it may result in legal problems. Suppose that 200 boxes of cereal are weighed and that the mean and standard deviation of this sample are $\bar{x} = 14.85$ oz and $s = .5$ oz, respectively. Major Mills wishes to test statistically whether the sample indicates that the mean filling weight (μ) differs from 15 oz.

Solution

As the problem involves a question of average weight, we use a test involving the arithmetic mean. Our null hypothesis is that the entire population (universe) of cereal boxes in this weight category has a mean weight of 15 oz, and any observed differences are a result of random sampling error. The null and alternative hypotheses are

$$H_o: \mu = 15 \text{ oz}$$

$$H_1: \mu \neq 15 \text{ oz}$$

This is a two-tailed (two-sided) statistical test because we wish to detect whether $\mu > 15$ oz or $\mu < 15$ oz. Therefore, we have two rejection regions, one on either tail.

If Major Mills wishes to avoid committing Type I error (rejecting the null hypothesis when it is true), it should choose a very small α. Suppose that the management of Major Mills chooses a level of significance of 1% (risk of Type I error); then $\alpha = .01$.

Because the sample size is greater than 30, the appropriate test involves use of the normal curve. Figure 6.2 illustrates the acceptance and rejection regions. The z value from Appendix C for the given level of significance is 2.57. If our sample test ratio has a z value less than

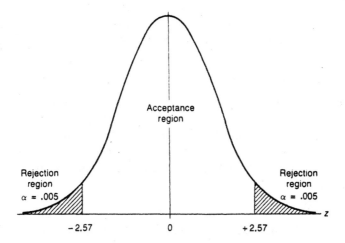

Figure 6.2 Acceptance and rejection regions for a two-tailed test of the mean at the 1% level.

–2.57 or greater than +2.57, we shall reject our hypothesis; otherwise, we shall accept it. The statistical test is performed as follows:

$$z = \frac{\bar{x} - \mu_0}{s/\sqrt{n}}$$

$$z = \frac{14.85 - 15}{5/\sqrt{200}}$$

$$= -4.2$$

The null hypothesis is rejected because –4.2 is well into the lower tail rejection region. It can be concluded that there is sufficient evidence to indicate that the machine is underfilling. Given the observed value of $z = -4.2$, any value of z less than –4.2 or greater than +4.2 would even more strongly contradict the null hypothesis. The observed significance level for the test is

$$p \text{ value} = P \ (z < -4.2 \text{ or } z > +4.2)$$

In Appendix B, we find that $P \ (z > 4.2) = .5 - .4999 = .0001$. Therefore, the p value for the test is

$$2(.0001) = .0002$$

$$p = .0002$$

Consequently, we can say that the test results disagree strongly with the null hypothesis H_o: $\mu = 15$. The probability of observing a z value as large as 4.2 or as small as –4.2 is only .0002, if in fact H_o is true.

6.3.2   One-tailed test of mean

In the previous example we performed a two-tailed test because we were concerned about the possibility that the population mean would be either larger or smaller than the hypothesized value of the mean. In a one-tailed test, we are only concerned about the possible deviation in one direction from the hypothesized value of the mean.

Example 6.2

The Environmental Protection Agency requires that potable water should contain no more than .025 mg/l of lead. An environmental scientist wishes to determine if government regulations have contributed to reduced lead content in the drinking water. A random sample of 150 water specimens from different parts of the country shows an average lead content of .038 mg/l and a standard deviation of .10 mg/l. The scientist wishes to test her results at the .05 level of significance.

Solution

To determine whether the sampled specimen contains no more than .038 mg/l of lead with a risk of 5%, one proceeds as follows. The hypotheses are

$$H_o: \mu \leq .025 \text{ mg/l}$$

$$H_1: \mu > .025 \text{ mg/l}$$

This is a one-tailed test because the null hypothesis may be rejected if the sample mean is too high. Using a significance level of $\alpha = .05$, we reject the null hypothesis for this test if:

$$z > 1.64$$

(See Figure 6.3.)

Calculating the value of the test statistic, we obtain:

$$z = \frac{\bar{x} - \mu_o}{s_{\bar{x}}}$$

or

$$z = \frac{\bar{x} - \mu_0}{s/\sqrt{n}}$$

$$z = \frac{.038 - .025}{.10/\sqrt{150}}$$

$$= \frac{.013}{.008}$$

$$= 1.63$$

Because this value falls to the left of 1.64, the null hypothesis is not rejected and the environmental scientist concludes that the true mean lead content in water is less than .025 mg/l, and that the difference may be attributable to chance sampling error.

6.4 Test of a mean: small sample

The discussion in Section 6.3 deals with a simple significance test when the underlying distribution is normal and we assume the presence of large samples. However, researchers often face situations where large samples are not possible, and a decision on small samples of data must be judged and interpreted largely on their internal evidence.

Some argue that small samples are inherently unreliable and that one should not draw conclusions from them at all. A little reflection shows that this attitude is unable to deal with a number of common difficulties. First, it is often not possible to increase the size of the data

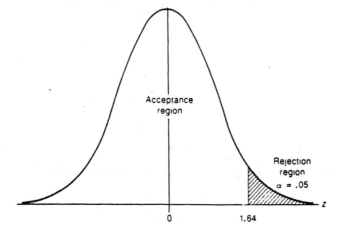

Figure 6.3   Illustration of the rejection region for a lower tailed test.

to any appreciable extent. For example, the number of animals or birds with some rare disease that are available for experimentation may well be quite small, and we have no choice but to make the best of the situation. Second, environmental conditions may be so variable in time and space that we can obtain reasonable homogeneity only within the compass of a comparatively small setup. Third, whether we like it or not, much scientifically relevant evidence that is available from the work of ourselves or others is, in fact, scanty. Given these difficulties, the most sensible procedure is to use methods of analysis that tell us objectively what conclusions, however vague, can justifiably be drawn from the data. Small sample tests of significance are used widely and their utility is recognized in a variety of experimental conditions. Thus, it is worthwhile to become familiar with the small sample or t test. The t test is the appropriate procedure when sample size is less than 30. In the following example, we use the t *distribution* in testing a hypothesis.

Example 6.3

A farmer, considered to be a price taker, is always interested in the maximum price commodity futures will attain over the next year. One way of trying to predict the ceiling price is to solicit the opinions of all futures market experts. Because of the cost and time factors involved, only a small sample ($n < 30$) can be used to draw conclusions about the maximum price of commodity futures. Suppose the *Wall Street Journal* reports that in years of economic conditions such as these, the average price of cotton futures over the next year is typically no more than \$0.70/lb. A cotton producer obtains the opinion of 15 futures market experts. According to these experts, the mean projected ceiling price of cotton futures is \$0.75 with a standard deviation of $s = \$0.05$. Using $\alpha = .01$, test the hypothesis that the population mean is \$0.70.

Solution

The elements of the test are

$$H_o: \mu \leq \$0.70$$

$$H_1: \mu > \$0.70$$

To test at $\alpha = .01$ level, the rejection region is found in the t distribution table to be

$$t > 2.624$$

where degrees of freedom $(d.f.) = n - 1 = 14$. The rejection region is shown in Figure 6.4.

To obtain the value of t:

$$t = \frac{\bar{x} - \mu_o}{s/\sqrt{n}}$$

$$t = \frac{.075 - .070}{0.05/\sqrt{15}}$$

$$= 3.87$$

Because the obtained t value falls in the rejection region, as illustrated in Figure 6.4, the farmer concludes that the predicted mean price for cotton futures is significantly greater than $0.70.

6.5 Test of a proportion

Many statistical surveys are performed by plant scientists to make an inference about the proportion (fraction) of new hybrid plants that can tolerate high doses of fertilizers. Such a scientist selects a random sample from the total number of new hybrid plants, and treats each with a high dose of plant nutrient. The number of plants that shows tolerance (no fertilizer burns), divided by the total (n), represents the sample proportion. This quantity should be close to the population proportion. We should note that the preceding sampling procedure is a binomial experiment if there are a large number of plants in the surveyed group.

The testing procedure for a proportion is analogous to the procedure for testing a hypothesis concerning the population mean, as

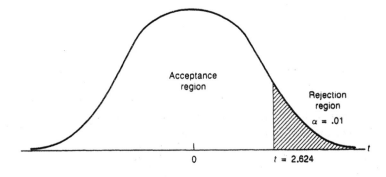

Figure 6.4 Illustration of the rejection region for an upper tailed test.

described in Section 6.3. The only difference is that a hypothesized value of a proportion instead of a mean is involved. The test statistic converts the sample proportion into a z value as follows:

$$z = \frac{p - \pi_0}{\sigma_{\bar{p}}} \qquad [6\text{--}5]$$

where π_0 = hypothesized value of the population proportion.

To compute the standard error of the proportion ($\sigma_{\bar{p}}$), we use the hypothesized value of π as follows:

$$\sigma_{\bar{p}} = \sqrt{\frac{\pi_0(1 - \pi_0)}{n}} \qquad [6\text{--}6]$$

To determine the testing procedure for proportions, let us follow the steps discussed in Section 6.2 on hypothesis testing in the following examples.

Example 6.4

Recent studies indicate that to supplement their incomes, Georgia farmers look for off-farm employment. The percentage of off-farm employment in recent years is shown to be 60%. A random sample of 600 Georgia farmers indicates 354 (59%) to have off-farm jobs. To determine if the sample provides evidence that the true proportion of Georgia farmers holding off-farm employment is no longer 60%, using $\alpha = .01$, the test is performed as follows.

Solution

The hypotheses are

$$H_0: \pi = 60\% = .60$$
$$H_1: \pi \neq 60\% \neq .60$$

This is a two-tailed test because H_0 may be rejected if the sample percentage is too high or too low. Using a significance level of $\alpha = .01$, we reject the null hypothesis for this test if

$$z < -2.57 \text{ or } z > +2.57$$

Figure 6.2 shows the acceptance and the rejection region for this problem.

The test statistic is

$$z = \frac{p - \pi_o}{\sigma_{\bar{p}}}$$

$$z = \frac{p - \pi_o}{\sqrt{\dfrac{\pi_o(1 - \pi_o)}{n}}}$$

$$= \frac{.59 - .60}{\sqrt{\dfrac{.60(1 - .60)}{600}}}$$

$$= -.50$$

Because the computed z value falls in the acceptance region, there is not enough evidence to refute the assertion that 60% of Georgia farmers hold off-farm employment.

Example 6.5

A seed company is being monitored by a regulatory agency to check on the percentage of impurities found in its Red Fescue seed. Certified seed can contain no more than 2% impurities. The regulatory agency analyzed a random sample of 80 bags of Red Fescue seed and found the sample to have 4% impurities. To determine whether the sampled bags contain no more than the allowable percentage of impurities with a risk of 5%, one would proceed as follows. The hypotheses are

$$H_o: \pi \leq 2\% \text{ impurities}$$

$$H_1: \pi > 2\% \text{ impurities}$$

This is a one-tailed test because H_o may be rejected if the sample percentage is too high. Using a significance level of $\alpha = .05$, we reject the null hypothesis for this test if:

$$z > 1.64$$

(See Figure 6.3)

The test statistic is computed as follows:

$$z = \frac{p - \pi_0}{\sqrt{\dfrac{\pi_0(1-\pi_0)}{n}}} \qquad [6\text{--}7]$$

$$= \frac{.04 - .02}{\sqrt{\dfrac{.02(1-.02)}{80}}}$$

$$= 1.28$$

Because the computed z value falls within the acceptance region, the regulatory agency concludes that there is not enough evidence to indicate greater seed impurities than the 2% allowed.

6.6   An alternative approach to hypothesis testing

The previous discussion is based on the procedure whereby acceptance and rejection regions are separated by a critical value that is a certain number of standard errors from the hypothesized value. Identical results can be obtained using an algebraic equivalent approach.

Instead of working with the number of standard errors that a number lies from the hypothesized value, the critical value of the parameter estimate can be used, for we are only making minor alterations in the procedure. Let us follow several examples to see how the procedures differ.

Example 6.6

The Orange Advisory Board has the power to regulate somewhat the rate at which oranges are sent to market, to ensure that there is an adequate supply through the marketing season for shipping the product during the whole season. To correctly accomplish this objective, the board must know what level of production to expect in a given year. It is known that in a particular area production over several years has amounted to 487 boxes per acre with a standard deviation of 37.3 ($\sigma = 37.3$) boxes per acre. A sample of 25 acres has been taken during the past two weeks with the result that $\bar{x} = 470$ boxes per acre. Does this indicate, at $\alpha = .05$, that this year's production is not the same as the average for previous years?

Solution

Because previous production normally amounts to 487 boxes per acre, we use this yield as the population mean ($\mu = 487$). The question becomes, would it be possible to take a sample of 25 acres and find a sample mean of only 470? By recalling earlier sections on sampling distributions, we know the upper and lower values forming the boundary for 95% probability can be found as:

$$\text{Upper boundary} = \mu + z_{\alpha/2}\frac{\sigma}{\sqrt{n}} \qquad [6\text{--}8]$$

$$\text{Lower boundary} = \mu - z_{\alpha/2}\frac{\sigma}{\sqrt{n}} \qquad [6\text{--}9]$$

The population mean as well as the upper and lower boundaries are shown in Figure 6.5.

Note that z is used here even though $n < 30$ because the population standard deviation, σ, is known. By inserting the numbers for the problem into the two equations, we get:

$$\text{Upper boundary} = \mu + z_{\alpha/2}\frac{\sigma}{\sqrt{n}}$$

$$= 487 + (1.96)\frac{37.3}{\sqrt{25}}$$

$$= 487 + 14.62$$

$$= 501.62$$

$$\text{Lower boundary} = \mu - z_{\alpha/2}\frac{\sigma}{\sqrt{n}}$$

$$= 487 - 14.62$$

$$= 472.38$$

In Figure 6.7, it can be seen that if the population mean is actually 487, then 95% of the time (or 95 chances out of 100) a sample of 25 acres from that population has a mean no lower than 472.38 and no higher than 501.62.

If the mean of a sample of 25 acres is outside the boundaries, there are a couple of possibilities. First, because the sample should fall within the boundaries 95% of the time, then a value outside these boundaries might indicate that an extremely unusual group of trees had been selected, or it might indicate that the population mean is no longer 487 as in previous years.

The preceding procedure can be put into steps appropriate for testing hypotheses.

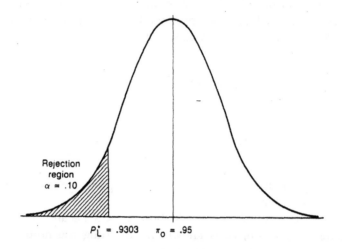

Figure 6.5   Illustration of the mean and the upper and lower boundaries.

Step one — State the null and alternative hypotheses.

$$H_0: \mu = 487$$
$$H_1: \mu \neq 487$$

Step two — Determine the upper and lower boundaries.

Upper boundary $\quad = \bar{x}_U^* = \mu + z_{\alpha/2}\dfrac{\sigma}{\sqrt{n}}$

$$= 487 + (1.96)\dfrac{37.3}{\sqrt{25}}$$

$$= 501.62$$

Lower boundary $\quad = \bar{x}_L^* = \mu - z_{\alpha/2}\dfrac{\sigma}{\sqrt{n}}$

$$= 487 - (1.96)\dfrac{37.3}{\sqrt{25}}$$

$$= 472.38$$

Step three — State the decision rule: reject H_0 if $\bar{x} < 472.38$ or if $\bar{x} > 501.62$; otherwise do not reject H_0.

Step four — Draw conclusions. Because $\bar{x} = 470$, we reject H_o and conclude that the population mean for this year is no longer 487. Although the test shows only that the population mean is no longer 487, the fact that the sample mean is in the lower region gives reason to suspect that the actual mean is somewhat lower than 487. It would be unwise to assume that the new population mean is 470, because that was based on a very small sample in relation to the total number of acres.

Example 6.7

By using the Orange Advisory Board from Example 6.6, we assume that the population mean of 487 is known but that the population standard deviation is not. Again, a sample of 25 acres is taken with a mean $\bar{x} = 470$ and a sample standard deviation of $s = 27.2$. This time it is desired to determine, at $\alpha = .05$, if the average yield per acre is less than the average for past years.

Solution

A one-tailed test is being used here because, in this example, it is being assumed that a hard freeze throughout the production area had caused considerable damage and that production may have dropped.

It should be noted that in the absence of a factor such as a freeze that causes one to suspect that production is either lower (as in this case) or higher, a two-tailed test would be more appropriate. To determine if production has dropped, the following hypothesis is used:

$$H_o: \mu \geq 487$$

$$H_1: \mu < 487$$

The boundary or critical value for this problem is

$$\bar{x}_L^* = \mu - t_\alpha \frac{s}{\sqrt{n}}$$

$$= 487 - (1.711)\frac{27.2}{\sqrt{25}}$$

$$= 477.69$$

t test is being used here because $n < 30$ and σ is not known. The value of $t = 1.711$ is found by using $\alpha = .05$ and the degrees of freedom ($d.f.$) which is $n - 1$. For this problem we have

$$d.f. = 25 - 1 = 24$$

The decision rule is

Reject H_o if $\bar{x} < 477.69$

The rejection region is illustrated in Figure 6.6.

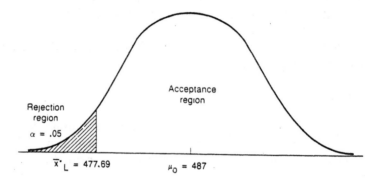

Figure 6.6   Illustration of the rejection region.

Because $\bar{x} = 470$, we reject the null hypothesis and conclude that the population mean has dropped and is now less than 487.

The basis for this conclusion is that because $\bar{x} = 470$, which puts it in the region such that it would occur less than 5 times out of 100 if $\mu = 487$, we are willing to run the risk that we may be reaching a false conclusion.

Example 6.8

A researcher has developed a new insecticide that is believed to kill more than 95% of cotton borers on contact and is environmentally safe. In a separate laboratory test, the insecticide is applied to 200 cotton borers. Although all 200 eventually die, only 180 die immediately on contact. Do the data contradict at $\alpha = .10$ the claim that the insecticide can kill 95% of cotton borers on contact?

Solution

Note that this situation involves a percentage instead of a mean. It follows a binomial distribution. The steps in the test are

$$H_o : \pi_o \geq .95$$

$$H_1 : \pi_o < .95$$

$$p_L^* = \pi_o - z_\alpha \sqrt{\frac{\pi_o(1-\pi_o)}{n}} \qquad \text{[6–10]}$$

$$= .95 - (1.28) \sqrt{\frac{.95(1-.95)}{200}}$$

$$= .95 - .0197$$

$$= .9303$$

The rejection region is illustrated in Figure 6.7.

The decision rule means rejecting H_o if $p < .9303$. Otherwise, do not reject H_o. Because $p = 180/200 = .90$, which is less than .9303, we do reject H_o. We conclude that there is sufficient evidence that the claim made by the developer of the insecticide is incorrect.

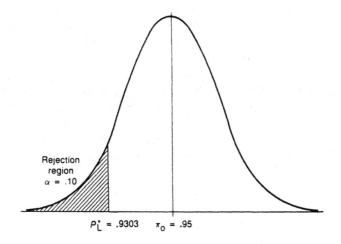

Figure 6.7 Illustration of the rejection region.

6.7 *Relationship of α and β risk*

Recall from Section 6.1 that the probability of accepting H_o when H_o is false is referred to as the β risk or a Type II error. It is also mentioned that there is an inverse relationship between the α and β risk. For instance, if a researcher wishes to be quite certain to minimize the risk of committing a Type I error (rejecting H_o when H_o is true), then he or she must choose a very small α. By doing so, he or she increases the risk of committing a Type II error.

In this section we examine the β risk and how β is calculated. Additionally, we look at:

1. The *power* of a test $(1 - \beta)$ which gives the probability of rejecting a false null hypothesis
2. The *power function*
3. The *power curve*

As was pointed out earlier, the α risk in a given problem is determined by the level of significance one chooses. The β risk, on the other hand, is not as easily determined because it is dependent on:

1. The α risk
2. The sample size
3. The alternative hypothesis being considered

As an illustration, consider Example 6.6, and observe how the risk of a Type II error increases as we consider the α risk, the sample size, and the alternative hypothesis being tested.

Assume that the α in the preceding example is reduced from 5 to 1%. The new z value of ± 2.57 changes our boundaries to 467.83 and 506.17. The new acceptance region is larger than the old acceptance region (472.38 to 501.62). This change therefore increases the chances of accepting a false null hypothesis or the β risk. To reduce the chances of accepting a false H_o (the β risk), we can increase the sample size. A larger sample means a smaller standard error. As the standard error gets smaller, the acceptance region for any given α becomes smaller. This reduces the risk of a Type II error. For the preceding example, let us increase the sample size from 25 to 100 acres. The standard error then becomes:

$$\sigma_{\bar{x}} = \frac{s}{\sqrt{n}}$$

$$= \frac{37.3}{\sqrt{100}}$$

$$= 3.73$$

As can be seen from the preceding equation, the standard error is reduced from 7.46 (when $n = 25$) to 3.73 (when $n = 100$). To use this new standard error with the original α of 5%, the boundaries will be 479.69 and 494.31. As you observe, the new acceptance region is smaller. The smaller the acceptance region, the less the likelihood of accepting a false null hypothesis. This in turn means a reduction in the risk of committing a Type II error.

To make a Type II error, we have in fact accepted a false null hypothesis. This implies that some alternative hypothesis must be true. As can be seen in Figure 6.8, the numerical value of β varies according to possible true values of μ. If the true alternative μ is close to μ_o, the β risk is large, whereas if the true alternative is very different from μ_o, the β risk is small. As mentioned earlier, specific values for β can be computed for particular alternative values of μ.

As an illustration, consider again Example 6.6, where the population standard deviation is 37.3 and the hypothesis H_o: $\mu = 487$ is to be tested with a 5% level of significance and a sample of $n = 25$ observations. Assume the true mean is 471. Notice that changing μ changes only the

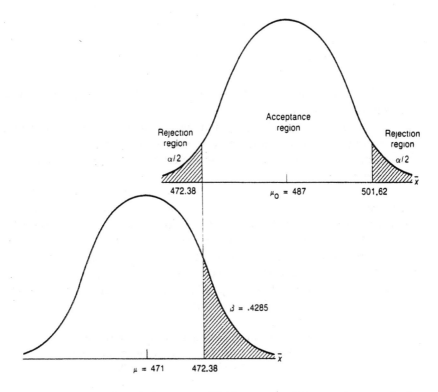

Figure 6.8 Sampling distribution of $\bar{x}$ if $\mu = 487$ and if $\mu = 471$, respectively, for $n = 25$ and $\sigma = 37.3$.

location of our random sampling distribution of $\bar{x}$. Figure 6.8 illustrates the values of β obtained when $\mu = 471$. The probability of accepting the null hypothesis (H_o: $\mu = 487$) when the true mean is 471 is calculated as follows:

$$z = \frac{\bar{x}^* - \mu}{\sigma_{\bar{x}}}$$

[6–11]

where $\bar{x}^* $ = critical value.

The new z value for the preceding problem is

$$z = \frac{472.38 - 471}{37.3 / \sqrt{25}}$$

$$= .1850$$

We now use this new z value to determine the probability of falsely accepting H_o. Appendix B is used to determine the probability as follows:

$$\beta = P(z - 0.1850)$$
$$= .5000 - .0714$$
$$= .4286$$

This probability of 42.86% is the β risk for the particular alternative mean of 471 ($\mu = 471$).

By using the preceding method, we can determine the probabilities of a Type II error for different values of μ. Table 6.2 presents several possible values of μ, and the associated probabilities of accepting a false null hypothesis.

The third column of Table 6.2 shows the set of $1 - \beta$ probabilities. We refer to these probabilities as the *power function*. The graph of a power function is called a *power curve* or *operating characteristic (OC) curve*. Because the power curve shows the probability of rejecting the null hypothesis when the null hypothesis is false for all possible values of the parameter tested, it is used to evaluate a hypothesis test. The power curve is illustrated in Figure 6.9. This curve simply shows graphically the probability of not committing a Type II error.

Table 6.2 **Probabilities of a Type II Error and Power Functions When $\mu = 487$ Boxes of Oranges and Selected Alternative Means, with $\alpha = .05$**

Selected alternative means (μ) boxes	Probability of Type II error (β)	Probability of not making a Type II error ($1 - \beta$)
457	.0197	.9803
464	.1314	.8686
467	.2358	.7642
471	.4286	.5714
497	.2709	.7291
507	.2358	.7642
510	.1314	.8686
517	.0197	.9803

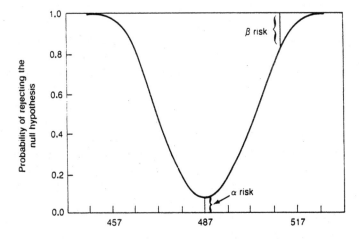

Figure 6.9 Power curve for testing the mean production level of oranges.

Chapter Summary

This chapter covered the basic concepts of hypothesis testing. Procedurally, you were introduced to the six systematic steps in testing hypotheses, namely, (1) stating the null and an alternative hypothesis, (2) selecting the level of significance, (3) determining the test distribution, (4) defining the rejection or critical region, (5) performing the statistical test, and (6) drawing the statistical conclusion.

Each of the tests given in this chapter concerned testing to determine if the true population mean or proportion is equal to some particular value. In the next chapter, we expand the discussion to include tests of one mean against another or one proportion against another. This greatly expands our ability to make decisions regarding real situations.

The techniques in the chapter are based on the assumptions of *large samples* as well as Central Limit Theorem.

Case Study

Testing of Hypotheses on Groundwater Pollution from Fertilizers

Increased agricultural productivity through plant and animal breeding, mechanization, and other management technologies has helped to maintain the balance between food supply and demand around the globe. The critical link between food production to meet today's demands and long-term agricultural sustainability seems to be the adequate supply of essential nutrients.

With the increasing use of fertilizers, scientists and the general public are concerned that high-input farming is causing considerable damage to the environment. Air, water, and soil pollution have led to loss of genetic diversity and ecosystem disturbances. Excessive use of nitrogen fertilizers and animal manure have been linked to increase of nitrates in the drinking water (Keeney, 1989). The significance of fertilizers in polluting the air is highlighted in reports that nitrous oxide originating from agricultural soils can damage the ozone layer in the stratosphere (Bouwman, 1990; Smith et al., 1990).

Groundwater pollution occurs when excess N from fertilizers not taken up by crop plants finds its way below the root zone and leaches into the groundwater. Excess rainfall, poor quality soils with low water-holding capacity, and heavy fertilizer application practices contribute to excess amounts of nitrate–N in the soil. Once the groundwater is polluted, it remains in this state for extended periods of time. The World Health Organization (WHO) of the United Nations has set a standard of 10 mg NO_3–N/l of water for safe drinking water (Singh and Kansal, 1994). It is important to know if the level of nitrate concentration exceeds the safety standards, and if there are significant changes over time.

Water samples collected from deep wells situated on cultivated farms of the Ludhiana district in Punjab, India, were analyzed for nitrate–N content for two different years. The results are as follows.

Nitrate Concentration (mg NO_3–N/l) in Water Samples
from Wells Located in Cultivated Areas of Ludhiana
District (central Punjab, India) from 1982 to 1988

	1982	1988
Number of observations	26	28
Mean	2.13	2.29

Adapted from Singh, B. et al., *Indian J. Environ. Health*, 33, 516, 1991.

By assuming that the standard deviation for the samples in 1982 and 1988 were .23 and .31, respectively:

1. Test the hypothesis that the nitrate concentration found in these wells in 1982 is less than the international standard set by the WHO. Use $\alpha = .05$.

2. Test the hypothesis that the nitrate concentration in 1988 exceeded the 1982 level. Use $\alpha = .05$. Would this hypothesis be accepted at $\alpha = .01$?

References

Bouwman, A. F., *Soils and Greenhouse Effect,* John Wiley & Sons, New York, 1990.

Keeney, D. R., Sources of nitrate groundwater, in Follett, R. F., Ed., *Nitrogen Management and Groundwater Protection,* Elsevier, Amsterdam, 1989, 23.

Singh, B. and Kansal, B. D., Role of fertilizers in environmental pollution, in Dahliwal, G. S. and Kansal, B. D., *Management of Agricultural Pollution in India,* Commonwealth, New Delhi, India, 1994, 138.

Smith, S. J., Schepers, J. S. and Porter, L. K., Assessing and managing nitrogen losses to the environment, *Adv. Soil Sci.,* 14, 1, 1990.

Review questions

1. An agronomist states that the average increase in height of a new variety of avocado (Hass-CP84) developed by her research institution is 2.5 in./week. To establish this fact, a random sample of 36 new seedlings is taken and the average weekly growth rate is recorded as 2.1 in. with a standard deviation of $s = .2$. At $\alpha = .01$, do the data present significantly different evidence than that of the research institution?

2. A random sample of 40 acres of rice treated with granular insecticide (Azodrin®) for the control of brown plant hoppers and stem borers yields the following results:

$$\bar{x} = 2250 \text{ lb/acre}$$
$$s = 140 \text{ lb/acre}$$

   a. Test the null hypothesis that $\mu = 2300$ lb against the alternative hypothesis that $\mu < 2300$, at $\alpha = .10$. Interpret the results.
   b. Calculate the p value, and interpret its meaning.

3. A fertilizer currently being used on corn is known to increase yield by 30 lb/acre. To compare a new fertilizer with the one currently being used, the new fertilizer is applied to a random sample of 500 acres throughout Kansas. The average increase in yield for the sampled acres is 45 lb, and the standard deviation is 15 lb. Do the data provide sufficient evidence, at $\alpha = .05$, to conclude that the new fertilizer is effective in increasing yield?

4. A citrus cooperative wishes to determine whether the mean sugar content per orange shipped from a particular grove is less than 0.025 g. A random sample of 150 oranges produces a mean sugar content of .021. By using a level of significance of $\alpha = .01$, and assuming $\sigma = .005$ g, what can be concluded?

5. An animal scientist conducts an experiment to determine the effect of benzedrine on the heart rate of sheep. A random sample of 60 sheep is included in the study. At the beginning of the experiment, the heart beat per minute is measured for each sheep. After the administration of a measured dose of benzedrine, the heart beat is recorded again. If the sample mean increase in beats per minute is 5.2 with a standard deviation of 4.0, test the research hypothesis that the mean increase in heart rate for sheep after the use of benzedrine is greater than zero. Use $\alpha = .10$.

6. An ecologist concerned with the amount of industrial effluents that are found in the streams and rivers wants to determine their impact on the amount of dissolved oxygen available for fish and other aquatic organisms. Recent studies suggest that a minimum of 6 ppm of dissolved oxygen is required for support of aquatic life. The scientist takes a random sample of 36 specimens from a river at specific locations. The sample yields a mean of 5.2 and a standard deviation of .3 of dissolved oxygen. Do the data provide sufficient evidence to indicate that the average dissolved oxygen content is less than six? Test at $\alpha = .05$.

7. Yecora Rojo-16, a new variety of wheat, is believed to contain approximately 25% protein by weight. In a recent study of 200 random samples, 1 kg each, analyzed for protein content, a researcher observes the mean protein content of Yecora Rojo-16 to be 24% protein by weight with a standard deviation of 3%. Do the data provide sufficient evidence to contradict the original claim? Test at $\alpha = .05$.

8. Suppose an agricultural lender routinely analyzes its accounts concerning default rates so as to spot any developing trends, up or down. In this case, the lender is interested in any change, positive or negative. Because this is only to spot trends, the lender uses a sample of $n = 16$ accounts. The most recent sample shows 3% to be in default. Does this indicate any change in the default rate? Test at $\alpha = .10$.

9. Ecotourism has become a major source of revenue for Costa Rica. To remain a profitable venture, the hotel occupancy rates must average at least 72%. A survey of all hotels for 90 d shows a

mean occupancy rate of 70% and a standard deviation of 8%. Test using $\alpha = .05$.

10. Since 1970 there has been a decline in the health of forests in large areas of Germany (FRG) (Urfer, 1993). A forest inventory done in 1991 indicates that 60% of forested areas shows a decline in health. To determine whether the percentage of forested areas has remained healthy or not, an environmentalist takes a random sample of 300 trees in a forest. The sample produces 165 trees that are diseased.

   a. State the null and alternative hypotheses.

   b. Conduct a statistical test at $\alpha = .10$ and state your conclusions.

11. An animal scientist is interested in learning whether a new ration will result in a mean weight gain in steers of 20 lb/week. In this particular instance, the scientist does not want to reject the null hypothesis that the mean weight gain is 20 lb/week if the population mean is in fact greater than 20 lb. The scientist believes that the new ration is superior. Only if the population mean is less than 20 lb does the scientist want to reject the superiority of the new feed ration.

   a. State the null and alternative hypotheses.

   b. Conduct a statistical test at $\alpha = .10$ and state your conclusions.

References and suggested reading

Dixon, W. J. and Massey, F. J., Jr., *Introduction to Statistical Analysis*, 4th ed., McGraw-Hill, New York, 1983, chap. 8.

Feinberg, W. E., Teaching the Type I and Type II errors: the judicial process, *Am. Statist.*, 25, 30, 1971.

Gibbons, J. D. and Pratt, J. W., P values: interpretation and methodology, *Am. Stat.*, 29, 20, 1975.

Gill, J. L., *Design and Analysis of Experiments: In the Animal and Medical Sciences.*, Vol. 1, Iowa State University Press, Ames, IA, 1978, chap. 1.

Hamburg, M., *Statistical Analysis for Decision Making*, 4th ed., Harcourt Brace Jovanovich, New York, 1987, chap. 7.

Lehmann, E. L., *Testing Statistical Hypotheses*, John Wiley & Sons, New York, 1959.

Mason, R. D., *Statistical Techniques in Business and Economics*, 5th ed., Homewood, Duxbury, Boston, 1982, chaps. 13 and 14.

Mendenhall, W. and Beaver, R. J., *Introduction to Probability and Statistics*, 8th ed., PWS-KENT, Boston, 1991, chap. 8.

Mendenhall, W. and Reinmuth, J. E., *Statistics for Management and Economics*, 4th ed., Duxbury, Boston, 1982, chaps. 8 and 9.

Mendenhall, W., Wackerly, D., and Scheaffer, R. L., *Mathematical Statistics with Applications*, 4th ed., PWS-KENT, Boston, 1990.

Snedecor, G. W. and Cochran, W. G., *Statistical Methods*, 7th ed., Iowa State University Press, Ames, IA, 1980, chap. 5.

Urfer, W., Statistical analysis of climatological and ecological factors in forestry; spatial and temporal aspects, in Barnett, V. and Turkman, K. F., Eds., *Statistics for the Environment*, John Wiley & Sons, New York, 1993.

chapter seven

Hypothesis Testing: Two-Sample

Learning objectives

After you have mastered this chapter and worked through the review questions, you should be able to do the following:

- Explain the purpose of two-sample test of hypothesis for the mean and the percentage.
- Explain the methodology involved in conducting these tests.
- Conduct tests of hypotheses about the differences between two population parameters such as the mean and the percentage.

7.1   Introduction: questions about differences

Environmental and agricultural scientists, and managers are often interested in determining whether two populations of a crop or an animal group are different with respect to some characteristic. For instance, an environmental scientist may want to know if logging in an old growth forest has impacted the diet of a bird species, or an agronomist may want to determine if one variety of a crop responds differently to a chemical fertilizer than another variety. An animal scientist may be interested in the differences that may exist in milk production of two different genetic breeds. An agricultural economist may hypothesize that the mean farm income in a certain region is some specific value μ_o. In all these situations two random samples, one from each population, provide the necessary data on which to make inferences about the population parameter.

In the previous chapter, you learned how to perform a test of hypotheses with a single sample. Our interest was to see whether sample results supported a hypothesis about a population characteristic or not. In this chapter, we adopt hypothesis-testing procedures that specifically make relative comparison of the two population means and percentages. The two-sample hypothesis tests of means assist us in making inferences about the population by using sample data to see if there is a statistically significant difference between the means of two populations.

As in the case of the single-sample test of a hypothesis, the null and alternative hypothesis for differences between two population means can be stated mathematically as:

$$H_o: \mu_1 - \mu_2 = 0$$
$$H_1: \mu_1 - \mu_2 \neq 0 \qquad \qquad [7\text{-}1]$$

$$H_o: \mu_1 - \mu_2 \geq 0$$
$$H_1: \mu_1 - \mu_2 < 0 \qquad \qquad [7\text{-}2]$$

or

$$H_o: \mu_1 - \mu_2 \leq 0$$
$$H_1: \mu_1 - \mu_2 > 0 \qquad \qquad [7\text{-}3]$$

The null hypothesis in Equation 7–1 states that there is no difference between the two population means. That is, the true mean of the first and the second populations are equal to each other. It is not required that the null hypothesis always be stated as equal. More generally, we can hypothesize that μ_1 and μ_2 differ by any amount D_o:

$$H_o: \mu_1 - \mu_2 = D_o$$
$$H_1: \mu_1 - \mu_2 \neq D_o \qquad \qquad [7\text{-}4]$$

The one-tailed tests are stated with appropriate inequalities as shown in Equation 7–2 and Equation 7–3. The determination of whether to use the left-tailed or the right-tailed test depends on the nature of the problem. If the problem states that H_o should be rejected only if μ_1 is significantly greater than μ_2, we use the *right-tailed* test. On the other hand, if the problem indicates that μ_1 is significantly less than μ_2, a *left-tailed* test is used. As you recall from the previous chapter, the null hypothesis is the one for which we can compute the probabilities of outcomes. A sample outcome may either support or deny the null hypothesis.

In the following sections of this chapter we discuss hypothesis tests involving the differences between two population means and percentages when we have large samples and small samples.

7.2 Two-sample test of means: large sample

The most common test of comparison between two populations is the test for comparing two means. If the samples are selected independently of each other, the procedure is similar to that used when testing a hypothesis for a single sample mean.

The *test of difference between means* basically asks the question, is the mean (μ_1) of a random sample (n_1) drawn from a population equal to the mean (μ_2) of another sample (n_2)? If we continue to select samples from the population and observe the difference between means, we could theoretically construct a sampling distribution of these differences. Such a sampling distribution is called the *sampling distribution of the differences between sample means*. Figure 7.1 shows the sampling distribution of differences between sample means.

If we have a large sample from both populations, the shape of the distribution approximates the normal distribution. Recall that when we assume a normal distribution, the middle 68.3% of the differences in the sampling distribution are found within one standard deviation of the mean, and 95.4% of the differences in sampling distribution are found within two standard deviations of the mean and so on. With the assumption that we have drawn randomly large samples from the population, the z test can be applied as a test of significance. With the use of the z table and the level of significance (α), the boundaries of the rejection region for the null hypothesis may be established. The test between means must use the standardized difference between the mean of sample one (μ_1) and the mean of sample two (μ_2). This means that we have to calculate the *critical ratio* of the differences as follows:

$$z = \frac{\bar{x}_1 - \bar{x}_2}{\sigma_{\bar{x}_1 - \bar{x}_2}} \qquad [7\text{--}5]$$

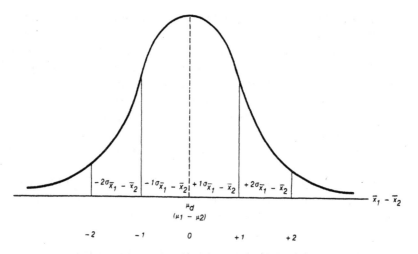

Figure 7.1 **The sampling distribution of differences between two sample means.**

and the standard error of the difference is computed as follows:

$$\sigma_{\bar{x}_1-\bar{x}_2} = \sqrt{\frac{\sigma_1^2}{n_1} + \frac{\sigma_2^2}{n_2}} \qquad [7\text{--}6]$$

We see from Equation 7–6 that we must know the population standard deviation to calculate the standard error. In the absence of knowledge about the population standard deviation, we may use the sample standard deviation (if we have a large sample) to estimate the standard error as follows:

$$s_{\bar{x}_1-\bar{x}_2} = \sqrt{\frac{s_1^2}{n_1} + \frac{s_2^2}{n_2}} \qquad [7\text{--}7]$$

The following illustrations provide a better understanding of how to perform a test of means between two samples for a two-tailed as well as a single-tailed condition.

Example 7.1

An environmental scientist is concerned that increased logging of the old growth forests in Canada has impacted the diet of a bird species. He notices that the birds in the logged forests appear to be living on a diet that is different than usual. Various questions can be asked about the possible effect of a different diet on physiological function. The scientist wants to know if there is any real difference in body weight between the birds in the logged forest and those in forests that have not been logged. A sample of 45 birds each from the logged and unlogged forests produces the following results:

Unlogged forest	Logged forest
$n_1 = 45$	$n_2 = 45$
$\bar{x}_1 = 95.50$ g	$\bar{x}_2 = 92.33$ g
$s = 9.25$ g	$s = 8.50$ g

Determine for the scientist whether at $\alpha = .05$ there is a significant difference between the two mean body weights.

Solution

We follow the procedural steps outlined in Chapter 6 in testing a hypothesis.

Step one — Because the environmental scientist is only interested in determining if there is a significant difference between the two means, this is a two-tailed test. Therefore, we state the null and the alternative hypothesis as follows:

$$H_o: \mu_1 = \mu_2$$

$$H_1: \mu_1 \neq \mu_2$$

Step two — Given that the environmental scientist would like to use the .05 level of significance, and that we have a large sample, we are able to use the z distribution. Thus, the rejection boundaries are $z = \pm 1.96$.

Step three — The decision rule is

Accept H_o if calculated z (critical ratio) falls between ± 1.96

or

Reject H_o and accept H_1 if critical ratio < -1.96 or critical ratio $> +1.96$.

Step four — The test is

$$z = \frac{\bar{x}_1 - \bar{x}_2}{\sigma_{\bar{x}_1 - \bar{x}_2}}$$

To calculate for z, we have to first determine the value of $\sigma_{\bar{x}_1 - \bar{x}_2}$ from the sample data as follows:

$$S_{\bar{x}_1 - \bar{x}_2} = \sqrt{\frac{s_1^2}{n_1} + \frac{s_2^2}{n_2}}$$

$$= \sqrt{\frac{(9.25)^2}{45} + \frac{(8.50)^2}{45}}$$

$$= \sqrt{\frac{85.56}{45} + \frac{72.25}{45}}$$

$$= \sqrt{1.90 + 1.61}$$

$$= 1.87$$

Now that we have the estimated value for the $\sigma_{\bar{x}_1 - \bar{x}_2}$, the critical ratio is

$$z = \frac{95.50 - 92.33}{1.87}$$

$$= \frac{3.17}{1.87}$$

$$= 1.70$$

Graphically, the acceptance and rejection regions are shown in Figure 7.2.

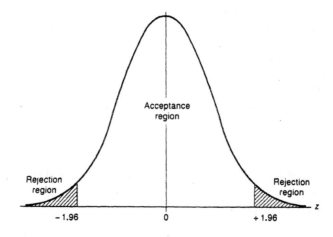

Figure 7.2   Two-tailed test, areas of acceptance and rejection, with a level of risk of .05.

In Figure 7.2 there are two rejection regions, one above +1.96 and the other below —1.96. The implication of a two-tailed test is that the null hypothesis that $\mu_1 = \mu_2$ would be rejected at .05 level of significance if the critical ratio is less than 1.96 or greater than 1.96.

Step five — Because 1.70 is less than 1.96, we cannot reject the null hypothesis and conclude that there is no significant difference between the mean bird weight of the two groups.

Example 7.2

A plant scientist is experimenting with a new chemical, Borkill, for corn borers. She claims that Borkill will reduce corn borer damage and increase yield. Alternate acres are treated with Borkill during the growing season. To determine its effectiveness, a random sample of 50 ears of corn from the treated acres is selected. The mean weight for the corn ear is computed as 12 oz, with a standard deviation of 2.0 oz. Similarly, a random sample of 50 corn ears from the untreated acres is selected; the mean weight and the standard deviation were 10 and 1.4 oz, respectively. At the .01 level of significance, should the plant scientist's claim be accepted?

Solution

This is a right-tailed test because the scientist's claim of the treated corn being improved over the untreated corn is to be tested. The hypotheses are

$$H_o: \mu_1 \leq \mu_2$$
$$H_1: \mu_1 > \mu_2$$

In testing the validity of the scientist's claim, we are asked to use .01 level of significance. The z value that bounds the rejection region is 2.33, as shown in Figure 7.3. Therefore, the decision rule is

Accept H_o if critical ratio ≤ 2.33

or

Reject H_o if critical ratio > 2.33

The statistical test ratio is

$$z = \frac{\bar{x}_1 - \bar{x}_2}{\sigma_{\bar{x}_1 - \bar{x}_2}} = \frac{\bar{x}_1 - \bar{x}_2}{\sqrt{\dfrac{s_1^2}{n_1} + \dfrac{s_2^2}{n_2}}}$$

$$z = \frac{12 - 10}{\sqrt{\dfrac{(2.0)^2}{50} + \dfrac{(1.4)^2}{50}}}$$

$$= \frac{2}{\sqrt{0.12}}$$

$$= \frac{2}{.35}$$

$$= 5.71$$

The statistical conclusion for this problem is to reject the null hypothesis because the computed ratio 5.71 is greater than the critical value of 2.33. We conclude that the mean corn yield from the treated area is greater than the untreated area.

7.3   Two-sample test of means: small sample

In many instances it is economically and procedurally not feasible to take a large sample for a study. When we are limited to a small sample, we use the t distribution to make inferences about population means. In using the t statistic, we assume that both sampled populations are

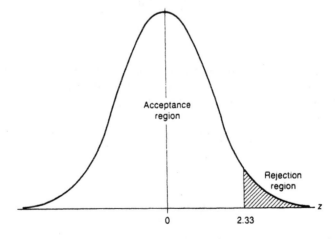

Figure 7.3   Acceptance and rejection regions for a right-tailed test of difference between means with .01 level of significance.

approximately normally distributed with equal population standard deviations ($\sigma_1 = \sigma_2 = \sigma_3$). Given the assumption of equal standard deviations, it is reasonable to combine the sum of deviations from the two samples to construct a *pooled* sample estimator of σ for use in the t statistic as follows:

$$t = \frac{\bar{x}_1 - \bar{x}_2}{s_p \sqrt{\dfrac{1}{n_1} + \dfrac{1}{n_2}}} \qquad [7\text{--}8]$$

The pooled standard deviation (s_p) is computed by averaging the two sample standard deviations as follows:

$$s_p = \sqrt{\frac{(n_1 - 1)s_1^2 + (n_2 - 1)s_2^2}{n_1 + n_2 - 2}} \qquad [7\text{--}9]$$

Examples 7.3 and 7.4 provide, respectively, the illustration for a two-tailed and a single-tailed test of means for small samples.

Example 7.3

Two different heating systems (natural gas and cogeneration) are offered for greenhouse use. An agricultural engineer is interested in determining if there is a difference in cost of operation between the two systems. A sample of 16 greenhouses using natural gas produces an average annual cost of $35,000 with a standard deviation of $800. Another sample of 14 greenhouses with space capacity equal to sample one produces an average annual cost of $32,800, with a standard deviation of $1,000. At $\alpha = .05$, determine if the average cost of the two systems are significantly different.

Solution

Because it was not predicted as to which system is best, we use a two-tailed test. The null and alternative hypotheses are

$$H_o: \mu_1 = \mu_2$$
$$H_1: \mu_1 \neq \mu_2$$

The critical value for the test is determined by the level of significance and the sample size adjusted for loss of degrees of freedom. In this instance, the degrees of freedom (*d.f.*) are

$$d.f. = n_1 + n_2 - 2$$
$$= 16 + 14 - 2$$
$$= 28$$

The critical values of t from Appendix E for 28 *d.f.* and an α of .05 are $\pm$ 2.048. In computing the degrees of freedom in this case, we subtract 2 from the sum of the two sample sizes. This means that *1 d.f. is lost* for each sample. For a detailed explanation of the computation of the degrees of freedom, see Mason (1982).

The test statistic is

$$t = \frac{\bar{x}_1 - \bar{x}_2}{S_p \sqrt{\dfrac{1}{n_1} + \dfrac{1}{n_2}}} \qquad\qquad [7\text{–}8]$$

To compute the t ratio, we have to first calculate the *pooled* standard deviation, in this case by substituting the sample values into Equation 7–9 as follows:

$$S_p = \sqrt{\frac{(16-1)(800)^2 + (14-1)(1000)^2}{16+14-2}}$$

$$= \sqrt{\frac{(15)(640,000) + (13)(1,000,000)}{28}}$$

$$= \sqrt{\frac{9,600,000 + 13,000,000}{28}}$$

$$= \sqrt{807,142.9}$$

$$= 898.4$$

This pooled standard deviation is now substituted into Equation 7–8 to determine the t ratio:

$$t = \frac{35,000 - 32,800}{898.4 \sqrt{\dfrac{1}{16} + \dfrac{1}{14}}}$$

$$= \frac{2,200}{898.4\,(0.37)}$$

$$= 6.62$$

The computed t value falls in the rejection region as shown in Figure 7.4. The null hypothesis is rejected, and we conclude that there is a difference in the heating cost of the two systems.

Example 7.4

An animal researcher is conducting an experiment on rice bran as a feedstuff for swine. A rice bran diet is more expensive than the regular diet. Farmers would prefer the less expensive diets. The animal researcher performs a cost analysis and finds that the greater expense incurred by using the rice bran diet is justified if it yields an average of at least 5 kg more in final weight gain than the regular diet. The researcher believes that the extra cost is justified because the rice bran diet meets this criterion. In a pilot study, the performance of pigs fed the rice bran diet and the regular diet is examined. The results are as follows. Do these data provide sufficient evidence to indicate that pigs on the rice bran diet gain more weight than on the regular diet? Let $\alpha = .01$.

Results

Rice bran diet	Regular diet
$n_1 = 10$	$n_2 = 8$
$\bar{x}_1 = 99.5$ kg	$\bar{x}_2 = 95.8$ kg
$s_1 = 3.8$ kg	$s_2 = 2.5$ kg

Because the rice bran diet is predicted to be superior to the regular diet, a one-tailed test is involved. The null and alternative hypotheses are

$$H_o: \mu_1 - \mu_2 \geq 5$$

$$H_1: \mu_1 - \mu_2 < 5$$

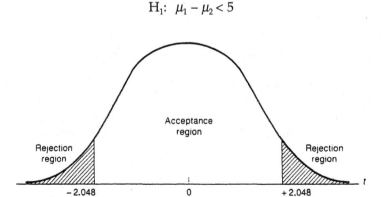

Figure 7.4 Region of acceptance and rejection for 28 *d.f.* and $\alpha = .05$.

At the .01 level of significance, the critical value with 16 *d.f.* is −2.583.

The test statistic is performed as follows by first computing the pooled standard deviation:

$$s_p = \sqrt{\frac{(10-1)(3.8)^2 + (8-1)(2.5)^2}{10+8-2}}$$

$$= \sqrt{\frac{9(14.4) + 7(6.25)}{16}}$$

$$= \sqrt{\frac{173.35}{16}}$$

$$= 3.29$$

The *t* statistic for this problem is

$$t = \frac{(\bar{x}_1 - \bar{x}_2) - 5}{s_p \sqrt{\dfrac{1}{n_1} + \dfrac{1}{n_2}}}$$

$$= \frac{(99.5 - 95.8) - 5}{3.29 \sqrt{\dfrac{1}{10} + \dfrac{1}{8}}}$$

$$= -.83$$

The calculated *t* ratio of −.83 is less than the critical value of −2.583. Therefore, we accept the null hypothesis and conclude that there is not sufficient evidence to indicate that the increase in final weight gain is at least 5 kg higher than the regular diet.

7.4   *Two-sample test of percentage: large sample*

When we make the assumption that we have taken a large random sample from the population of interest, the sample percentage is considered to be an unbiased estimator of the population percentage π. If we hypothesize that $\pi_1 = \pi_2$, the null hypothesis is written as:

$$H_o: \pi_1 - \pi_2 = 0$$

If the null hypothesis is true, the sample percentages p_1 and p_2 are the unbiased estimators of the population percentage. To obtain the best estimate of the value of π, we pool the weighted mean of the two sample proportions as follows:

$$\bar{p} = \frac{x_1 + x_2}{n_1 + n_2} \qquad [7\text{--}10]$$

where $\bar{p}$ = weighted mean of two sample proportions
x_1 = number of desired characteristic observed in sample one
x_2 = number of desired characteristic observed in sample two
n_1 = size of sample one
n_2 = size of sample two

For a large sample, the sampling distribution is approximately normal, and the test statistic is:

$$z = \frac{(p_1 - p_2) - (\pi_1 + \pi_2)}{s_{p_1 - p_2}} \qquad [7\text{--}11]$$

where p_1 = percentage of sample one with desired characteristic
p_2 = percentage of sample two with the desired characteristic
π_1 = true percentage of population one
π_2 = true percentage of population two
$s_{p_1 - p_2}$ = standard error of the percentage

Because we are hypothesizing that $\pi_1 - \pi_2 = 0$, the test statistic is

$$z = \frac{(p_1 - p_2) - 0}{s_{p_1 - p_2}} \qquad [7\text{--}12]$$

Based on our hypothesis that $\pi_1 = \pi_2 = 0$, the procedure for computing the standard error is to pool the sample percentages to obtain a best estimate of the common value of π.

The standard error of the percentage using the pooled estimate of p_1 and p_2 is computed as follows:

$$s_{p_1 - p_2} = \sqrt{\frac{\bar{p}(1-\bar{p})}{n_1} + \frac{\bar{p}(1-\bar{p})}{n_2}} \qquad [7\text{--}13]$$

This standard error is substituted for the standard error of percentage ($s_{p_1 - p_2}$) into Equation 7–12 to determine the z statistic. When the hypothesized difference between population percentages is other than zero, it is not correct to pool the sample data. In such a situation the test statistic is

$$z = \frac{(p_1 - p_2) - D_o}{\sqrt{\dfrac{p_1(1 - p_1)}{n_1} + \dfrac{p_2(1 - p_2)}{n_2}}}$$ [7-14]

where D_o is the difference between $\pi_1 - \pi_2$. The following example illustrates how we use Equation 7–14.

Example 7.5

An ornamental horticulturist is interested in determining the effects of PP333, a growth retardant, on stem elongation of beans and chrysanthemums. He believes that both plants respond equally to the growth retardant. A random sample of 36 bean plants and 40 chrysanthemums is selected from plots drenched with PP333 during the early stages of growth. The numbers of bean and chrysanthemum plants that show reduced stem elongation were 23 and 20, respectively. Test the validity of the researcher's statement, using a significance level of .05.

Solution

The null and alternative hypotheses are:

$$H_o: \pi_1 = \pi_2$$
$$H_1: \pi_1 \neq \pi_2$$

This is a two-tailed test because we are interested in determining whether the two percentages are equal or not. Because we have a large sample, the z distribution is used. The boundaries of the rejection region are ±1.96.

The decision rule is to accept H_o if the critical ratio falls between ±1.96, or to reject H_o if the critical ratio is less than –1.96 or greater than 1.96.

Because we are hypothesizing that $\pi_1 = \pi_2$, the pooled estimate of population percentage ($\bar{p}$) according to Equation 7–10 is

$$\bar{p} = \frac{23 + 20}{36 + 40}$$

$$= 0.57$$

The test statistic is

$$z = \frac{(p_1 - p_2)}{\sqrt{\frac{\bar{p}(1-\bar{p})}{n_1} + \frac{\bar{p}(1-\bar{p})}{n_2}}}$$

$$z = \frac{0.64 - 0.50}{\sqrt{\frac{0.57(1-0.57)}{36} + \frac{0.57(1-0.57)}{40}}}$$

$$= \frac{0.14}{0.11}$$

$$= 1.27$$

Because 1.27 falls within the acceptance region, we conclude that both beans and chrysanthemums respond equally to the growth retardant.

Example 7.6

An agricultural engineer believes that a new attachment designed by his firm for plowing reduces friction and is preferred by more farmers in those regions where the soil is of heavy clay than those regions where the soil is clay in nature.

To test this claim, the manufacturer asked 35 farmers in each of the two regions of the country to use the new attachment. It is observed that 22 farmers in the heavy-clay region and 18 of the clay region prefer the new attachment. At .01 level of significance, should we accept the agricultural engineer's assertion?

Solution

This is a right-tailed test because we are interested in determining whether the percentage of farmers who prefer the new attachment in the heavy-clay region is significantly greater than the percentage of farmers in the clay region. The null and alternative hypotheses are

$$H_o: \pi_1 \leq \pi_2$$

$$H_1: \pi_1 > \pi_2$$

Because we are asked to use .01 level of significance, and this is a single-tailed test, the z value from Appendix C is 2.33. The decision rule is

Accept H_o if the critical ratio ≤ 2.33.

or

Reject H_o if the critical ratio > 2.33.

The test statistic is given by Equation 7–12 as:

$$z = \frac{(p_1 - p_2)}{\sqrt{\dfrac{\bar{p}(1-\bar{p})}{n_1} + \dfrac{\bar{p}(1-\bar{p})}{n_2}}}$$

To calculate the critical ratio, we have to first compute the standard error of the percentage by pooling the percentage estimates as follows:

$$\bar{p} = \frac{22 + 18}{35 + 35}$$

$$= \frac{40}{70}$$

$$= 0.57$$

Next, the z statistic is computed:

$$z = \frac{0.63 - 0.51}{\sqrt{\dfrac{0.57(1-0.57)}{35} + \dfrac{0.57(1-0.57)}{35}}}$$

$$= \frac{0.12}{\sqrt{\dfrac{0.245}{35} + \dfrac{0.245}{35}}}$$

$$= \frac{0.12}{0.118}$$

$$= 1.02$$

Because the critical ratio falls within the acceptance region, the null hypothesis is accepted and we conclude that there is not a significant difference between the preference of the farmers in the two regions.

Example 7.7

Researchers of the International Wheat Center have developed a new variety of Mexican and Pakistani wheat called the Mexi-Pak 420. It is predicted that the percentage increase in yield of more than 2 bushels/acre will exceed the yield of the first generation of the high yielding variety Mexi-Pak 400 by more than 5%. To see whether the facts support this hypothesis, researchers collect data on the yield of Mexi-Pak 420 and Mexi-Pak 400 with the following results:

Variety	Sample size (acres)	Number of acres with 2 or more bushels/acre increase in yield
Mexi-Pak 420	50	40
Mexi-Pak 400	50	36

Test the hypothesis at .05 level of significance.

Solution

The null and alternative hypotheses are

$$H_o: \ p_1 - p_2 \le .05$$

$$H_1: \ p_1 - p_2 > .05$$

Given that we are using a level of significance of .05, and we have a large sample, the critical value of z is 1.64. The decision rule is

Accept H_o if the critical ratio is ≤ 1.64

or

Reject H_o if the critical ratio is > 1.64

Because it is hypothesized that the difference between the population percentages is not zero, we do not use the pooled estimate of the sample percentages, but instead use the formula given in Equation 7–14 to calculate the z value as follows:

$$z = \frac{(p_1 - p_2) - 0.05}{\sqrt{\dfrac{p_1(1-p_1)}{n_1} + \dfrac{p_2(1-p_2)}{n_2}}}$$

$$= \frac{(0.8 - 0.72) - 0.05}{\sqrt{\dfrac{(0.80)(0.20)}{50} + \dfrac{(0.72)(0.28)}{50}}}$$

$$= \frac{0.03}{0.08}$$

$$= 0.38$$

Because the computed z of 0.38 is less than 1.64, we do not reject the null hypothesis. It is concluded that the researchers' hypothesis is not true.

Chapter summary

The emphasis in this chapter is on the testing of hypotheses about differences between means and percentages. The procedure for testing a hypothesis to make relative comparisons between two population means or percentages is similar to the *single-sample hypothesis tests,* except for the computation of the critical ratio that uses the formula:

$$z = \frac{\bar{x}_1 - \bar{x}_2}{\sigma_{\bar{x}_1 - \bar{x}_2}}$$

Procedurally, we follow the same six steps in testing a hypothesis as used in Chapter 6. The basic assumption in hypothesis testing between two means or percentages is that there is no difference between the two populations. The alternative hypothesis may be either a *single-tailed* or a *two-tailed* test. This of course depends on the nature of the problem.

When we have small samples to test the differences between means, we *pool* the standard deviations of the samples to obtain the best estimate of the standard error or the difference between means.

Similarly, in the two-sample test of the percentage, if we hypothesize that $\pi_1 = \pi_2$ and if the null hypothesis is true, then p_1 and p_2 are each an unbiased estimator of the respective population percentage. To obtain the best estimate of the value of π, we pool the weighted mean of the two sample proportions. You should keep in mind that when the difference between the population percentages is not zero, we do not pool the sample percentage in the calculation of the critical ratio.

Case Study

Comparing Relative Loss of Tropical Forests from Logging

Tropical forests play an important role in climate regulation by storing carbon that would have otherwise been released to the atmosphere. In addition, these forests provide about a fifth of all wood used worldwide (Miller and Tangley, 1991). Their importance in the pharmaceutical industry cannot be overstated. A quarter of the world's drugs uses plant extracts from the tropical forest as ingredients (World Wide Fund for Nature, 1988). Unfortunately, tropical forests are faced with constant danger from a variety of sources. Economic activities such as logging, cattle ranching, growing crops, and constructing dams and highways have all contributed to the decline in tropical forests.

Of the world's estimated 3.4 billion hectares of forests in 1990, tropical forests covered 1.76 billion hectares. Net annual deforestation between 1981 and 1990 exceeded 2% in ten tropical countries, all located in either Asia or the Americas. Brazil and Indonesia accounted for the largest extent of forest area lost annually, with deforestation of 3.7 and 1.2 million hectares respectively (World Resources Institute, 1995). In other words, during the 1980s alone, the world lost 8% of its tropical forests — a decline from 1,910 million hectares in 1980 to 1,756 million hectares in 1990 (U.N. Food and Agriculture Organization, 1993).

The primary causes of deforestation vary among the countries and regions of the world. In Asia, the forests are threatened by commercial logging. In Africa, the rising demand for fuel wood, overgrazing by cattle, and logging all contribute to deforestation. In Latin America, much of the destruction is a result of cattle ranching, resettlement schemes, and major development projects, with some commercial logging (Park, 1992).

The world community has to recognize logging to be one of the major causes of the loss of tropical forests (Acharya, 1995). New and strengthened efforts are needed to arrest tropical deforestation from logging. To assess the amount of deforestation, we need to know if there are significant changes in the extent of forests and woodlands over time. The following table shows the data from tropical South America on the extent of forests and woodlands in 1980 and 1990.

Extent of Forest and Woodland (000 ha)

Countries	1980	1990
Bolivia	55,564	49,317
Brazil	597,816	561,107
Colombia	57,734	54,064
Ecuador	14,342	11,962
Paraguay	16,884	12,858
Peru	70,618	67,906
Venezuela	51,681	45,690

From World Resources Institute, *World Resources 1994 to 1995*, Oxford University Press, New York, 1995, Table 19.1.

Using the preceding data:

1. Compute the mean and the standard deviation for each time period.
2. Test the hypothesis that there is a significant difference in the mean of the forests and woodlands between the two time periods. Use $\alpha = .05$.

References

Acharya, A., Tropical forests vanishing, in Brown, L., Lenseen, N., and Kane, H., *Vital Signs 1995*, W. W. Norton, New York, 1995, 116.

Miller, K. and Tangley, L., *Trees of Life: Saving Tropical Forests and Their Biological Wealth*, Beacon Press, Boston, 1991.

Park, C., *Tropical Rainforests*, Routledge, New York, 1992.

U.N. Food and Agriculture Organization, Forest Resource Assessment 1990: Tropical Countries, Forestry Paper 112, Rome, 1993.

World Resources Institute, *World Resources 1994–1995*, Oxford University Press, New York, 1995.

World Wide Fund for Nature, Conservation of Tropical Forests, Special Report 1, Gland, Switzerland, 1988.

Review questions

1. An agronomist is conducting an experiment to determine the yield differences of two types of rice plants (varieties one and two). Both varieties are grown under identical conditions, and the data (tons/hectare) from the experiment on the two varieties are as follows. Test at $\alpha = .05$ if variety one has higher yield than variety two.

$$n_1 = 40 \text{ ha} \qquad n_2 = 35 \text{ ha}$$
$$\bar{x}_1 = 3.6 \text{ ton/ha} \qquad \bar{x}_2 = 3.2 \text{ ton/ha}$$
$$s_1 = 0.3 \text{ ton/ha} \qquad s_2 = 0.1 \text{ ton/ha}$$

2. In a recent study to determine the influence of various lactational treatments on the incidence of estrus and subsequent fertility of sows between two breeds, animal scientists took a random sample and found the following results.

Breed	Sample size	Number of sows showing incidence of estrus
A	59	22
B	65	35

Do these data provide sufficient evidence to indicate that the percentage of sows showing incidence of lactational estrus for the two breeds are different? Test at $\alpha = .01$.

3. Two new methods of processing tomatoes are proposed for use at a tomato processing plant. Method one requires new technology and is more expensive than method two. Managers of the plant favor the less costly method of processing tomatoes. An agricultural economist has done a cost benefit analysis of the new procedure. She believes that if the new process can handle an additional 2 ton/d of tomatoes, the adoption of the new process is economically justified. A study of the new and the old processes produces the following results. Do the data support the claim made by the agricultural economist that process one is justified? Test at $\alpha = .05$.

$$n_1 = 10 \qquad n_2 = 8$$
$$\bar{x}_1 = 22 \qquad \bar{x}_2 = 18$$
$$s_1 = 2.5 \qquad s_2 = 3.2$$

4. The Verdelli process of growing lemons has been practiced in Europe for over 50 years. Researchers are interested in learning whether the process would result in higher yields if grown on clay loam or loam types of soils in Florida. The results from a random sample of acres show the following:

Soil type	Sample size (acres)	Yield (field boxes/acre)	Standard deviation
Clay loam	100	340	20
Loam	120	320	15

Test at $\alpha = .05$ whether the yield from clay loam soil is greater than from the loam soil.

5. Animal researchers have been concerned with the negative effects of high-energy, low-fiber rations on milk production. To minimize the negative effects of such rations that produce high acidity in the rumen, it is hypothesized that sodium bicarbonate ($NaHCO_3$) supplements will reduce the acidity and increase milk production. In a recent study, the following results are obtained from two random samples of cows: one is fed the high-energy, low fiber ration; and the other is fed the same, but also containing a sodium bicarbonate supplement.

$$n_1 = 60 \qquad\qquad n_2 = 90$$
$$\bar{x}_1 = 16{,}150 \text{ lb} \qquad \bar{x}_2 = 17{,}030 \text{ lb}$$
$$s_1 = 485 \text{ lb} \qquad\quad s_2 = 572 \text{ lb}$$

By testing at $\alpha = .01$, is there any difference between the mean milk production of the two rations?

6. Gypsy moth defoliation of deciduous trees centers mostly in the northeastern parts of the U.S., but now the moth is moving westward. To eradicate the gypsy moth, a number of techniques are being adopted. Applications of chemical insecticide carbaryl and microbial insecticide *Bacillus thuringiensis* (Bt) are part of the eradication strategy. In a laboratory experiment, 160 larvae exposed to Bt produce a mortality rate of 35%. In a similar test of 150 larvae exposed to carbaryl, a mortality rate of 40% is observed.
   a. Formulate the null and alternative hypotheses.
   b. What is the test statistic for this test?
   c. Determine the standard error of the test statistic.
   d. At the .05 level of significance, what decision should be made concerning the hypothesis?

7. Researchers test the effects of two different antibiotics, penicillin and tetracycline, for a bacterial study on ruminants. Bacterial cultures from a random sample of 45 cows treated with penicillin and 35 cows with tetracycline produce 35 and 20 cultures with antibacterial activity, respectively. By testing at $\alpha = .01$, is there evidence to indicate a difference between the percentage of cows showing antibacterial activity for the two antibiotics?

8. Black vine weevil is a problem in nurseries throughout the country. Two different insecticides are used to measure the effectiveness of each on the weevil. Mortality rate is recorded after 72 h of exposure to the insecticide. The results of the experiment are given as follows:

Insecticide	Sample size	Number of weevils dead
Carbofuran	250	200
Diazinon	185	160

At the .05 level of significance, is there a difference between the mortality rates of the two different insecticides?

9. Chloroform, which in its gaseous form is suspected of being a cancer-causing agent, is reported to be present in minute quantities in wells and canal water sources in California. A study of wells and canals in two separate counties in California produce the following results:

$$n_1 = 40 \qquad n_2 = 35$$
$$\bar{x}_1 = 36 \ \mu g/l \qquad \bar{x}_2 = 42 \ \mu g/l$$
$$s_1 = 10 \ \mu g/l \qquad s_2 = 8 \ \mu g/l$$

At the .01 level of significance, is there a significant difference in the amount of chloroform found in the two water sources?

10. An animal scientist is interested in efficiency of production in sheep. It is hypothesized that there is no difference between the mean birth weight of purebred (Polled Dorset) sheep and a crossbreed (one-quarter Finn). A random sample of 35 purebred and 45 crossbred is tested. The mean birth weight for Dorset dams is 3.8 kg, with a standard deviation of 0.8 kg. The mean birth weight of the one-quarter Finn is 3.5 kg, with a standard deviation of 0.5 kg. At the .05 level of significance, is there a significant difference in the birth weight of the two groups?

References and suggested reading

Cochran, W. G. and Cox, G. M., *Experimental Designs*, 2nd ed., John Wiley & Sons, New York, 1957.

Dixon, W. J. and Massey, F. J., Jr., *Introduction to Statistical Analysis*, 4th ed, McGraw-Hill, New York, 1983, chap. 8.

Gibbons, J. D. and Pratt, J. W., P values: interpretation and methodology, *Am. Stat.*, 29, 20, 1975.

Gill, J. L., *Design and Analysis of Experiments: In the Animal and Medical Sciences*, Vol. 1, Iowa State University Press, Ames, IA, 1978, chap. 1.

Hamburg, M., *Statistical Analysis for Decision Making*, 4th ed., Harcourt Brace Jovanovich, New York, 1987, chap. 7.

Lehmann, E. L., *Testing Statistical Hypothesis*, John Wiley & Sons, New York, 1959.

Mason, R. D., *Statistical Techniques in Business and Economics*, 5th ed., Richard D. Irwin, Homewood, IL, 1982, chaps. 13 and 14.

Mendenhall, W., *Introduction to Probability and Statistics*, 6th ed., Duxbury, Boston, 1983, chap. 8.

Netter, J., Wasserman, W., and Whitmore, G. A., *Applied Statistics*, 3rd ed., Allyn & Bacon, Boston, 1987, chap. 13.

Snedecor, G. W. and Cochran, W. B., *Statistical Methods*, 7th ed., Iowa State University Press, Ames, IA, 1980, chap. 6.

Triola, M. F., *Elementary Statistics*, 4th ed., Benjamin/Cummings, Redwood City, CA, 1989.

chapter eight

Analysis of Variance

Learning objectives

The basic purpose of this chapter is to introduce you to analysis of variance, also known as ANOVA. The ANOVA is used to compare several populations in terms of their means and variances. After mastering this chapter and working through the review questions, you should be able to:

- Explain when and how ANOVA is used, and the underlying assumptions.
- Describe the techniques used in the one-factor and two-factor ANOVA with equal or unequal observations.
- Describe the Latin square design in agricultural research.
- Use the ANOVA techniques in testing hypotheses concerning the means of three or more populations.

8.1 Introduction

In the previous chapter you learned the techniques for evaluating differences between two populations. In this chapter you are introduced to procedures that allow you to compare three or more populations with ease. Although the procedures we developed in Chapter 7 could be adopted to test hypotheses about differences between two or more populations, they are not best suited for that purpose. For instance, if you were to compare 15 different varieties of rice to see if there are differences between the yield of these varieties, you would have to compute the mean and the standard deviation of each variety from a random sample, and then compare them with each other, two at a time. This approach is time consuming. To avoid such a cumbersome procedure for comparing differences between populations, a technique called the analysis of variance (ANOVA) is used. The techniques of ANOVA were developed by R. A. Fisher in the 1920s and have many uses in agriculture and other applied fields.

Before examining the techniques of ANOVA, it is important to be familiar with the specific terminology used. In ANOVA we use sample data to compare several levels of a variable of interest called a *treatment* to see if they achieve different results. A treatment in a broad sense refers to any variable that the researcher controls. Thus, for an agronomist several fertilizer concentrations applied to a particular crop is a treatment. Similarly, an animal scientist looks on several concentrations of a drug given to an animal species as a treatment. In agribusiness a treatment may refer to a particular advertising technique for a cereal product.

Another term you encounter in this context is the term *experimental unit*. An experimental unit is an entity that receives a treatment. For example, for an environmental scientist a river or a forest is an experimental unit. For an animal scientist the experimental unit may be a cow or a goat. For an agronomist it may be a plot of land, and for an agricultural engineer it may be a manufactured item.

ANOVA permits us to see whether or not the degree of variability in the sample data has contributed to the differences in the means of the populations of interest. In ANOVA we compare variances σ^2 (recall from Chapter 4 that the variance is the square of the standard deviation) of more than two populations to determine if the means are equal. This is accomplished by the use of a distribution called the *F distribution*. Like the *t*, the *F* distribution depends on the degrees of freedom (*d.f.*). When graphed, the *F* distribution is positively skewed, and its values range from zero to infinity. The shape of the curve, however, depends on the degrees of freedom. The *F* test is a ratio of the variance explained by treatments divided by the error or unexplained variance.

Because of the way in which the *F* statistic is defined, we have two kinds of degrees of freedom to consider. One is associated with the numerator and the other, with the denominator.

The *F* distribution is a continuous probability distribution, and therefore the values of *F* are provided by the areas under the curves. In Figure 8.1 some typical *F* curves are shown.

In general, we can say that the ANOVA techniques divide the total variance of a set of data into component parts. Each component part

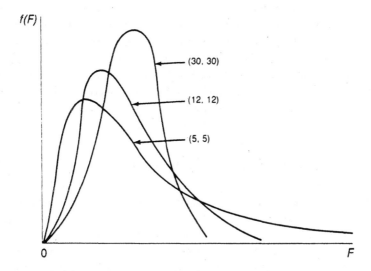

f(F)

(30, 30)

(12, 12)

(5, 5)

0   F

Figure 8.1 Various *F*-distribution probability curves.

has its own source of variation that the ANOVA procedure identifies and locates. Additionally, the magnitude of contribution of each source of variation is delineated by this procedure. Because this procedure divides the total variance into component parts, we may view ANOVA as a special kind of multiple regression (see Chapter 12).

The theoretical concepts of ANOVA are based on a set of assumptions:

1. The samples are drawn independently of each other and are random.
2. The populations of interest have distributions that approximate the normal distribution.
3. The variances (σ^2) of the populations are equal.

It is important to remember that not all assumptions are met perfectly in a situation. Departures from these basic assumptions should be carefully understood because they have serious implications. Cochran (1947) suggested that ANOVA results should be looked on as approximate instead of exact, because we seldom encounter experiments in which all the assumptions are met completely.

In the following sections, you can see ANOVA techniques applied to three experimental designs. In Section 8.2, one-factor ANOVA with the use of *completely randomized design* is discussed. We use this design when the experimental material or experimental units are relatively homogeneous. In Section 8.3, the two-factor ANOVA or the *randomized block design* experiments are presented. The two-factor designs are used when there is heterogeneity in experimental material. A way to accomplish the homogeneity of the experimental material is to divide the experimental material into various smaller subsets, called *blocks*, in such a way that variability within a block is small. In such a design the comparisons among the treatments are made within homogeneous blocks of material. In Section 8.4, the *Latin square* design is covered. This design is used when there are two sources of variation to control, and can be viewed as a method of assessing treatment effects when a double type of blocking is imposed on the experimental unit.

8.2   One-factor analysis of variance

Equal sample size. As the name implies, in this type of an experiment only a single factor varies, whereas others are kept constant. Additionally, we have equal sample size or equal replications in the experiment.

For example, an agronomist may vary the application rate of a fertilizer while keeping other production management factors such as insect control and water distribution constant.

In the one-factor ANOVA we partition the total variation as measured by the *total sum of squares (SST)* of the entire experiment into two components. One is the treatment variation, and the other is the variation as a result of experimental error. The *treatment variation (SSTr)* is due to variability between sample results, whereas variability within sample results is measured by the *error sum of squares (SSE)*. The methods of calculating each of the sums of squares are given next.

Total sum of squares (SST). In comparing several groups we add each score in the groups to compute a total mean (grand mean) for the group as a whole. To determine the variability of each score from the mean, we can calculate the total sum of squares that is the difference between each score and the grand mean. The total sum of squares is calculated as follows:

$$SST = \sum_j \sum_i \left(x_{ij} - \bar{\bar{x}} \right)^2 \qquad [8\text{--}1]$$

where SST = total sum of squares
x_{ij} = value of observation in *i*th row (replications) and *j*th column (treatment)
$\bar{\bar{x}}$ = grand mean (read as *x double bar*)
Σ = summation sign

Mathematically it can be shown that *SST* is the summation of the *SSTr* and *SSE*. That is,

$$SST = SSTr + SSE \qquad [8\text{--}2]$$

Treatment sum of squares (SSTr). This term refers to the difference between each group mean and the grand mean, and is calculated as follows:

$$SSTr = \sum_j r\left(\bar{x}_j - \bar{\bar{x}}\right)^2 \qquad [8\text{--}3]$$

where $SSTr$ = sum of squares of treatments
$\qquad r$ = number of replications (rows) per treatment
$\qquad \bar{x}_j$ = mean of the jth column or treatment
$\qquad \bar{\bar{x}}$ = the grand mean

Error sum of squares (SSE). This sum of squares refers to the variability within samples. It measures the difference between each individual score and the mean of the group from which the score comes. The error sum of squares is noted mathematically as:

$$SSE = \sum_j \sum_i \left(x_{ij} - \bar{x}_j\right)^2 \qquad [8\text{--}4]$$

where $\quad x_{ij}$ = value of observation in the ith row and jth column (treatment)
$\qquad \bar{x}_j$ = mean of the jth column (treatment)

After we have computed the sums of squares, we are then able to perform the F *test* to decide whether to reject H_o: $\mu_1 \ldots = \mu_k$. The F test in mathematical notation is:

$$F = \frac{SSTr / k - 1}{SSE / N - k} \qquad [8\text{--}5]$$

As seen in Equation 8–5, this statistic has an F distribution with $k - 1$ and $N - k$ degrees of freedom, $(d.f.)^*$

where $\quad k$ = number of treatments
$\qquad N$ = number of observations

---

* As mentioned in the discussion on the t distribution, degrees of freedom may be interpreted as the number of independent deviations of the form $x - \bar{x}$ present in the calculation of the standard deviation. In this particular situation, the k differences $x_j - \bar{x}$ must sum to zero so only $k - 1$ of them are able to vary *freely*. With N number of observations, $x_{ij} - \bar{x}_j$ in each of the k samples must add to zero. Therefore, this leaves us with $N - k$ differences that vary *freely*.

Notice that in Equation 8–5 we divide the numerator and the denominator by their respective degrees of freedom. In the ANOVA, we refer to a sum of square divided by the appropriate degree of freedom as a *mean square*. The F statistic may then be written as:

$$F = \frac{\text{treatment mean square}}{\text{error mean square}} \qquad [8\text{--}6]$$

If the numerator and the denominator are nearly equal, the variance ratio is close to one. This supports the null hypothesis of equal group means. If the treatment mean square is greater than the error mean square, the variance ratio is going to be much larger than one. In such a situation, the hypothesis of equal group means is not supported. The size of the observed F that serves as the basis for rejecting the null hypothesis of equal population variances depends on the critical value selected. The critical value of course depends on the level of significance chosen by the researcher or the manager. The one-way ANOVA technique is summarized in Table 8.1.

Table 8.1   **Table for One-Way ANOVA**

Variation	Degrees of freedom	Sum of square	Mean square	F
Treatment	$k - 1$	SSTr	$MSTr = SSTr/k - 1$	MSTr/MSE
Error	$N - k$	SSE	$MSE = SSE/N - k$	
Total	$N - 1$	SST		

Let us illustrate the one-factor ANOVA technique with the following example.

Example 8.1

Rivers around the world serve a number of vital functions. Their water is used not only for drinking but also for irrigation, transportation, recreation, and sewage disposal. With an increase in the population and rapid industrialization of many countries, these rivers are being polluted at an alarming rate. To find out the pollution level in a river, an environmental scientist takes four random samples of water at three different locations. Samples from location one are taken upstream from an industrial plant located on the shore of the river. Location two samples are from waters just below the industrial plant. Location three samples are taken two miles outside the city where its treated sewage is released into the river. Test at $\alpha = .05$ to determine if there is a difference in the quantity of dissolved oxygen in the samples.

Location	Mean dissolved oxygen content in parts per million			
1	6.8	6.5	6.3	6.6
2	3.9	4.8	4.2	5.0
3	4.7	4.9	4.1	3.9

Solution

We follow a step-by-step procedure in performing the test again.

Step one — State the null and alternative hypothesis as follows:

$$H_o: \mu_1 = \mu_2 = \ldots = \mu_k$$

$$H_1: \text{ not all } \mu_j \text{ are equal}$$

Step two — Calculate the treatment totals and their respective means, and the grand mean.

The grand mean is

$$\bar{\bar{x}} = \frac{26.2 + 17.9 + 16.9}{4(3)}$$

$$= \frac{61}{12}$$

$$= 5.08$$

Mean Dissolved Oxygen Content in the Different Locations

	Location		
Sample	1	2	3
1	6.8	3.9	4.2
2	6.5	4.8	4.7
3	6.3	4.2	4.1
4	6.6	5.0	3.9
Total	26.2	17.9	16.9
Mean	$\bar{x}_1 = 6.55$	$\bar{x}_2 = 4.48$	$\bar{x}_3 = 4.23$

Step three — Calculate the sum of squares for the treatment, error, and total as follows:

$$SSTr = \sum_j r\left(x_j - \overline{\overline{x}}\right)^2$$

$$= 4(6.55 - 5.08)^2 + 4(4.48 - 5.08)^2 +$$

$$4(4.23 - 5.08)^2$$

$$= 13.03$$

The procedure to compute the *SSE* is illustrated in Table 8.2.

Table 8.2 **Calculation of the Error Sum of Squares (SSE)**

i	$(x_{i1} - \overline{x}_1)^2$	$(x_{i2} - \overline{x}_2)^2$	$(x_{i3} - \overline{x}_3)^2$
1	$(6.8 - 6.55)^2 = .0625$	$(3.9 - 4.48)^2 = .3364$	$(4.2 - 4.23)^2 = 0.009$
2	$(6.5 - 6.55)^2 = .0025$	$(4.8 - 4.48)^2 = .1024$	$(4.7 - 4.23)^2 = .2209$
3	$(6.3 - 6.55)^2 = .0625$	$(4.2 - 4.48)^2 = .0784$	$(4.1 - 4.23)^2 = .0169$
4	$(6.6 - 6.55)^2 = .0025$	$(5.0 - 4.48)^2 = .2704$	$(3.9 - 4.23)^2 = .1089$
Total	.13	.7876	.3476

$$SSE = \sum_j \sum_i \left(x_{ij} - \overline{x}_j\right)^2$$

$$= 0.13 + 0.786 + 0.3476$$

$$= 1.27$$

As pointed out earlier, the *SST* can be calculated by two methods. Both procedures are illustrated:

$$SST = \sum_j \sum_i \left(x_{ij} - \overline{x}_j\right)^2$$

$$= (6.8 - 5.08)^2 + (6.5 - 5.08)^2 + (6.3 - 5.08)^2 + \ldots +$$

$$(3.9 - 5.08)^2$$

$$= 14.29$$

On the other hand, *SST* is equal to the *SSTr* and *SSE*:

$$SST = SSTr + SSE \qquad [8\text{–}7]$$

$$SST = 13.03 + 1.27$$

$$= 14.30$$

The small difference in the value of the *SST* calculated using Equation 8–7 results from rounding.

Step four — Set up an ANOVA table as shown in Table 8.3.

Table 8.3 **ANOVA Table for Mean Dissolved Oxygen Content**

Variation	Degrees of freedom	Sum of squares	Mean square	F
Treatment	3 – 1 = 2	12.97	6.49	6.49/0.14 = 46.35
Error	12 – 3 = 9	1.27	0.14	
Total	12 – 1 = 11	14.24		

Step five — We are now able to make a statistical decision: should we reject the null hypothesis? The critical value of F for $\alpha = .05$, and 2 and 9 *d.f.* from Appendix F is 4.26. Because 46.35 is greater than 4.26, the null hypothesis is rejected, and the environmentalist may conclude that the three water samples from the river have different quantities of dissolved oxygen.

Shortcut formulas for computation of sum of squares. The method of calculating the sum of squares in the previous example provides the rationale for the ANOVA procedure. In this section you are introduced to a shortcut method of calculating the sum of squares. The computational formula for the different sums of squares is given as follows:
The *SSTr* in mathematical notation is

$$SSTr = \sum_j \frac{T_j^2}{r} - C \qquad [8\text{--}8]$$

where T_j = total of r observations (replications) in jth column
C = correction term that in mathematical notation is

$$C = \frac{T^2}{rc} \qquad [8\text{--}9]$$

where T = grand total of all rows (r) and columns (c) observations.
The *SSE* in mathematical notation is

$$SSE = \sum_j \sum_i x_{ij}^2 - \sum_j \frac{T_j^2}{r} \qquad [8\text{--}10]$$

Finally the *SST* in mathematical notation is

$$SST = \sum_j \sum_i x_{ij}^2 - C \qquad\qquad [8\text{–}11]$$

where x_{ij} = value of observation in *i*th row of *j*th column.

The use of the shortcut computation method is shown in Example 8.1. Table 8.4 displays the data used in this example.

Table 8.4   **Calculation Needed for the F Ratio Using the Shortcut Method**

Samples	Location 1	Location 2	Location 3
1	6.8	3.9	4.2
2	6.5	4.8	4.7
3	6.3	4.2	4.1
4	6.6	5.0	3.9
Total	26.2	17.9	16.9
Mean	$\bar{x}_1 = 6.55$	$\bar{x}_2 = 4.48$	$\bar{x}_3 = 4.23$

Note: Grand mean = (6.55 + 4.48 + 4.23)/3 = 5.08.
Sum of totals (*T*) = 26.2 + 17.9 + 16.9 = 61.0

$$C = \frac{(61.0)^2}{(3)(4)}$$
$$= 310.08$$

$$SSTr = \frac{[(26.2)^2 + (17.9)^2 + (16.9)^2]}{4} - 310.08$$
$$= 13.04$$

$$SSE = (6.8)^2 + \ldots + (3.9)^2 - \frac{[(26.2)^2 + (17.9)^2 + (16.9)^2]}{4}$$
$$= 324.38 - 323.11$$
$$= 1.27$$

$$SST = [(6.8)^2 + (6.5)^2 + \ldots + (3.9)^2] - 310.08$$
$$= 324.38 - 310.08$$
$$= 14.3$$

After we have determined the sum of squares for the total, treatment, and error using this shortcut formula, we set up an ANOVA table and make a statistical decision as was shown previously in Example 8.1.

Unequal sample size. The preceding section deals with the one-factor ANOVA when there are samples of equal size. Sometimes it is not possible to have samples of equal size.

For instance, animal scientists may be performing experiments on different breeds of animals where the number of a particular breed may be limited to a few. In such situations we may not have a sample of equal size.

In another situation, a researcher may start with an experiment that has equal samples. However, for previously unseen reasons, elements of the experimental unit may be destroyed or lost in the process. If such conditions prevail, how do we perform an ANOVA test? Although the general format of one-factor ANOVA of equal sample size is the same as those for unequal sample size, the shortcut computational formulas are slightly different. The shortcut computational formulas are given next.

The *correction term* as well as the sum of squares for treatment, error, and total in mathematical notation are

$$C = \frac{T^2}{\sum_j r_j} \qquad\qquad [8\text{--}12]$$

where T = grand total
         r_j = number of observations in jth column

$$SSTr = \sum_j \frac{T_j^2}{r_j} - C \qquad\qquad [8\text{--}13]$$

$$SSE = \sum_j \sum_i x_{ij}^2 - \sum_j \frac{T_j^2}{r_j} \qquad\qquad [8\text{--}14]$$

$$SST = \sum_j \sum_i x_{ij}^2 - C \qquad\qquad [8\text{--}15]$$

T_j is the total of observations in the *j*th column. All other terms are as previously defined in conjunction with experiments of equal-sized samples. The computational differences are illustrated in the following example.

Example 8.2

An agricultural economist is interested in the profit rate of three different sizes farms (small, medium, and large). A random sample of 15 farms in a region produces the following profit-to-investment ratios. Use a .01 level of significance to test whether there is a difference in the profit rates of the small, medium, and large farms.

Profit to Investment Ratios of Three
Different Size Farms (%)

Small	Medium	Large
4	5	7
6	6	4
5	7	4
4	6	6
	5	7
		5

Solution

The null and alternative hypotheses are

$$H_o: \mu_1 = \mu_2 = \mu_3$$

$$H_1: \text{ not all } \mu_j \text{ are equal}$$

Before we compute the sum of squares, the totals for the different size farms are computed as shown in Table 8.5.

Table 8.5   **Profit-to-Investment Ratio of Three
Different Size Farm (%)**

	Small	Medium	Large
	4	5	7
	6	6	4
	5	7	4
	4	6	6
		5	7
			5
Total	19	29	33

Note: Grand total = 19 + 29 + 33 = 81.

The treatment, error, and total sums of squares are computed by using Equations 8–11 to 8–14.

$$C = \frac{T^2}{\sum_j r_j}$$

$$= \frac{(81)^2}{15} = 437.40$$

$$SSTr = \sum \frac{T_j^2}{r_j} - C$$

$$= \frac{(19)^2}{4} + \frac{(29)^2}{5} + \frac{(33)^2}{6} - 437.40$$

$$= 439.95 - 437.40$$

$$= 2.55$$

$$SSE = \sum_j \sum_i x_{ij}^2 - \sum_j \frac{T_j^2}{r_j}$$

$$= (4)^2 + (6)^2 + \ldots + (5)^2 - \left[\frac{(19)^2}{4} + \frac{(29)^2}{5} + \frac{(33)^2}{6} \right]$$

$$= 455 - 439.95$$

$$= 15.05$$

$$SST = \sum_j \sum_i x_{ij}^2 - C$$

$$= (4)^2 + (6)^2 + \ldots + (5)^2 - 437.40$$

$$= 455 - 437.40$$

$$= 17.60$$

As shown earlier, the *SST* is the sum of treatment and error sums of squares. Therefore:

$$SST = SSTr + SSE$$

$$= 2.55 + 15.05$$

$$= 17.60$$

After having determined the sums of squares, we now can determine the degrees of freedom associated with treatment and error sources of variation. They are determined as follows:

$$\text{Treatment } d.f. = k - 1 = 3 - 1 = 2$$

$$\text{Error } d.f. = N - k = 15 - 3 = 12$$

Given our calculations, we are now able to construct the ANOVA table as shown in Table 8.6.

Table 8.6   **ANOVA Table for the Rate of Profit of Three
Different Size Farms**

Source of variation	Degrees of freedom	Sum of squares	Mean square	F
Treatment	2	2.55	1.28	1.28/1.25 = 1.02
Error	12	15.05	1.25	
Total	14	17.60		

The F ratio of 1.02 computed in Table 8.6 is compared with the critical value of F given in Appendix F. The critical value for $\alpha = .01$, and 2 $d.f.$ for the treatment and 12 $d.f.$ for the error is

$$F_{.05} (2,12) = 6.93$$

Because 1.02 is less than 6.93, we accept the null hypothesis (H_o) and conclude that there is no difference in the profit rate of the three farms of different size.

Before we move to the discussion of two-factor ANOVA, you should be aware that the one-factor ANOVA is conventionally known as a *completely randomized experimental design*. This expression is based on the fact that treatments are applied at random to individual experimental units.

8.3   *Two-factor analysis of variance*

The objective of the two-factor ANOVA is to isolate the effect of not one, but two variables of interest in an experiment. There are two types of two-factor ANOVA. The first one is the *randomized block design* and the second is the *completely randomized design*. We shall first elaborate on the randomized block design where inferences are made with respect to the factor that is central to the experiment; and then discuss the use of the completely randomized design where inferences are made about both factors, namely, the factor in which the experimenter is interested and one additional factor that may play an important role in the results.

In agriculture we use the term block to mean parcels of land, breed of animal, or any other factor in the analysis that may account for some of the variation in the SST and reduce the unexplained or error variation. When we use randomized block design, the treatments are randomly assigned to units within each block. Therefore, when an agronomist performs a fertilizer trial test with block design, he or she makes sure that the best fertilizer is applied to the best as well as the worst soils. Further, the design isolates through blocking the effect of one other extraneous factor such as the irrigation method, pest control practice, or soil quality.

The experimental layout of such a design is simple. An agronomist who performs a test on the yield of corn resulting from different rates of application of fertilizer may consider the quality of soil present on the farm as an additional factor in the test. He divides a field into several blocks of equal size according to soil quality, and then divides each block into equal plots as shown in Figure 8.2. One plot from each block is then randomly selected for fertilizer application. The corn yield data from the plots receiving different rates of application are then recorded. With the randomized block design we may test the hypothesis that there is no difference in the population mean yield from different fertilizer treatments. The null hypothesis for this problem is the same as in the case of one-factor ANOVA, and is written as:

$$H_o: \ \mu_1 = \mu_2 = \mu_3$$

Similarly, an animal researcher in testing feed rations, for example, may use randomized block design to test the hypothesis that the mean weight gain from different rations is the same. Suppose the animal

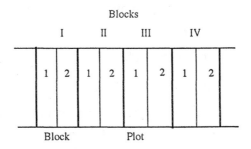

1 = Fertilizer applied
2 = No fertilizer

Figure 8.2 Field map of a fertilizer trial test.

researcher experiments with four different rations and 40 steers on which to perform the experiment. Using this design, she would first group (block) the animals in such a way that the steers within each group (block) are as homogeneous with respect to age, initial weight, etc. as possible. Assume that she is able only to separate the steers into four blocks according to their initial weight; she would then have ten steers in each block. The steers in a block are then randomly assigned to the four different rations. The data from such blocks are then collected and analyzed.

Table 8.7 generalizes the display of data generated from a two-factor ANOVA experiment.

Table 8.7 **Two-Factor ANOVA for Sample Data**

Blocks	Treatment 1	2	3	....	k	Total	Mean
1	x_{11}	x_{12}	x_{13}		x_{1k}	T_1	$\bar{x}_{1.}$
2	x_{21}	x_{22}	x_{23}		x_{2k}	T_2	$\bar{x}_{2.}$
3	x_{31}	x_{32}	x_{33}		x_{3k}	T_3	$\bar{x}_{3.}$
4	x_{41}	x_{42}	x_{43}		x_{4k}	T_4	$\bar{x}_{4.}$
.	.	.	.		.	.	.
.	.	.	.		.	.	.
n	x_{n1}	x_{n2}	x_{n3}		x_{nk}	$T_{n.}$	$\bar{x}_{n.}$
Total	$T_{.1}$	$T_{.2}$	$T_{.3}$		$T_{.k}$	$T_{..}$	
Mean	$\bar{x}_{.1}$	$\bar{x}_{.2}$	$\bar{x}_{.3}$		$\bar{x}_{.k}$		$\bar{x}_{..}$

The notation used in two-factor ANOVA is similar to that used in one-factor ANOVA:

k = number of treatments

n = number of blocks

T_i = total of the ith block

$\bar{x}_i$ = mean of the ith block

$T_{..}$ = grand total that is found by adding either row totals or column totals

In the two-factor ANOVA, we partition the total variation as measured by the *SST* into three components, namely, the *SSTr*, replication or block (*SSB*), and *SSE*. The formula for computing each of the total sum of squares is

Total sum of squares:

$$SST = \sum_j \sum_i x_{ij}^2 - C \qquad [8\text{--}15]$$

or we may compute *SST* by:

$$SST = \sum_j \sum_i (x_{ij} - \bar{x}..)^2 \qquad [8\text{--}16]$$

where x_{ij} = value of observation of treatment i in block j
$\bar{x}_{..}$ = grand mean

C = correction term that is $\dfrac{T^2_{...}}{bk}$

$T_{..}$ = total of all observations
b = number of blocks similar to rows
k = number of treatments similar to number of columns

Treatment sum of squares:

$$SSTr = \sum \frac{T^2_{.j}}{b} - C \qquad [8\text{--}17]$$

or we may compute the *SSTr* using Equation 8–18 given as shown:

$$SSTr = \sum_j \sum_i (\bar{x}_{.j} - \bar{x}_{..})^2 \qquad [8\text{--}18]$$

where $T_{.j}$ = total of all observations in jth treatment
b = number of blocks
$\bar{x}_{.j}$ = sample mean of jth treatment
$\bar{x}_{..}$ = grand mean

Block sum of squares:

$$SSB = \sum \frac{T^2_{i.}}{k} - C \qquad [8\text{--}19]$$

or we may compute the *SSB* as follows:

$$SSB = \sum_j \sum_i (\bar{x}_{i.} - \bar{x}_{..})^2 \qquad [8\text{--}20]$$

where $T_{i.}$ = total of all observations in block i
k = number of treatments
C = correction factor
$\bar{x}_{i.}$ = sample mean of ith block
$\bar{x}_{..}$ = grand mean

Error sum of squares:

$$SSE = SST - SSTr - SSB$$

After we have computed the sum of squares, we are then able to compute the mean squares and the ratio as shown in Table 8.8.

Table 8.8   **ANOVA Table for Randomized Block Design with *k* Treatments and *b* Blocks**

Source of variation	Degrees of freedom	Sum of squares	Mean square	F
Between treatments	$k - 1$	SSTr	$SSTr/(k - 1)$	$\dfrac{SSTr/(k-1)}{SSE/(k-1)(b-1)}$
Between blocks	$b - 1$	SSB	$SSB/(b - 1)$	
Error	$(k - 1)(b - 1)$	SSE	$SSE/(k - 1)(b - 1)$	
Total	$bk - 1$	SST		

The following example illustrates the step-by-step computational procedure for a randomized block design.

Example 8.3

An agricultural researcher wishes to determine the effect of three different irrigation methods ($k = 3$) on the yield of tomatoes. An experimental field is divided into five blocks and a random sample of 5 acres each under different irrigation methods. The following results in boxes of tomatoes per acre were produced. Test at $\alpha = .05$ to see if there is any significant difference in the mean yield from different irrigation methods.

Yield in Boxes per Acre from Three Different
Irrigation Methods

Blocks	Irrigation method		
	Drip	Sprinkler	Flood
Parcel I	22	19	16
Parcel II	19	19	18
Parcel III	20	16	19
Parcel IV	23	20	13
Parcel V	21	17	15

Solution

Given the preceding information, we are able to test the hypothesis that the mean yield from the three different methods of irrigation are equal. The null and alternative hypotheses are

$$H_o: \mu_1 = \mu_2 = \mu_3$$

$$H_1: \text{ not all } \mu\text{'s are equal}$$

The sums of squares for treatment, block, error, and total are calculated from the information presented in Table 8.9.

Table 8.9 **Yield in Boxes from 15 Acres Irrigated by Three Different Methods on Five Different Parcels of Land**

	Irrigation method			Row sum
Blocks	Drip	Sprinkler	Flood	$(T_{i.})$
1	22	19	16	57
2	19	19	18	56
3	20	16	19	55
4	23	20	13	56
5	21	17	15	53
Column sum $(T_{.j})$	105	91	81	
Grand total $(T_{..})$				277

The corrected term that is needed in the calculation of the sum of squares is computed as follows:

$$C = \frac{T_{..}^2}{bk}$$

$$= \frac{(277)^2}{15}$$

$$= 5115.26$$

$$SST = \sum_j \sum_i x_{ij}^2 - C$$

$$= (22)^2 + (19)^2 + \ldots + (15)^2 - 5115.26$$

$$= 5217 - 5115.26$$

$$= 101.74$$

$$SSTr = \sum \frac{T_{\cdot j}^2}{b} - C$$

$$= \frac{[(105)^2 + (91)^2 + (81)^2]}{5} - 5115.26$$

$$= 5173.40 - 5115.26$$

$$= 58.14$$

$$SSB = \sum \frac{T_{i\cdot}^2}{k} - C$$

$$= \frac{[(57)^2 + (56)^2 + (55)^2 + (56)^2 + (53)^2]}{3} - 5115.26$$

$$= 5118.33 - 5115.26$$

$$= 3.07$$

$$SSE = SST - SSTr - SSB$$

$$= 101.74 - 58.14 - 3.07$$

$$= 40.5$$

Table 8.10   **ANOVA Table for Two-Factor Irrigation Problem**

Source of variation	Degrees of freedom	Sum of squares	Mean square	F
Irrigation method	$3 - 1 = 2$	58.14	29.07	$29.07/5.07 = 5.73$
Parcels of land	$5 - 1 = 4$	3.07	0.77	
Error	$(3-1)(5-1) = 8$	40.53	5.07	
Total	14	101.74		

The computed F ratio of Table 8.10 is compared with the critical value of F given in Appendix F. The critical value at $\alpha = .05$ (2 numerator *d.f.* and 8 denominator *d.f.*) is

$$F_{.05}(2, 8) = 4.46$$

Because the computed F value of 5.73 is greater than 4.46, the null hypothesis (H_o) is rejected, and we conclude that there is a significant difference in mean yield from three different irrigation methods.

Analysis of a completely randomized design. The second type of two-factor ANOVA is the *completely randomized design*. In this design the researcher is able to make inferences about both factors. For instance, in Example 8.3 we were only interested in the effect different irrigation methods had on the yield. The randomized experimental design allows us to test two different null hypotheses:

1. H_o: No difference in the mean yield among different irrigation methods.
2. H_o: No difference in the mean yield among parcels of land with different soil quality.

Because we have already tested the first hypothesis using the randomized block design, we can use the same set of experimental data to construct the ANOVA table for the second null hypothesis. Table 8.11 shows the sum of square, and the F ratio for testing the hypothesis of no difference in the mean yield among parcels of land with different quality soils. The computed F value of 0.15 is compared with the critical value of F given in Appendix F. The critical F at $\alpha = .05$ and 4 numerator *d.f.* and 8 denominator *d.f.* is

$$F_{.05} (4, 8) = 3.84$$

Table 8.11 ANOVA Table for a Two-Factor Randomized Design and Sample Results Relating the Yield Differences due to Irrigation Methods and Soil Quality

Source of variation	Degrees of freedom	Sum of squares	Mean square	F
Irrigation method	$3 - 1 = 2$	58.14	29.07	$0.77/5.07 = 0.15$
Parcels of land	$5 - 1 = 4$	3.07	0.77	
Error	$(3-1)(5-1) = 8$	40.53	5.07	
Total	14	101.74		

Because the computed F ratio is less than the critical F, the null hypothesis is accepted; and we conclude from these data that there is no significant difference in yield among the soils of different quality.

There are some distinct advantages to using the randomized block design. First, if a researcher is working with experimental material that is heterogeneous, this design offers the ability to study the interaction, or joint effects, of the two variables of interest as well as their separate effects. Second, the variance in the randomized block design is smaller than the completely randomized design, especially when there is more variation among the replications than within replications. Third, the randomized block design reduces experimental error that leads to increase in power. Increase in power refers to the likelihood of rejecting the false hypothesis. The increase in power results from a larger F value in the randomized block design.

This has been a relatively simple explanation of this experimental design. For further elaboration of the technique and design, refer to Fisher (1950), Cochran and Cox (1968), Gomez and Gomez (1984), and Hoshmand (1994).

8.4   Latin square design

In the previous two sections we elaborated on one-factor and two-factor experimental designs. When several factors are to be included in an experiment, and each factor has several categories, we use what is known as the *factorial experiment*.

To conduct a factorial experiment we may need to have a great number of test subjects. For example, an animal scientist conducting an experiment to evaluate the effects of four different rations on weight gain may choose a design that takes into consideration other factors that might explain weight-gain variability, such as the initial weight and age of animals. Thus, the animal scientist might block the sample subjects into four initial weight and four age categories. In such a situation where the scientist wants replications of the four treatments with two blocking factors, each also having four categories, $4 \times 4 \times 4 \times n = 64n$ subjects are required. In experiments with a large number of factors and numerous categories of each factor, the number of subjects required is very large indeed.

The *Latin square design* permits us to reduce the number of subjects required, and yet assesses the relative effects of various treatments when a two-directional blocking restriction is imposed on the experimental units. Basically, this design requires that each factor have the same number of categories. In the case of the feed ration experiment, we need only 16 instead of 64 subjects.

The design layout of this experiment is such that every treatment occurs once in each category of both blocking factors. The blocking factors are commonly referred to as *row blocking* and *column blocking*. The randomized layout of a Latin square design with four treatments, A, B, C, and D, is shown as follows:

	Blocking factor			
	I	II	III	IV
1	A	B	C	D
2	B	A	D	C
3	C	D	A	B
4	D	C	B	A

Notice that in the preceding illustration we have two blocking factors, each identified by either Roman numerals or arabic numbers. The letters A, B, C, and D are the treatments and appear in each row and column only once. The computational procedures of the Latin square design are illustrated in the following example.

Example 8.4

An agronomist tests the effect of four different fertilizers, A, B, C, and D, on the yield of wheat. He arranges plants in an experimental field in a Latin square design. Thus, rows and columns in the table are rows and columns in the field. The yields in tons per hectare are as shown. Test at the 5% level that the mean yields are not equal for the four fertilizers.

	Column			
Row	I	II	III	IV
1	A 4.8	B 5.2	C 4.9	D 4.7
2	B 4.6	A 5.0	D 4.8	C 5.1
3	C 4.5	D 3.8	A 5.0	B 5.3
4	D 3.9	C 5.4	B 4.9	A 5.0

Solution

The null and alternative hypotheses are

$$H_o: \mu_1 = \mu_2 = \mu_3 = \mu_4$$

$$H_1: \text{ not all } \mu\text{'s are equal}$$

The calculation of the sum of squares is the same as before. In Latin square design we have to compute the totals for the row, column, and treatment as shown in Table 8.12.

Table 8.12 **Row, Column, and Treatment Totals for the Wheat Yield Experiment**

	Column				Row	Treatment
Row	I	II	III	IV	total	total
1	A 4.8	B 5.2	C 4.9	D 4.7	19.6	A 20.3
2	B 4.6	C 5.0	D 4.8	A 5.1	19.5	B 20.0
3	C 4.5	D 3.8	A 5.0	B 5.3	18.6	C 19.4
4	D 3.9	A 5.4	B 4.9	C 5.0	19.2	D 17.2
Column total	17.8	19.4	19.6	20.1		
Grand total						76.9

From Table 8.12, we compute the correction factor, and the various sums of squares as follows:

$$C = \frac{T_{..}^2}{rc}$$

$$= \frac{(76.9)^2}{16}$$

$$= 369.60$$

Total sum of squares:

$$SST = \sum x_{ijk}^2 - C$$

$$= [(4.8)^2 + (5.2)^2 + \ldots + (5.0)^2] - 369.60$$

$$= 2.95$$

Treatment sum of squares:

$$SSTr = \frac{\sum T_r^2}{k} - C$$

$$= \frac{[(20.3)^2 + (20.0)^2 + (19.4)^2 + (17.2)^2]}{4} - 369.60$$

$$= \frac{1484.29}{4} - 369.60$$

$$= 1.47$$

Row sum of squares:

$$Row\ SS = \frac{\sum T_{i.}^2}{r} - C$$

$$= \frac{[(19.6)^2 + (19.5)^2 + (18.6)^2 + (19.2)^2]}{4} - 369.60$$

$$= \frac{1479.01}{4} - 369.60$$

$$= 369.75 - 369.60$$

$$= 0.15$$

Column sum of squares:

$$\text{Column } SS = \frac{\sum T_{\cdot j}^2}{c} - C$$

$$= \frac{[(17.8)^2 + (19.4)^2 + (19.6)^2 + (20.1)^2]}{4} - 369.60$$

$$= \frac{1481.37}{4} - 369.60$$

$$= 369.75 - 369.60$$

$$= 0.74$$

Error sum of squares:

$$SSE = TSS - SSTr - \text{Row } SS - \text{Column } SS$$

$$= 2.95 - 1.47 - 0.15 - 0.74$$

$$= 0.59$$

The mean square for each source of variation is computed by dividing the sum of squares with the corresponding degrees of freedom, shown in Table 8.13 as follows.

Table 8.13 **ANOVA Table of the Effect of Four Different Fertilizers on Yield of Wheat Using a Latin Square Design**

Source of variation	Degrees of freedom	Sum of squares	Mean square	F
Row	3	0.15	0.050	
Column	3	0.74	0.247	
Treatment	3	1.47	0.490	5.0
Error	6	0.59	0.098	
Total	15	2.95		

From Table 8.13 we can see that the computed F value is 5.0 at the .05 level of significance. The critical value of F given in Appendix F is

$$F_{.05}(3, 6) = 4.76$$

Because the computed F is greater than the critical F, the null hypothesis of equal mean yield from four different fertilizers is rejected.

We conclude that there is a significant difference among the mean yields of wheat from different fertilizers. Notice once again that, while the F test indicates a significant difference in yield, this test does not tell us which pair may have caused the difference. For such an analysis we use the procedure for mean comparisons as discussed in Chapter 7.

Chapter summary

In this chapter you were introduced to the ANOVA techniques to compare three or more sample means simultaneously. This comparison is done to determine if any statistically significant differences exist between the means of the populations. The ANOVA uses the F distribution. The underlying assumptions of ANOVA are

1. The samples are drawn randomly from the population and are independent of each other
2. The population from which the samples are drawn has a normal distribution
3. The variance of all populations are equal.

In comparing several means, we follow the six-step hypothesis testing procedure of Chapter 6, namely:

1. Stating the null (H_o) and alternative (H_1) hypotheses
2. Deciding on the level of significance (α)
3. Selecting the appropriate test statistic (in this case the F test)
4. Formulating the decision rule
5. Performing the statistical test using the sample data
6. Finally, reaching a statistical conclusion

Three different ANOVA designs often used in environmental and agricultural sciences were introduced to you. Many other experimental designs are used in environmental and agricultural sciences, but are beyond the scope of this text. You are encouraged to refer to the suggested readings at the end of this chapter for more elaborate ANOVA designs and techniques.

The one-factor ANOVA considers the effect of one variable of interest, called *treatment*, on an experimental unit. One factor or variable of interest may have several levels or categories. For instance, an agronomist may use several application rates (treatment) of a particular fertilizer on the yield of a crop. The different application rates are considered as different levels or categories of treatments. In this situation, the agronomist is only interested in the effect that the fertilizer would have on yield.

To determine if any differences exist in the mean yield of several populations, the ANOVA technique identifies the sources of variability within and between the sample observations. In the one-factor ANOVA, there are three sources of variation:

1. *Treatment variation*, which is the variation between treatments
2. *Error variation*, which is the variation within each sample
3. *Total variation*, which is the combination of the between-treatment and within-treatment variation

The F test that uses the F ratio is computed by dividing the between-treatment mean square by the within-treatment mean square. If the F ratio is equal to one, it supports the null hypothesis that the population treatment means are equal, and we accept the null hypothesis. However, if the F ratio is greater than one, the rejection of the null hypothesis depends on comparing the computed F ratio with the tabular F distribution given in the Appendix F. *Completely randomized design* was illustrated, using one-factor ANOVA.

The two-factor ANOVA takes into account the effects of two variables of interest in an experimental unit. For instance, an animal scientist may be interested in the effect that a new formulation of a feed ration and the breed of an animal have on the weight gain of the animal. There are two basic types of two-factor ANOVA. The first is known as the *randomized block design* where blocks refer to the second factor in an experiment that may account for some of the differences in the results, and treatment is the major factor in which the experimenter is interested. In this design the treatments are randomly assigned to units within each block.

The second is known as the *completely randomized design* where sample units are randomly assigned to each factor combination. For instance, the agronomist in testing the fertilizer effect on yield may use the soil quality present on a farm as a second or block factor. The experimental design allows us to test two different null hypotheses:

1. H_o: No difference in population mean yield from different fertilizer treatments
2. H_o: No difference in population mean yield among the different parcels of land

In the *Latin square design* we have a method for analyzing three or more factors. If we are experimenting with three factors, for instance, then we are interested in one treatment factor and two blocking variables. The treatments are assigned in such a way that they occur once and only once in each column and each row. Therefore, the number of rows, columns, and treatments are all equal.

Latin square design permits us to perform a factorial design without requiring large sample units, which is considered to be the primary advantage of the design. Its disadvantage is that the number of degrees of freedom for the error mean square is small. This makes it difficult to reject the null hypothesis unless the total sum of squares is large in relation to the error sum of squares.

Case Study

Comparing Drought-Resistant Varieties of Cowpea in Senegal

Plant breeders and crop physiologists have adopted many different strategies to cope with drought conditions around the world. These strategies have included supplying irrigation water to traditional crop varieties, transferring drought resistance genes, and improving on those crops that are well adapted to dry areas.

In Senegal, scientists have used the transfer of drought resistance genes in cowpea (a legume) to improve on the yield of this crop and to cope with the drought conditions that prevail in the country. Prior to 1980, most cowpea crops that were grown in the Sahel required a growing season of 80 to 120 d (Chrispeels and Sadava, 1994). The aim of the scientists has been to develop new varieties of cowpea that have rapid initial growth allowing for early closure of the canopy. When the plants have grown enough to touch each other (well-developed canopy), water evaporation from the soil is minimized and weed growth is reduced. Both of these conditions improve the yield of the crop.

After years of experimentation, scientists have developed a new variety of cowpea that requires only 65 d to produce a crop. This is considered to be more effective not only in Senegal but also in other drier areas of the Sahel. Breeders from the University of California crossed a variety that flowers early under the hot long-day conditions prevalent in the Imperial Valley of California with a day-neutral variety from Senegal. They also selected from the progeny a variety that does well under conditions of low rainfall (short rain season). How these two new varieties compare in performance to the local varieties is of interest to dry area farming.

The table following shows the comparison of yield among different varieties of cowpea under different conditions of rainfall.

Comparison of the Yields of Different Varieties of Cowpea under
Different Conditions of Rainfall in Senegal

Cowpea variety	Grain yields (kg/ha)		
	452 mm Rain	315 mm Rain	135 mm Rain
Local variety	2260	1300	50
Senegalese parent	—	1070	130
U.S. parent	2350	1350	195
New variety	2400	1800	200

Adapted from Chrispeels, M. and Sadava, D., 1994. *Plants, Genes, and Agriculture*,
Jones & Bartlett, Boston, 1994, 193.

As can be seen in the table, the new variety does not appear
to perform that much better than the parents or the local variety
when there is a sufficient (425 mm) amount of rainfall. Furthermore, it appears that even though the yield drops with the
decrease in rainfall, the relative performance of the new variety
is superior to the local variety or the Senegalese parent. However,
these apparent differences need to be statistically determined.

1. What type of ANOVA design is this?
2. State the null and alternative hypotheses.
3. Test the results at $\alpha = .01$.

References

Chrispeels, M. and Sadava, D., *Plants, Genes, and Agriculture*, Jones &
Bartlett, Boston, 1994.

Review questions

1. An agronomist uses a randomized block design to test the null
hypothesis that mean responses of sorghum are identical under
four treatments. Using three levels for blocking factor, he obtains
the following data:

$$SSTr = 68$$

$$SSB = 124$$

$$SST = 248$$

a. Construct the ANOVA table.

b. Should the null hypothesis (H_o) of identical population means be accepted or rejected at the .05 level of significance?

2. An agricultural economist is assessing the effects of education (factor A) and practical experience on the farm (factor B) on annual earnings of farmers. He chooses five levels or categories of education, and four categories of practical experience. The following data are obtained in each category for 20 randomly chosen farmers:

$$SSTr = 600,000$$

$$SSB = 700,000$$

$$SST = 1,800,000$$

a. By using education as the only factor, construct the ANOVA table.
b. At the $\alpha = .01$ level of significance, can the agricultural economist conclude that treatment means differ?
c. By using practical experience as the only treatment, construct the ANOVA table.
d. Construct a two-factor ANOVA table, using both education and practical experience as treatments.
e. What can the agricultural economist conclude from these data regarding the null hypotheses of identical mean incomes for education levels and practical experience at the .05 level of significance?

3. Crop scientists are interested in the effect temperature has on seedling development of perennial warm-season grasses. In a study (Hsu et al., 1985), the investigators obtain the following data:

Species	Emergence to first tiller (d)		
	20°C	25°C	30°C
Crabgrass	9.0	8.0	7.0
Big bluestem	17.5	18.5	19.8
Caucasian bluestem	24.6	17.3	12.1
Indiangrass	32.0	19.7	20.2
Switchgrass	28.4	16.4	14.6

Adapted from Hsu, F.H., Nelson, C.T., and Mateches, A.G., *Crop Sci.*, 25(2), March-April, 249, 1985.

From these data, can the investigators conclude that different temperatures have different effects on the warm-season grass development? Let $\alpha = .01$.

4. Environmental scientists are concerned that deep tilled soils and irrigation methods are major contributors to soil erosion. Some scientists have argued that the increased yield compensates for the erosion that may be minimal. A group of scientists are interested in the effect that tillage and low-pressure center pivot irrigation have on corn (*Zea mays* L.) yield. In a recent study the following data are gathered on yield of corn:

Mean Yield (Mg/ha) of Corn Grown under Three
Methods of Irrigation and Three Tillage Practices

	Tillage method		
Irrigation method	Till plant	Chisel	Disk
High-pressure impact	8.82	8.54	8.55
Low-pressure impact	8.36	8.13	8.32
Low-pressure spray	8.80	8.22	8.56

Adapted from Wilhelm, L.N., Milken, L.N., and Gilley, J.R., *Agr. J.*, 77, 258, 1985.

5. A researcher is interested in the effect of four different insecticides on four different crops in four different regions. By using a Latin square design, the following table shows the rate of kill of the insecticide in percentages. Do these data provide sufficient evidence to indicate a difference in the rate of kill among the different insecticides? Let $\alpha = .05$.

	Crops			
Region	1	2	3	4
I	A (45)	B (63)	C (68)	D (72)
II	B (52)	A (64)	D (82)	C (76)
III	C (48)	D (75)	A (59)	B (81)
IV	D (62)	C (58)	B (73)	A (66)

6. A researcher is interested in testing the impact of different rates of application of nitrogen on a high-yielding variety of rice called IR-8. All other management factors such as pest control and irrigation programs are held constant, whereas the fertilizer treatment rates vary. The results of a random sample from the experiment follow. At $\alpha = .05$, is there a difference in yield between the three treatments?

| Sample observation | Fertilizer treatment nitrogen rate, kg/ha | | |
number	0	60	90
1	4118	4560	4600
2	4205	5000	4800
3	3985	4980	4900
4	4320	4490	5100

References and suggested reading

Cochran, W. G., Some consequences when the assumptions for the analysis of variance are not satisfied, *Biometrics*, 3, 22, 1947.

Cochran, W. G. and Cox, G. M., *Experimental Designs*, 2nd ed., John Wiley & Sons, New York, 1968.

Dunn, O. J. and Clark, V. A., *Applied Statistics: Analysis of Variance and Regression*, 2nd ed., John Wiley & Sons, New York, 1987.

Fisher, R. A., *Contributions to Mathematical Statistics*, John Wiley & Sons, New York, 1950.

Fisher, R.A., *The Design Experiments*, 8th ed., Oliver and Boyed, Edinburgh, 1966.

Fisher, R.A., *Statistical Methods for Research Workers*, 14th ed., Hefner, New York, 1973.

Gomez, K. A. and Gomez, A. A., *Statistical Procedures for Agricultural Research*, 2nd ed., John Wiley & Sons, New York, Chap. 2–4, 1984.

Hamburg, M., *Statistical Analysis for Decision Making*, 4th ed., Harcourt Brace Jovanovich, New York, 332, 1987.

Hoshmand, A. R., *Experimental Research Design and Analysis: A Practical Approach for Agricultural and Natural Sciences*, CRC Press, Boca Raton, FL, 1994.

Hsu, F. H., Nelson, C. J., and Mateches, A. G., Temperature effects on seedling development of perennial warm-season forage grasses, *Crop Science*, 25, 249, 1985.

Mendenhall, W., *An Introduction to Linear Models and the Design and Analysis of Experiments*, Wadsworth, Belmont, CA, 1968.

Mendenhall, W. and Beaver, R. J., *Introduction to Probability and Statistics*, 8th ed., PWS-KENT, Boston, MA, 1991.

Wilhelm, L.N., Milken, L.N., and Gilley, J.R., Tillage and low pressure center-pivot irrigation effects on corn yield, *Agronomy Journal*, 77, 258, 1985.

PART IV

Nonparametric Methods

chapter nine

Chi-Square Analysis

Learning objectives

In this chapter, you are introduced to another statistical concept of hypothesis testing, namely, that concerning the frequency distribution of the population or populations. After mastering this chapter and working through the review questions, you should be able to:

- Explain why the chi-square (χ^2) distribution can be used to approximate the sampling distribution when comparing observed frequencies with expected frequencies.
- Conduct statistical tests of hypotheses for goodness-of-fit tests, test of differences among proportions, tests of independence, and tests of homogeneity.
- Calculate the expected frequencies for these tests, and determine the degrees of freedom (*d.f.*) for each.

9.1   Introduction

The methods discussed in Chapters 1 through 7 center around the use of the normal or the t distribution. However, there are many instances when these distributions are not applicable. We covered some of these in Chapter 8. In this chapter we discuss the use of the chi-square (χ^2) distribution in environmental and agricultural research. The symbol χ is the lower case Greek letter pronounced "kie" and written as "chi." This distribution is similar to the binomial, t, and F probability distributions in that it is not a single probability curve, but instead an entire family of curves. The shape of these curves depends on the number of degrees of freedom ($d.f.$) that exist in a particular problem. Figure 9.1 shows several χ^2 probability distributions with different degrees of freedom.

Chi-square analysis is used to test hypotheses about the significance of the differences that may exist between three or more sample percentages. Research shows that nonparametric statistical tests are as powerful in detecting differences among populations as the parametric methods discussed in the previous chapters. In those instances where the normality and other assumptions are not satisfied, the nonparametric methods may yield better results. Additionally, χ^2 is used to test whether two variables are independent of each other. When data are derived from several trials in the same experiment, the χ^2 test of homogeneity is used to determine how to pool the sample data from all trials.

The procedure for performing a χ^2 test of homogeneity is similar to the χ^2 test of independence, as will be shown in a later section of this chapter.

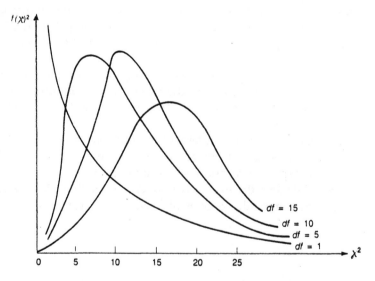

Figure 9.1 χ^2 **Distributions for different degrees of freedom.**

In situations where collected data on population variables are qualitative in nature (such as sex: male or female; state of health: diseased or healthy; marital status: single or married; response: yes or no), the absence of independence between variables has significant implications for researchers. For instance, a veterinarian who knows that preventative health measures (vaccinated, unvaccinated) and resistance (diseased, healthy) are dependent can use the χ^2 test of independence to make conclusions about the effectiveness of a vaccine on an animal.

In the following sections, we use the χ^2 distribution to test hypotheses about three or more population percentages, *test of independence*, *goodness-of-fit*, and *homogeneity*.

9.2   Test of differences among proportions

The procedure for χ^2 testing follows the same steps as those outlined in Chapter 6 on hypothesis testing.

Step one — State the null and alternative hypotheses. The null hypothesis may be so stated that there is no significant difference between the percentages of several populations under study. The alternative hypothesis, on the other hand, may state that not all population percentages are equal.

Step two — Select the level of significance for the test.

Step three — Record the observed frequencies actually obtained from random samples taken from the populations under study.

Step four — Compute the expected (theoretical) frequencies or percentages if H_o is true.

Step five — Compute χ^2 using the observed and expected frequencies with the formula:

$$\chi^2 = \Sigma \frac{(f_o - f_e)^2}{f_e} \qquad [9\text{--}1]$$

where f_o = observed sample frequency
   f_e = expected frequency if H_o is true

Step six — Make the statistical decision. The decision is based on comparing the computed χ^2 with the χ^2 table value given in Appendix G. If the computed value is less than or equal to the critical value of χ^2 given in the appendix, the null hypothesis is accepted. On the other hand, if the computed value is greater than the critical value, the null hypothesis is rejected. Figure 9.2 shows the area of acceptance and rejection in a χ^2 distribution.

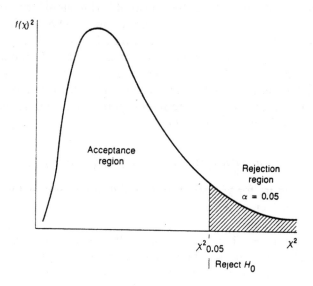

Figure 9.2   Acceptance and rejection regions for a 5% level.

The following example illustrates the procedure in testing a hypothesis using the χ^2 analysis method.

Example 9.1

An environmental scientist is interested in enhancing the germination rate of Indian rice grass seed. Two cultivars, 'Paloma' and 'Nezpar,' show high dormancy rates. Research shows that scarification of seed with acid increases germination rates. In this experiment 200 seeds of each cultivar are treated with three different concentrations of sulfuric acid. By testing at $\alpha = .01$, is there a significant difference in the percentage germination rate of the three acid concentrations?

Actual Results of Acid Scarification of
Indian Ricegrass

Cultivar	Number of germinated seeds under different H_2SO_4 concentration		
	5%	10%	15%
Paloma	48	52	45
Nezpar	62	55	48

Solution

The null and alternative hypotheses are

H_o: The population percentage of germinated seed is the same under different acid concentration.

H_1: The population percentage of germinated seed is *not* the same under different acid concentration.

Table 9.1 **Contingency Table for Actual Results of Acid Scarification of Indian Ricegrass**

Cultivar	Number of germinated seeds under different H_2SO_4 concentration			
	5%	10%	15%	Row totals
Paloma	48	52	45	145
Nezpar	62	55	48	165
Column totals	110	107	93	
Grand total				310

The sample observed frequencies (f_o) are the actual data collected from the experiment.

The expected frequencies (f_e) for each cell of the contingency table are computed using the following formula:

$$f_e = \frac{(\text{row total})(\text{column total})}{\text{grand total}} \qquad [9\text{-}2]$$

Given the information in Table 9.1, we compute the expected frequencies for the different cells in the contingency table. Thus, the f_e value for the cultivar 'Paloma' treated with 5% solution of sulfuric acid is

$$f_e = \frac{(145)(110)}{310}$$

$$f_e = 51.45$$

The same procedure is used to compute the f_e for other cells, and results are displayed in Table 9.2. Table 9.2 displays the computational results for χ^2 using Equation 9–1. The computed χ^2 is compared with the table χ^2 given in Appendix G. If the computed value is significantly different from zero, we do not accept the null hypothesis. However, if the computed value falls within the accepted region, the null hypothesis is accepted.

Table 9.2   **Computation of χ^2**

Row/column (cell)	f_o	f_e	$f_o - f_e$	$(f_o - f_e)^2$	$\dfrac{(f_o - f_e)^2}{f_e}$
1–1	48	51.45	−3.45	11.90	0.231
1–2	52	50.05	1.95	3.80	0.076
1–3	45	43.50	1.50	2.25	0.052
2–1	62	58.55	3.45	11.90	0.203
2–2	55	56.95	−1.95	3.80	0.067
2–3	48	49.50	−1.50	2.25	0.045
Total	310	310.00	0.00		0.674

To make the comparison between the computed χ^2 and the table value, we must locate the table value in Appendix G. To locate the table value, we have to know the level of significance and the degrees of freedom associated with the problem. The level of significance in this problem was set at $\alpha = .01$. The degrees of freedom for a contingency table that contains several rows and columns are found using the following formula:

$$d.f. = (r - 1)\,(c - 1) \qquad\qquad [9\text{--}3]$$

where r = number of rows in table
        c = number of columns in table

For the present example in which we have a contingency table with two rows and three columns, the degrees of freedom are

$$d.f. = (2 - 1)\,(3 - 1) = 2$$

Given the .01 level of significance and 2 *d.f.*, the critical χ^2 table value is 9.210.

Because the computed value is less than the critical value, the null hypothesis is accepted, and we conclude that the percentage of germinated seed scarified using different acid concentration is the same. Note that the table value of 9.210 means that if the null hypothesis is true, the probability of obtaining a computed χ^2 value as large as 9.210 is only .01.

9.3   Chi-square test of independence

In this test, our interest is focused on *independence* between two qualitative variables or population characteristics. Independence is tested for when one set of observations taken under one particular set of conditions is compared with a similar set of observations under a different set of conditions. Under such circumstances there are no definite expected values. Therefore, the question is whether the results of the experiment are dependent (contingent on) or independent of the conditions under which they are observed. This test is referred to as the *test of independence* or *contingency test*.

When results are classified according to two criteria, the results of a sample are presented in a 2 × 2 contingency table. A contingency table is a frequency distribution that displays the cross classification in tabular form. Table 9.3 shows the general format of a contingency table with data values in which:

O_{11} = cell in contingency table, first row and first column
r_1 = total of row frequencies
c_1 = total of column frequencies

Table 9.3   General Format of a Contingency Table

Second criterion of classification	First criterion of classification						
Level	1	2	3	4	.....	c	Total
1	O_{11}	O_{12}	O_{13}	O_{14}	....	O_{1c}	r_1
2	O_{21}	O_{22}	O_{23}	O_{24}	....	O_{2c}	r_2
3	O_{31}	O_{32}	O_{33}	O_{34}	....	O_{3c}	r_3
.	.	.	.	.		.	.
r	O_{r1}	O_{r2}	O_{r3}	O_{r4}	....	O_{rc}	r_k
Total	c_1	c_2	c_3	c_4	....	c_m	n

When there are 3 rows and 3 columns in a contingency table, we have a 3 × 3 contingency table.

The χ^2 test of independence involves comparing the observed frequencies with the expected frequencies. We compute the expected frequencies under the hypothesis of independence between two sets of events. The expected frequencies are calculated in the same manner as in Section 9.2, using Equation 9–3 as follows:

$$f_e = \frac{(\text{row total})(\text{column total})}{\text{grand total}}$$

After we have computed the expected frequencies for each cell, the χ^2 is calculated, using Equation 9–1. When we have a 2 × 2 contingency table, where a, b, c, and d are the observed frequencies, the following shortcut method is used to compute the χ^2:

	Categories of observations		
	I	II	Total
A	a	b	$a + b$
B	c	d	$c + d$
Total	$a + c$	$b + d$	$a + b + c + d = n$

$$\chi^2 = \frac{[(ad - bc)^2]n}{(a+b)(c+d)(a+c)(b+d)} \qquad [9\text{--}4]$$

The following example illustrates the χ^2 test of independence.

Example 9.2

A plant geneticist performs two different experiments, A and B, to determine the dominant:recessive ratio of crosses between individuals heterozygous for the same single gene. The following results are obtained:

	Phenotype of offspring	
	Dominant	Recessive
A	70	30
B	75	25

By testing at $\alpha = .05$, are the observed results independent of experimental conditions?

Solution

The dominant:recessive ratio for experiment A is 2.33:1; and for experiment B, it is 3.0:1. When such differences between ratios of different experiments exist, one may ask the question, are the observed results independent of experimental conditions? In environmental and agricultural sciences, problems of this kind where the results of experimental observations are affected by different environments or genetic constitutions are common. The test of independence provides statistical answers to such problems.

The step-by-step procedure using the shortcut formula is given as follows.

Step one — The null and alternative hypotheses are

H_o: The observed results are independent of the experimental conditions.

H_1: The observed results are dependent of the experimental conditions.

Step two — When we use Equation 9–4 to compute the χ^2, there is no need to compute the expected frequencies as was done in the previous section. Therefore, the χ^2 is computed as follows:

	Phenotype of offspring		
	Dominant	Recessive	Total
A	70	30	100
B	75	25	100
Total	145	55	200

$$\chi^2 = \frac{[(70)(25)-(30)(75)]^2(200)}{(100)(100)(145)(55)}$$

$$= \frac{[(1,750)-(2,250)]^2(200)}{79,750,000}$$

$$= \frac{50,000,000}{79,750,000}$$

$$= 0.627$$

Step three — The decision criterion is to reject H_o if the computed χ^2 is greater than the critical χ^2 given in Appendix G. Given the level of significance of $\alpha = .05$, and $k-1$ or $2-1 = 1$ *d.f.*, the critical χ^2 is 3.841.

Step four — Because the computed χ^2 is less than the critical χ^2, the null hypothesis is accepted, and we conclude that the observed results are independent of experimental conditions.

9.4 Test of goodness-of-fit

Another important χ^2 analysis procedure is the *test of goodness-of-fit*. The purpose of this test is to determine if an observed set of data fits an expected set of data. The test helps us determine whether sample results are consistent with the hypothesis that they were drawn from a population with a known distribution — the *uniform distribution* or the *normal distribution*. The uniform distribution is a continuous distribution that assumes all possible values are equally likely. A more extensive discussion of this distribution is given in Ostle and Mensing (1975) and Lapin (1987).

We can also use the test of goodness-of-fit to validate the assumption that a particular population follows a normal distribution. The following examples illustrate the test of goodness-of-fit for both uniform and normal distributions.

Example 9.3

Uniform distribution. An animal scientist is interested in determining beef cattle preferences to various rations. A random sample of 60 steers was introduced to five different rations. The steers had free access to all five rations. Preference for a ration is measured by the number of steers eating from a particular bin at any one feeding time. The observed results are displayed as follows. By testing at $\alpha = .05$, are the rations equally preferred?

Ration	A	B	C	D	E
Number of steers	9	13	14	11	13

Solution

If the animals do not have a preference for the rations, we would expect the same number of steers to feed at each of the five ration bins. This means that we would expect the number of steers to be uniformly distributed among the five rations. By keeping this in mind, we set the null and alternative hypotheses as:

H_o: The rations are equally preferred.
H_1: The rations are not equally preferred.

Under the null hypothesis, the expected number of steers favoring each ration is $60/5 = 12$. The χ^2 test is computed using the following formula:

$$\chi^2 = \Sigma \frac{(f_o - f_e)^2}{f_e}$$

Table 9.4 shows the computed χ^2 to be 1.332.

To make a statistical decision, we have to compare the computed χ^2 with the critical χ^2 given in Appendix G. The number of degrees of

Table 9.4   Computation of χ^2 Statistic from the Data in Example 9.3

Ration	Number preferring f_o	f_e	$f_o - f_e$	$(f_o - f_e)^2$	$\dfrac{(f_o - f_e)^2}{f_e}$
A	9	12	−3	9	0.750
B	13	12	1	1	0.083
C	14	12	2	4	0.333
D	11	12	−1	1	0.083
E	13	12	1	1	0.083
Total	60	60	0		1.332

freedom is equal to the number of categories minus 1, or $5 - 1 = 4$. The critical value of χ^2 for $\alpha = .05$ and 4 *d.f.* is 9.488.

Because 1.332 is less than 9.488, we cannot reject H_o. Therefore, we conclude that the rations are equally preferred, and that the distribution of steers is uniform.

Example 9.4

Normal distribution. In an attempt to incorporate the salt-tolerant qualities of *Elytrigia pontica* into wheat, a plant scientist has hybridized *E. pontica* with wheat. A simple random sample of 100 hybridized plants with different chromosome numbers is exposed to 20.5 g/l of NaCl. Table 9.5 displays the frequency distribution of the plants that survive after 30 d. We wish to determine whether these data provide sufficient evidence to indicate that the observations were not drawn from a normal population. Test at $\alpha = .01$.

Table 9.5   Frequency Distribution of Number of
Wheat Plants Surviving Exposure to 20.5 g/l of NaCl

Chromosome no.	No. of surviving plants
40 < 48	10
48 < 56	25
56 < 64	28
64 < 72	34
72 < 80	38
80 < 88	15

Solution

The null and alternative hypotheses are

H$_o$: The sample data come from a normally distributed population.

H$_1$: The sample data do not come from a normally distributed population.

The test statistic is computed using Equation 9–1.

$$\chi^2 = \sum \frac{(f_o - f_e)^2}{f_e}$$

where f_o = observed frequencies
       f_e = expected frequencies that need to be determined

Based on the principles of probability distributions, the relative frequency of occurrence of values of a given magnitude is found by finding the area under the curve defined by the distribution. For the normal distribution that is defined by its mean, μ, and its standard deviation, σ, the relative frequency of occurrence of values is found by the area under the normal curve. Recall from Chapter 5 that the table of areas under the normal curve shows the area between the mean and a given number of standard deviations from the mean. The numerical value corresponding to this area is found by transforming the normal distribution in a z, that is, converting the units of measurement into standard unit or z score. A z score is computed using the formula:

$$z = \frac{x - \mu}{\sigma}$$

Because the present example has not specified the mean and the variance (or the standard deviation) of the hypothesized normal distribution, we must use the sample mean and standard deviation as estimates of μ and σ to compute the area under the normal curve for each class interval.

The procedure for computing the area, and hence the relative frequency of occurrence of values for each class interval, is displayed in Table 9.6. The mean and the sample standard deviation for this example is computed as $\bar{x} = 65.87$ and $s = 11.40$ (see Chapter 2 for computation procedures). The expected frequency of occurrence of value for the interval 48 to 56 is found by the following steps.

Step one — Compute the z corresponding to the lower and upper limits of the class interval as follows:

Table 9.6 Calculation of the Expected Frequency and the χ^2 Test of the Goodness-of-Fit

1	2	3		4	5	6
Class interval	Observed frequency (f_o)	Standardized z		Probability (area under the normal curve)	Expected frequency (f_e)	$\dfrac{(f_o - f_e)^2}{f_e}$
		z_l	z_u			
40 < 48	10	$-\infty$	−1.57	.5000 − .4418 = .0582	8.73	0.18
48 < 56	25	−1.57	−0.87	.4418 − .3078 = .1340	20.10	1.19
56 < 64	28	−0.87	−0.16	.3078 − .0636 = .2442	36.63	2.03
64 < 72	34	−0.16	0.54	.0636 + .2054 = .2690	40.35	0.99
72 < 80	38	0.54	1.24	.3925 − .2054 = .1871	28.07	3.51
80 < 88	15	1.24	∞	.5000 − .3925 = .1075	16.12	0.08
	150			1.0000	150.00	$\chi^2 = 7.98$

Lower limit:

$$z_l = \frac{L_l - \bar{x}}{s}$$

$$= \frac{48 - 65.87}{11.40}$$

$$= -1.57$$

Upper Limit:

$$z_u = \frac{L_u - \bar{x}}{s}$$

$$= \frac{56 - 65.87}{11.40}$$

$$= -0.87$$

Column three of Table 9.6 displays the standardized z for the remaining class intervals.

Step two — Determine the probability of occurrence of the values in each class interval by determining the area under the normal curve. Because the true lower limit of the first class interval is $-\infty$, and the true upper limit of the last class interval is $+\infty$, the area under the normal curve for the first interval is simply the difference between the area from $-\infty$ to z. The area from 0 to 1.57 found in Appendix B is .4418. Therefore, the area between $-\infty$ to −1.57 is

$$.5000 - .4418 = .0582$$

232 Statistical Methods for Environmental and Agricultural Sciences

Column four of Table 9.6 shows the area under the normal curve for each class interval.

Procedurally, you should keep in mind that the area between any two z values that have the same sign, that is, the z_l lower and z_u upper with negative signs, is equal to the difference between areas from 0 to z_l and from 0 to z_u. For the z values carrying different signs, that is, either of the lower or upper z having a positive or a negative sign, the area between the z values is equal to the sum of the areas from 0 to z_l and the area from 0 to z_u. The class interval 64 to 72 shows the z values with different signs.

Step three — Compute the expected frequency for each class interval. The expected frequency is the product of the probability in each class and the total number of observations. That is

$$f_e = (P_j)(n) \qquad\qquad [9\text{--}5]$$

where f_e = expected frequency
P_j = probability of jth class
n = total number of observations

Column five of Table 9.6 shows the computed expected frequency for each class interval.

Step four — Compute the χ^2. By determining the expected frequency, you are then able to compute the χ^2 as shown in column six of Table 9.6. The computed χ^2 using Equation 9–1 is 7.98.

Step five — Make a statistical decision. The decision is based on the comparison of the computed χ^2 with the critical χ^2 value given in Appendix G.

Again, to find the critical value in the table, we must determine the degree of freedom. The degree of freedom is determined by the number of constraints that have been imposed on the data. In the present example, three constraints have been imposed on the data. The first restriction is that the expected frequencies must add to 150, the sample size. The second and third restrictions have to do with the estimation of the mean and the variance (or standard deviation). If the mean and the variance were given to us, only one restriction is imposed on the data. Given the three restrictions and the six classes or categories in which we have grouped the data, the degrees of freedom are:

$$d.f. = 6 - 3 = 3$$

The degree of freedom for the goodness-of-fit tests of this type is generally equal to the number of categories minus one. Additionally, we subtract one for each parameter that has to be estimated.

The critical value for $\alpha = .01$ and 3 *d.f.* is 11.34. Because the computed value is less than the critical value, the null hypothesis is accepted and we conclude that the sample data did come from a normally distributed population.

9.5 *Tests of homogeneity*

In environmental and agricultural research we often replicate experiments several times to make sure that error is minimized in experimentation. When the data from several replications are available for analysis, the χ^2 test of homogeneity is used to pool the data over all replications. In such a situation we are able to test the proposition that several populations are homogeneous with respect to some characteristic. We can also test the hypothesis that an observed ratio significantly deviates from a hypothesized ratio.

The χ^2 test of homogeneity is different from the χ^2 test of independence in two respects. First, the sampling procedure in the test of homogeneity is different from the χ^2 test of independence in that we identify two or more populations of interest in advance and draw a sample from each. In the test of independence, the sampling procedure is to select a single sample from a population, and then to cross classify the sample according to two criteria.

Second, the test of homogeneity is different from the test of independence in the rationale used in calculating the expected frequencies and how the results are interpreted. The rationale behind pooling the data from available samples is that best estimates are obtained from pooling instead of calculating the χ^2 from individual samples.

The following example illustrates the procedure for applying the test of homogeneity.

Example 9.5

An agricultural economist is asked to perform a marketing study to determine whether farmers of different age groups differ in the type of tractor they prefer. A random sample is selected from three different age groups. Each farmer is asked to specify which of the three types of tractor he or she prefers. The table below shows the results. By testing at $\alpha = .05$, are the farmers of different age groups not homogeneous with respect to preference in tractors?

Population age group	Type of tractor			Total
	A	B	C	
Under 35	140	100	60	300
35–45	90	100	40	230
45 and over	75	60	35	170
Total	305	260	135	700

Solution

To compute the χ^2, the steps used in testing a hypothesis are followed.

Step one — State the null and alternative hypotheses as:

H_o: The three age groups are homogeneous with respect to tractors preferred.
H_1: The three age groups are not homogeneous.

Step two — Compute the expected frequencies, applying the rationale underlying the test of homogeneity. This means the best estimate of the true proportion is found by pooling the sample observations. That is, the true proportion of the farmers who prefer type A tractor in each age group is

$$p = \frac{305}{700} = 0.44$$

To find the expected frequency for type A tractor, for example, we multiply each sample total by 0.44 as shown below.

$$f_e = (0.44)(300) = 132$$

Table 9.7 shows the expected frequencies for each type of tractor and age group.

Table 9.7 **Expected Frequencies of Types of Tractor Preferred by Different Age Group Farmers**

Population age group	Type of tractor		
	A	B	C
Under 35	132.0	111.0	57.0
35–45	101.2	85.1	43.7
45 and over	74.8	62.9	32.3

Step three — The χ^2 statistic is computed using Equation 9–1 as follows:

$$\chi^2 = \Sigma \frac{(f_o - f_e)^2}{f_e}$$

$$= \frac{(140 - 132)^2}{132} + \frac{(90 - 101.2)^2}{101.2} + \ldots + \frac{(35 - 32.3)^2}{32.3}$$

$$= 6.255$$

Step four — Determine the critical value from Appendix G using the level of significance and the degrees of freedom, and compare with the computed χ^2. Accept H_o if the computed value is less than or equal to the critical value; otherwise, reject H_o.

Step five — Make a statistical decision. The decision to accept or reject the null hypothesis is based on the critical value. To locate the critical value in Appendix G, we must know the level of significance and the number of degrees of freedom. The level of significance is given to us as $\alpha = .05$. The degrees of freedom for this problem are

$$d.f. = (r - 1)(c - 1)$$
$$= (3 - 1)\,(3 - 1)$$
$$= 4$$

The critical value of χ^2 for $\alpha = .05$ and 4 *d.f.* is 9.488.

Because the computed value 6.255 is less than 9.488, we accept the null hypothesis (H_o), and conclude that the populations are homogeneous with respect to the type of tractors preferred.

Chapter summary

Because relations between variables are an important part of statistics, various conceptual methods are used in measuring such relationships. The chi-square (χ^2) tests provide a means of measuring relations between nominal or qualitative variables. When a researcher is performing analyses on qualitative variables such as state of health of an animal or plant (healthy or diseased), sex (male or female), type of treatment or kind of response, the χ^2 test is used.

In this chapter you were introduced to four χ^2 tests: *tests for the equality of several proportions; tests of independence; tests of goodness-of-fit; and tests of homogeneity.*

In Chapter 7, you learned how to compare two population proportions using the normal distribution. When comparing several proportions, however, the χ^2 distribution and test is the appropriate measure.

When investigating whether two qualitative variables A and B are independent of each other, the χ^2 test of independence is used. Statistical independence means that the proportion of total population having any attributes of A and B are equal.

The test of goodness-of-fit is used to determine whether a variety of technical assumptions hold true with respect to the characteristics of the population under study.

Finally, you learned about the test of homogeneity where it is used to determine whether or not categories of a single variable are represented in the same proportions in two or more populations.

Case Study

The Relative Contribution of Men and Women to Agriculture in Africa

In every society, women play a critical role in its development. In many rural parts of the world they are responsible for farming, food provision, health care, and acquisition and stewardship of natural resources. In general, they perform the bulk of household subsistence work. Evidence indicates that in many societies, women have greater influence than men on rates of population growth and infant and child mortality, on health and nutrition, on children's education, and on natural resources management (World Resources Institute, 1995).

Although statistics often omit or obscure women's work, improved United Nations (U.N.) statistical methods show that, for example, 59% of women in eastern Asia, 60% in the former U.S.S.R., and 45 to 50% of women in Southeast Asia and sub-Saharan Africa are economically active (United Nations, 1991). In most developing countries, women's contribution to agricultural production has been significant, as was the case in the U.S. until farming was mechanized (Lewenhak, 1992, pp. 83-89; Davison, 1988). It is estimated that approximately one half of the world's food is grown by women, and that two thirds of women workers in developing countries are in the agricultural sectors (Buvinic & Yudelman, 1989). In many parts of the world, women are responsible for animal husbandry and poultry production. In Egypt, Chile, Pakistan, and Switzerland, between 80 and 100% of women are responsible for the care and production of poultry (Rhodda, 1991).

Agricultural production in developing countries tends to show a division of labor by gender, though women's responsibility varies from region to region. To successfully move the development agenda forward, policy makers and planners must incorporate the role of women in the development strategies. In the case of Africa, women have traditionally hoed, planted, weeded, harvested, stored, and processed crops and food, while men have cleared the land and plowed the soil (Buvinic and Yudelman, 1989, pp. 16–17). The relative contributions of men and women to agriculture are shown as follows:

Male and Female Share of Agricultural Work in Africa

Task	Male (%)	Female (%)
Plowing	70	30
Planting	50	50
Hoeing and weeding	30	70
Transporting	20	80
Storing	20	80
Processing	10	90
Marketing	40	60

From Chrispeels, M. and Sadava, D., *Plants, Genes, and Agriculture*. Jones & Bartlett, Boston, 1994, 321.

By using the data in the table, perform a χ^2 test to determine if there is a significant difference between the male and female share of agricultural tasks. Use $\alpha = .05$.

References

Buvinic, M. and Yudelman, S., *Women, Poverty, and Progress in the Third World*, Headline Series no. 289, Foreign Policy Association, New York, 1989.

Chrispeels, M. and Sadava, D., *Plants, Genes, and Agriculture*, Jones & Bartlett, Boston, 1994.

Davison, J., Land and women's agricultural production: the context, in Davison, J., Ed., *Agriculture, Women, and Land: The African Experience*, Westview Press, Boulder, CO, 1988, 1.

Lewenhak, S., *The Revaluation of Women's Work*, Earthscan, London, 1992.

Rhodda, A., *Women and the Environment*, Zed, London, 1991.

United Nations, *The World's Women, 1970-90: Trends and Statistics*, Social Statistics and Indicator Series K, no. 8, United Nations, New York, 1991, 84.

World Resources Institute, *World Resources: 1994–1995*, Oxford University Press, New York, 1995.

Review questions

1. An agronomist is interested in studying the genetic makeup and the pattern of inheritance of chlorophyll in a new maize cultivar. A sample of 200 seedlings from self-fertilized heterozygous plants shows 100 green and 100 yellow seedlings. Does the observed ratio deviate significantly from the hypothesized ratio of 3:1? Test at $\alpha = .01$.

2. To test the effect of four different chemicals on germination rate, a crop scientist treats 100 of each randomly selected seeds with chemicals. The following table displays the number of germinated seeds. Using $\alpha = .05$, test the hypothesis that the percentage of seeds germinating is independent of the type of chemical used.

Chemical	No. germinated	No. not germinated
A	85	15
B	92	8
C	78	22
D	89	11

3. An agricultural economist interested in energy cost as a percentage of total cost of production finds the following distribution

from a random sample of 170 farms. Use a level of significance of $\alpha = .01$ to test the hypothesis that the universe distribution is normal.

Energy cost as percentage of total cost	Number of farms
5 and under 10	40
10 and under 15	52
15 and under 20	35
20 and under 25	28
25 and under 30	15

4. Salt is considered an essential element in the diet. However, excessive intake of salt can cause health problems. A food and nutrition researcher performs a test on a meal with four different quantities of salt added. The results based on a sample of 100 people are as follows:

Meal	Number preferring
A	20
B	28
C	25
D	27

Using a .05 level of significance, test the hypothesis that a uniform distribution describes this population.

References and suggested reading

Cochran, W. G., The test of goodness of fit, Ann. Math. Stat., 23, 315, 1952.

Conover, W. J., *Practical Nonparametric Statistics*, 2nd ed., John Wiley & Sons, New York, 1980.

Dahiya, R. C. and Gurland, J., Pearson chi-squared test of fit with random intervals, *Biometrika*, 59, 147, 1972.

Daniel, W. W., *Applied Nonparametric Statistics*, 2nd ed., PWS-KENT, Boston, 1990.

Feinberg, B. S., The *Analysis of Cross-Classified Categorical Data*, 2nd ed., MIT Press, Cambridge, MA, 1980.

Fleiss, J. L., *Statistical Methods for Rates and Proportions.* 2nd ed., John Wiley & Sons, New York, 1981.

Friedman, M., The use of ranks to avoid the assumption of normality implicit in the analysis of variance, *J. Am. Stat. Assoc.*, 32, 1937.

Gill, J. L., *Design and Analysis of Experiments in the Animal and Medical Sciences,* Iowa State University Press, Ames, IA, 1978.

Gomez, K. A. and Gomez, A. A., *Statistical Procedures for Agricultural Research,* 2nd ed., John Wiley & Sons, New York, 1984.

Lapin, L. L., *Statistics for Modern Business Decisions,* 4th ed., Harcourt Brace Jovanovich, New York, 1987.

Mann, H. B. and Whitney, D. R., On a test of whether one of two random variables is stochastically larger than the other, *Ann. Math. Stat.,* 18, 50, 1947.

Ostle, B. and Mensing, R. W., *Statistics in Research.* 3rd ed., Iowa State University Press, Ames, IA, 1975.

Seigel, S. and Castellan, N. J., *Nonparametric Statistics,* 2nd ed., McGraw-Hill, New York, 1988.

Watson, G. S., The goodness of fit test for normal distributions, *Biometrika,* 44, 336, 1957.

Wilcoxon, F., Individual comparisons by ranking methods, *Biometrics,* 1, 80, 1945.

chapter 10

Additional
Nonparametric Methods

Learning objectives

After mastering this chapter and working through the review questions, you should be able to:

- Recognize situations that require the use of nonparametric methods.
- Apply the sign test for sample data.
- Use the Wilcoxon signed-rank test for small samples.
- Apply the Mann–Whitney U test to determine if two independent samples come from identical populations.
- Compute Spearman rank correlation coefficient for ranked data.

10.1 Introduction

In the previous chapters we generally assumed (with the exception of the chapter on chi-square [χ^2] that our sample data comes from one or more normally distributed populations. Furthermore, inferential procedures of Chapters 6 through 8 are concerned with population *parameters*. Thus we have dealt only with statistical methods called *parametric methods*. The body of statistical methodology that is not concerned with population parameters and is not dependent on rigid assumptions about the population probability distributions is referred to as *nonparametric* or *distribution free statistics*. For example, we can use nonparametric methods to test hypotheses when the data used are ranks or ordinal. *Ordinal* refers to numbers that primarily express ranking, preferences, and hierarchy. For instance, an agronomist in ranking his or her preference for the vigor of three varieties (A, B, and C) of corn (*Zea mays*) may use a scale of 1 to 5 (poor to excellent). If variety A receives a rating of five over variety B, which receives a rating of two, it is implied that the agronomist prefers A to B. The numbers given as ranks in this example, or the mean of such numbers, are not meaningful. Therefore, any statistical test requiring the calculation of means is not appropriate. Under such circumstances we use the nonparametric methods. If we have nominal data such as male or female, or yes or no, where there is no implication that one item is higher or lower than another, we also use nonparametric methods. Additionally, if the sample data collected are from the population that is not approximately normal with equal variances, we use nonparametric methods.

There are several advantages to the use of nonparametric statistics:

1. The computations used in nonparametric statistics are easier to conduct and understand.
2. The assumptions concerning populations are less restrictive, and thus are applicable to a wider range of conditions.
3. We may use small sample data to obtain exact results, whereas a similar situation using parametric methods may require a large sample.

The disadvantages of nonparametric methods are

1. Some information is ignored because order or rank may be used instead of numerical values of observations. For this reason, the nonparametric methods have been labeled as less efficient.
2. Because of their simplicity and ease of computation using small samples, they are sometimes used as a matter of convenience to the researcher in situations where larger samples and parametric methods may be appropriate.
3. Computations become burdensome as sample size increases.

Nonparametric methods are used when the following conditions apply. First, nonparametric statistics is used in those situations when the data do not meet the assumptions required for a parametric test. For instance, when we know that the data gathered are from a population that is not normally distributed, it is appropriate to use a nonparametric test.

Second, if the question to be answered does not involve a parameter, we may use nonparametric methods. For example, we may use a nonparametric test to decide if a sample is a random sample.

Third, nonparametric tests are used to derive quick and approximate results.

In this chapter you are introduced to some of the commonly used nonparametric methods in environmental and agricultural research and business analyses.

10.2 Sign test

The *sign test* is performed when we wish to analyze two sets of data that are not gathered independently; that is, they are from the same sample. For instance, in a before-and-after herbicide test, we observe the effectiveness of the herbicide by the presence or absence of weeds in the same sample. We can also apply the sign test to the data gathered from two related samples. For example, an agronomist may apply herbicide on two plots of land in different ways.

To perform the sign test, we must have pairs of observations where the differences between observations are computed, and then determine the sign associated with each. For example, if we collect data on two different treatments such as X and Y, the difference is computed as $X - Y$. If $X > Y$, the difference is recorded as +. If $X < Y$, the difference is recorded as −. The sign test is based on the resulting pluses and minuses.

Situations in which we have equal observations are by definition excluded from the sign test.

The sign test is most commonly used to test the hypothesis that the plus and minus signs are equally likely. If we denote the probability of obtaining a plus sign as p, the statistical hypothesis that the plus and minus signs are equally likely may logically be stated as:

$$H_o: \ p = .5$$

$$H_1: \ p \neq .5$$

We can use the binomial probabilities directly in performing the test; however, it is more convenient to use the normal approximation which is satisfactory for any sample size. The test statistic for the sign test is

$$z = \frac{2R - n}{\sqrt{n}} \qquad \text{[10–1]}$$

where R = number of positive signs
n = number of relevant paired observations

The same steps in hypothesis testing as discussed in Chapter 6 are used with the sign test. The following example illustrates the use of the sign test in an experiment.

Example 10.1

An agricultural marketing firm has designed a new television advertising campaign to increase the sale of a particular brand of California wine. To determine the effectiveness of the new advertising campaign, 20 randomly selected individuals are asked to rate this new television commercial in comparison with an old commercial. The commercials are rated on a scale of one to five, where a rating of one is poor, and five is excellent. The following scores are recorded by each viewer. By testing at .05 level of significance, is the new commercial superior to the old?

Solution

Following the steps in testing a hypothesis, the null and alternative hypotheses may be stated as:

H_o: New commercial ≤ old commercial

H_1: New commercial > old commercial

The sign of differences between observations are shown in Table 10.1. By eliminating the 4 tied cases from the test, we observe 12 minuses and 4 pluses in Table 10.1. Thus, the number of relevant paired observations is 16.

If the null hypothesis is true, we expect 50% of the random sample of 16 to have plus signs. That is, there are eight samples with plus signs. However, in this sample we have observed 4 samples with plus signs ($R = 4$). To test whether the disparity between expected and observed is due to chance, the normal approximation to the binomial distribution is used to determine the probability of observing four or more samples with plus signs.

Table 10.1 Ranking Scores Assigned to Two Television Commercials by a Panel of Viewers

| | Ratings | | Sign of |
| | Old commercial | New commercial | difference |
Viewer	(y)	(x)	($x - y$)
1	4	3	−
2	5	2	−
3	2	4	+
4	3	5	+
5	5	1	−
6	4	3	−
7	5	4	−
8	5	4	−
9	2	2	0
10	1	4	+
11	5	5	0
12	4	3	−
13	5	4	−
14	5	3	−
15	4	4	0
16	4	3	−
17	1	2	+
18	3	2	−
19	4	3	−
20	2	2	0

The test statistic given in Equation 10–1 is based on the number of positive signs R, and is

$$z = \frac{2(4) - 16}{\sqrt{16}} = -2.0$$

Given the level of significance of $\alpha = .05$, the critical normal deviate is $z = 1.64$. This test is upper tailed, because a large value of R (of z) refutes the null hypothesis. Figure 10.1 illustrates the rejection and acceptance regions.

Because the computed z falls in the acceptance region, the null hypothesis that the new commercial is not superior to the old is accepted at the 5% level of significance.

10.3 Wilcoxon signed-rank test

In situations where it is not possible to use the z test because we may have a small sample known to be from a nonnormally distributed

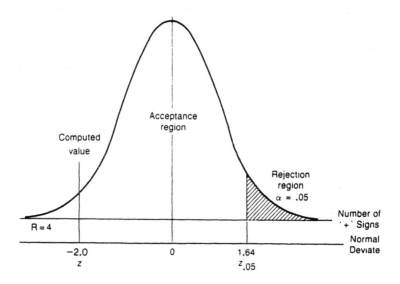

Figure 10.1   **Acceptance and injection regions with a significance level of .05.**

population, or the *t* test because the sampled population does not satisfy the normal distribution criteria, we may use the *Wilcoxon signed-rank test*. This test is employed when we have a pair of ordinal measurements taken from the same samples or matched samples, and we are interested in measuring the real differences in magnitude and direction that exist between those pairs.

The assumption underlying the Wilcoxon signed-rank test are

1. The sample is random.
2. The population is symmetrically distributed about its mean (recall, that a symmetrical distribution is one where the right half of the curve is a mirror image of the left half).
3. The variable of interest is continuous.
4. The measurement scale is at least interval. (This means that the unit of measurement not only indicates that one object is greater or less than another, but also indicates the magnitude.)

Seven steps are followed in calculating the Wilcoxon signed-rank test.

Step one — State the null and alternative hypotheses. The hypotheses about some unknown population mean μ may be stated as:

$$H_o: \ \mu_o = \mu_1$$

$$H_1: \ \mu_o \neq \mu_1$$

$$H_o: \ \mu_o \leq \mu_1$$

$$H_1: \ \mu_o > \mu_1$$

$$H_o: \ \mu_o \geq \mu_1$$

$$H_1: \ \mu_o < \mu_1$$

Step two — State the desired level of significance α.

Step three — Determine the size and sign differences between paired data as:

$$d_i = x_i - \mu$$

Once we have found the $d_i's$, eliminate those sets of data where $d_i = 0$ from the calculations, and reduce the sample size accordingly.

Step four — Rank the differences irrespective of the signs. This ranking is done from the smallest to the largest. In situations where we have tied rankings, the average of the two tied differences are given to each of the tied values.

Step five — Affix appropriate signs to the assigned ranks. That is, if the difference is positive between paired values, a plus (+) sign is affixed to the ranked value, and a minus (−) sign is affixed to those ranked values where the difference between paired values is negative.

Step six — Sum the ranks and determine the computed T value. After you sum the ranks, the smaller of the two sums ($T+$ or $T−$) is the computed T value.

Step seven — Draw a statistical conclusion. To draw a statistical conclusion, we compare the computed value with the appropriate table value given in Appendix H. If the computed T value is equal to or less than the table T value, we reject the null hypothesis.

The preceding steps are similar to those of the hypothesis testing procedure outlined in Chapter 6.

The following example illustrates the use of the Wilcoxon signed-rank test.

Example 10.2

An animal scientist formulates a new ration (*B*), and wishes to determine not only whether the new ration is superior to the old ration (*A*), but also the amount of daily weight gain. Ten heifers are fed the old and the new ration at two different time periods to determine the superiority of the new ration. Test at .05 level of significance.

Daily Weight Gain kg/d

Animal	Original (A)	New ration (B)
1	.50	.56
2	.62	.60
3	.70	.72
4	.60	.65
5	.71	.68
6	.82	.85
7	.65	.69
8	.59	.71
9	.80	.91
10	.70	.64

Solution

Step one — The null and alternative hypotheses are

H_o: The daily weight gains from the two rations are the same.

H_1: The daily weight gain of the new ration is greater.

Step two — The animal scientist would like to test this hypothesis at the .05 level of significance. Therefore $\alpha = .05$.

Step three — Determine the size and sign of differences between the paired data. The sign and size of the differences between the paired data are computed and shown in column four of Table 10.2.

Step four — Rank the differences irrespective of the signs. We assign a rank of one to the smallest difference, and a high rank to the largest difference. In this example, animals two and three are tied for first place with a difference of .02. We assign to each a rank of 1.5, which is the average of the ranks 1 and 2. The same procedure is used to assign the other ranks. You can observe that on three occasions we have tied ranks. The rankings are shown in column five of Table 10.2.

Step five — Affix the appropriate signs to the assigned ranks. In animal one, we observe a positive difference between rations A and B.

Table 10.2 **Computation for Wilcoxon Signed-Rank Test**

| | Daily weight gains | | | Rank | Signed rank | |
Animal	Ration (A)	Ration (B)	Differences (B − A)	irrespective of sign	Positive	Negative
1	.50	.56	+06	7.5	+7.5	
2	.62	.60	−.02	1.5		−1.5
3	.70	.72	+.02	1.5	+1.5	
4	.60	.65	+.05	6	+6	
5	.71	.68	−.03	3.5		−3.5
6	.82	.85	+.03	3.5	+3.5	
7	.65	.69	+.04	5	+5	
8	.59	.71	+.12	10	+10	
9	.80	.91	+.11	9	+9	
10	.70	.64	−.06	7.5		−7.5
Total					+42.5	−12.5

Note: n = Number of relevant observations = 10; T = the smaller of the two rank sums = 12.5.

Therefore, we affix a plus (+) sign to the rank given to animal one, that is, a +7.5.

Step six — Sum the ranks. We now sum the values of the positive ranks and the negative ranks. The sums are given in columns six and seven, respectively, of Table 10.2. The smaller of the two sums is the computed T value. In this problem, the sum of negative signed ranks is smaller than the sum of the positive signed ranks. Therefore the computed T value is 12.5.

Step seven — We are now in a position to draw a statistical conclusion. The computed T value given in step six is compared with the table value to determine whether the null hypothesis is accepted or rejected. Given that this is a one-tailed test, and the level of significance is $\alpha = .05$, the table value found in Appendix H is 10.

Because the computed T value of 12.5 is greater than the table value, we do not reject the null hypothesis, and conclude that the weight gain from the new ration is not an improvement over the original ration.

10.4 Mann–Whitney U test

The *Mann–Whitney U test* is a useful nonparametric technique that is employed in situations where the objective is to test whether two independent samples have been taken from populations with the same mean. We use this test in situations where selected sets of observations are at least ordinal; that is, they can be ranked from low to high or vice versa. This test is equivalent to the Wilcoxon rank sum test. However, instead of working directly with the sum of the ranks, Mann and Whitney have developed this test, also referred to as the U test. The U test uses the formula:

$$U = n_1 n_2 + \frac{n_1(n_1+1)}{2} - R_1 \qquad [10\text{--}2]$$

or

$$U = n_1 n_2 + \frac{n_2(n_2+1)}{2} - R_2 \qquad [10\text{--}3]$$

where R_1 = sum of ranks assigned to sample observation from population one

   R_2 = sum of ranks assigned to sample observation from population two

   Whichever of the two equations that results in a smaller value is used as the computed U value. This computed U value is compared with the table U value in Appendix I to determine whether the null hypothesis should be rejected. If the computed U value is less than the table value, reject the null hypothesis; otherwise do not reject the null hypothesis. The following example illustrates the use of the Mann–Whitney U test.

Example 10.3

An environmental scientist interested in the growth pattern of pine trees conducts an experiment at two locations in a forest. A random sample of ten 1-year-old pine trees' growth rates are observed and recorded. The following results are from the two sites:

Height of Trees in Centimeters at the End of First Growing Year

Site one:	95	97	86	83	98	102	85	120	110	94
Site two:	84	96	97	82	104	100	99	92	96	115

By testing at the .05 level of significance, is the median height of plants on site one greater than on site two?

Solution

We follow the same step-by-step procedure in testing as outlined in Chapter 6.

Let M_1 equal the median height of plants on site one and M_2 the median height of plants on site two.

Step one — The hypotheses are stated as:

$$H_o: \ M_1 \le M_2$$
$$H_1: \ M_1 > M_2$$

This is a right-tailed test, because the alternative hypothesis is that the height of plants on site one is greater than the height of plants on site two.

Step two — The data are pooled into a single group and are ranked as shown in Table 10.3. Because 82 cm is the lowest height observed among the sample of both sites, it is given a rank of one.

Step three — Sum the ranks as shown in columns two and three of Table 10.3.

Step four — Compute the U statistics using Equations 10–2 and 10–3 as follows:

$$U = 10(10) + \frac{10(10+1)}{2} - 104.5 = 50.5$$

$$U = 10(10) + \frac{10(10+1)}{2} - 105.5 = 49.5$$

The computed value is the lesser of the two values obtained from Equations 10–2 and 10–3. Thus, the computed value is 49.5.

Step five — Draw a statistical conclusion. Our conclusion depends on the comparison of the computed value with the table U value given in Appendix I. In this example, the sample sizes were equal, that is, $n_1 = 10$, and $n_2 = 10$. The desired level of significance was given as $\alpha = .05$, and this is a single-tailed test. Therefore, by looking in Appendix I, we find the appropriate U to be 27.

Because the computed value is greater than the table value, we cannot reject the null hypothesis, and conclude that there is no real difference in the height of pine trees grown on site one or two.

Table 10.3 **Rank Sums for Mann–Whitney Test**

Height (cm)	Rank Site one	Rank Site two
82		1
83	2	
84		3
85	4	
86	5	
92		6
94	7	
95	8	
96		9.5
96		9.5
97		11.5
97	11.5	
98	13	
99		14
100		15
102	16	
104		17
110	18	
115		19
120	20	
Rank sum	104.5	105.5

10.5 *Spearman rank correlation coefficient*

When we are interested in the degree of closeness of association between two ordinal variables — that is, the data are not available in numerical values but are only rank-order — we use a measure called *Spearman rank correlation coefficient*, r_s. In other words, r_s is a measure of the degree of correlation that exists between ranked data.

The Spearman rank correlation equation is

$$r_s = 1 - \frac{6 \sum D^2}{n(n^2 - 1)} \qquad [10\text{--}4]$$

where D = difference between ranks for paired observations
        n = number of paired observations

The Spearman rank correlation coefficient ranges in value from –1.0 to +1.0. A value of –1.0 (perfect negative correlation) means that a decreasing relationship exists between the two variables of interest in which a decrease in one variable is accompanied by an increase in the other variable. A value of +1.0 (perfect positive correlation) implies an increasing relationship between the two variables; that is, as one variable increases, so does the other. An r_s value of zero indicates no correlation between the two rankings.

Note this word of caution about the results from the Spearman rank correlation: they should not be interpreted as a measure of linear association between two variables, but instead as a measure of linear association between the ranks of the variables.

The step-by-step procedure for computing the Spearman rank correlation coefficient is illustrated with the following example.

Example 10.4

Two agronomists provide the following preference ranking for ten different preemergence herbicides that they use.

Herbicide	Ranking by agronomist I	Ranking by agronomist II
A	8	6
B	9	9
C	6	5
D	3	7
E	1	2
F	5	4
G	10	8
H	2	1
I	4	10
J	7	3

Can we conclude from these data that there is a significant rank correlation between the two agronomists? Let $\alpha = .01$.

Solution

Step one — The first step is to rank the data. This is usually done on a small-to-large basis. A ranking of one represents the highest ranking; two, the next highest, and so on. In the case of a tie ranking, the average of the two tied ranks is assigned to each rank as was done in the previous sections. In this particular example, the data have already been ranked. The ranks are shown in columns two and three of Table 10.4.

Table 10.4   Rank Correlation of Preference of Two Agronomist for Different Herbicides

Herbicide	Ranking I (X)	Ranking II (Y)	Difference in ranks $D = (X - Y)$	$D^2 = (X - Y)^2$
A	8	6	2	4
B	9	9	0	0
C	6	5	1	1
D	3	7	−4	16
E	1	2	−1	1
F	5	4	1	1
G	10	8	2	4
H	2	1	1	1
I	4	10	−6	36
J	7	3	4	16
Total			0	80

Step two — Compute the difference between ranks. The difference between ranks is labeled D and is shown in column four of Table 10.4.

Step three — Square the value of the difference between ranks and sum the values. This is shown in column five of Table 10.4.

Step four — Compute the Spearman rank correlation using Equation 10–4 as follows:

$$r_s = 1 - \frac{6 \sum D^2}{n(n^2 - 1)}$$

$$r_s = 1 - \frac{6(80)}{10(100 - 1)}$$

$$= 1 - 0.48 = 0.52$$

Step five — Draw a statistical conclusion. Our result of $r_s = 0.52$ indicates some degree of correlation but not very high.

There are many instances where rank correlation is a useful tool. When the data are ordered (i.e., no numerical value is attached to the data), we use the rank correlation. Ranking may be as a result of inadequate units of measurement, lack of time, lack of money, or lack of adequate measuring instruments. Rank correlation may be used to serve as a precursor to performing least squares correlation tests. Because correlation tests assume normality in the variables of interest, when there are substantial departures from normality in the variables of interest, we use rank correlation. Further elaboration on the concept of correlation is presented in Chapters 11 and 12.

Chapter summary

In this chapter you were introduced to a few of the nonparametric tests that are not dependent on the restrictive assumptions of the parametric methods. In the parametric methods we assumed that sample observations came from normally distributed populations. When it is apparent that the samples may not come from a normally distributed population, we use the distribution free methods of the *sign test, Wilcoxon signed-rank test, Mann–Whitney U test*, or the *Spearman rank correlation*.

In situations where one wishes to determine whether there are significant differences between paired observations drawn either from one sample or two related samples, the sign test or the Wilcoxon signed-rank test may be used. The sign test is often used when we are interested only in the direction of the differences.

When the data are drawn from two independent samples, and the researcher wishes to determine the difference between paired ranks, the Mann–Whitney *U* test should be used.

The Spearman rank correlation is employed when we are interested in determining correlation coefficient of ranked data. This test provides a measure of association between the ranks of two variables.

Case Study

Comparative Effects of Fungicides on Nodulation and Yield

Chemical pesticides are designed to eliminate pest problems in agricultural crops. They have played a major role in the agricultural productivity of many countries in the past several decades. Heavily promoted by manufacturers, international aid agencies, and governments of developing nations, pesticides until recently were viewed as the quickest path to food self-sufficiency and as a means to expand agricultural exports needed to fund cash-poor economies. Recently, however, attention has been focused on the negative consequences of pesticide use. It is argued that pesticides suppress the population and functions of indigenous organisms and disrupt the activities of diverse organisms (Rao et al., 1993).

The rates of application to soil of some pesticides such as certain fungicides are quite high, and microorganisms are exposed to levels that could seriously affect individual populations. Use of these fungicides can have additional unforeseen effects on nontarget organisms. This can influence crop productivity as profoundly or even more so than the pests they are intended to control (Smith, 1978; Rodriguez-Kabana and Curl, 1980). A group of nontarget organisms that can be affected by pesticides is nitrogen-fixing organisms. Nitrogen-fixing organisms such as *Rhizobium* bacterioids and certain fungi grow within the outside cell layers of the roots. *Rhizobium* bacteria that live in the root nodules of leguminous plants have the ability to convert atmospheric nitrogen through biological nitrogen fixation into usable nitrogen for easy uptake by the plants. However, other harmful organisms in the soil may prey on these nitrogen-fixing bacteria, reducing the amount of nitrogen introduced into the agricultural and natural ecosystem. There is a need to find pesticides that can control harmful predators or competitors and minimally affect the nitrogen-fixing ability of the *Rhizobium* bacteria.

Various studies have been conducted to determine the impact of fungicides on the nodulation and yield when *Rhizobium* inoculation is present or absent. Tewari and Chahal (1993) tested a number of strains of *Rhizobium* isolated from the nodules of mung bean that were sensitive to thiram (a fungicide). Thiram-resistant strains were effective in fixing nitrogen when seeds were simultaneously treated with thiram. Bandyopahyay (1986) found that captan at 0.3% concentration was inhibitory to five strains of cowpea *Rhizobium*. The next step is for researchers to determine if the differential effects of different uses of pesticides can be replicated over time.

The effects of fungicide on nodulation and yield of groundnut (peanut) were studied by Sridar et al. (1988), using data collected from separate samples in 1985 and 1986. They found that nodule numbers were not affected by the use of fungicides. As shown in the following table, the data suggest that treatment of Dithane M-45 followed by *Rhizobium* inoculation may increase the productivity of groundnut.

In comparing the yield data for the two time periods we can determine if the differences in yields from the different use of fungicides are consistent in the two samples.

1. What type of nonparametric test would you use?
2. Test at the .05 level of significance, if the null hypothesis will be rejected.

| | Pod yield (kg/ha) | |
Treatment	1985	1986
Captan + *Rhizobium*	1366	1244
Captan alone	1283	1170
Thiram + *Rhizobium*	1483	1280
Thiram alone	1150	1307
Bavistin + *Rhizobium*	1366	1413
Bavistin alone	1116	1364
Dithane M-45 + *Rhizobium*	1800	1520
Dithane M-45 alone	1283	1324
Rhizobium alone	1500	1289
Control (uninoculated)	950	1187

Adapted from Sridar, R., Ramalingam, R. S., and Sivaram, M. R., *Indian J. Microbiol.*, 28, 139, 1988.

References

Bandyopahyay, D., Induction of fungicide resistance to rhizobia, *Environ. Ecol.*, 4, 730, 1986.

Rao, V. R., Adhya, T. K., and Sethunathan, N., Effects of pesticides on soils health, in Dahliwal, G. S. and Singh, B., Eds., *Pesticides: Their Ecological Impact in Developing Countries*, Commonwealth, New Delhi, India, 112, 1993.

Rodriquez-Kabana, R. and Curl, E. A., Non-targeted effects of pesticides on soil borne pathogen and disease, *Annu. Rev. Phytopathol.*, 18, 311, 1980.

Smith, T. F., Some effects of crop protection chemicals on the distribution and abundance of vesicular-arbuscular endomycorrhizae, *J. Aust. Inst. Agric. Sci.*, 44, 82, 1978.

Sridar, R., Ramalingam, R. S., and Sivaram, M. R., Effect of fungicides on nodulation and yield of groundnut, *Indian J. Microbiol.*, 28, 139, 1988.

Tewari, N. and Chahal, V. P. S., Effects of fungicide resistant strains of *Rhizobium* spp. on N_2 fixation in mung bean (*Vigna radiata* L.), in *Proc. Natl. Symp. Integ. Pest Manage.*, 1993, 25.

Review questions

1. The marketing division of the Milk Producers Association develops a new promotional campaign to promote milk consumption. Ten different grocery stores are asked to participate in recording the milk sales before and after the commercial appeared on television. The sales for the ten grocery stores are recorded a week prior to and after the new commercial.

Store	Sales (thousand of $)	
	Before	After
1	16	20
2	22	18
3	26	30
4	20	28
5	34	35
6	20	20
7	36	34
8	29	31
9	18	22
10	20	26

Use the sign test to determine whether to accept or reject the null hypothesis that the sales are higher than before the new commercial. Use $\alpha = .01$.

2. Animal scientists formulate new rations to increase the maximum rate of lean tissue growth in steers. After feeding a sample of 15 steers this new ration, the following data show the maximum rate of lean tissue growth in grams per day at the beginning and end of trial test.

Steers	Before	After
1	420	600
2	480	720
3	360	350
4	470	760
5	560	530
6	670	850
7	550	690
8	580	540
9	480	445
10	590	740
11	580	730
12	650	600
13	640	620
14	690	790
15	520	500

Do these data provide sufficient evidence to indicate the new ration is effective in increasing the rate of lean tissue growth in steers? Use the Wilcoxon test. Let $\alpha = .05$.

3. An agronomist conducts an experiment comparing the effects of two different fertilizers on yield of wheat. The fertilizers are

applied at random to five different plots of uniform size. Yield in tons per hectare are given as follows:

Fertilizer A	3.2	4.5	5.6	4.4	4.6
Fertilizer B	3.8	5.2	5.6	5.8	5.0

Do these data provide sufficient evidence to indicate a difference in yield from the two different fertilizers? Let $\alpha = .10$.

4. An ornamental horticulturist experiments with a plant growth enhancer to determine its effectiveness on different varieties of ornamentals. Average plant height 4 weeks after treatment are as follows:

Plant	A	B	C	D	E	F	G	H	I	J
					Growth (cm)					
Growth enhancer	57	53	38	35	22	30	65	20	96	30
No growth enhancer	38	45	43	32	21	28	69	27	85	18

Conduct a Sign test at the .10 level.

5. A researcher is interested in the effectiveness of drip and flood irrigation on the yield of tomatoes. The yield from a random sample of ten experimental plots results in the following:

	Yield (in boxes)	
Plot	Drip irrigation X	Flood irrigation Y
1	20	20
2	22	18
3	24	19
4	30	25
5	21	19
6	20	20
7	24	25
8	23	26
9	25	22
10	18	20

Perform a Sign test at $\alpha = .05$.

6. Farm subsidy programs have been a center of controversy among the general public in recent years. A researcher gives a

random sample of ten rural residents and ten urban residents a test to measure their knowledge of the farm subsidy programs. The results of the test are given as follows. The researcher wishes to know whether she can conclude on the basis of these data that the two population scores are different in terms of their median score. Test at $\alpha = .05$.

Test scores on knowledge of farm subsidy

Rural residents	Urban residents
80	70
85	79
90	90
82	85
90	89
77	80
75	50
72	60
84	82
60	75

7. An animal science researcher is conducting an experiment comparing the effects of two different rations on ten Holsteins. The average milk yield, in pounds per cow per day, during the period of study are given as follows. Do these data provide sufficient evidence to indicate a difference in the rations? Let $\alpha = .01$.

Ration A	Ration B
65	45
60	50
62	60
55	52
46	51
75	48
68	58
70	65
63	68
66	61

8. Water samples at two different locations show the following bacterial counts.

Location A	22	25	30	28	24
Location B	21	26	36	37	31

Do the data present sufficient evidence to indicate a difference in the population distributions of bacteria count? Let $\alpha = .05$.

9. Using the data in Chapter 8, Example 8.1, suppose our environmentalist is interested in checking if there is a difference between the dissolved oxygen content at location (A) where the industrial plant is located and location (B) where the city releases its treated sewage water. The data are given as follows:

Location	Mean dissolved oxygen content (ppm)			
A	3.9	4.8	4.2	5.0
B	4.7	4.9	4.1	3.9

Perform a Mann–Whitney test with $\alpha = .01$.

References and suggested reading

Conover, W. J., *Practical Nonparametric Statistics,* 2nd ed., John Wiley & Sons, New York, 1980.

Daniel, W. W., *Applied Nonparametric Statistics,* 2nd ed., PWS-KENT, Boston, 1990.

Daniel, W. W. and Terrell, J. C., *Business Statistics: Basic Concepts and Methodology,* 4th ed., Houghton Mifflin, Dallas, 1986.

Friedman, M., The use of ranks to avoid the assumption of normality implicit in the analysis of variance, *J. Am. Stat. Assoc.,* 32, 1937.

Lapin, L. L., *Statistics for Modern Business Decision,* 4th ed., Harcourt Brace Jovanovich, New York, 1987.

Mann, H. B. and Whitney, D. R., On a test of whether one of two random variables is stochastically larger than the other, *Ann. Math. Stat.,* 18, 50, 1947.

Noether, G. E., *Introduction to Statistics: A Nonparametric Approach,* 2nd ed., Houghton Mifflin, Boston, 1976.

Seigel, S. and Castellan, N. J., *Nonparametric Statistics.* 2nd ed., McGraw-Hill, New York, 1988.

Wilcoxon, F., Individual comparisons by ranking methods, *Biometrics,* 1, 80, 1945.

PART V

Test of Association and Prediction

chapter eleven

Simple Regression and Correlation

Learning objectives

In this chapter you are introduced to two very important statistical techniques, regression analysis and correlation analysis. After mastering this chapter and working through the review questions, you should be able to:

- Explain the purpose of regression and correlation analysis in agricultural and environmental research.
- Compute the regression equation and the standard error of estimate.
- Use the computed equation to prepare interval estimates, and make predictions for forecasting purposes.
- Compute a measure of the strength of the correlation between two variables (coefficients of determination and correlation).

- Perform tests of hypothesis about the relationship of the two variables.
- Clearly state the assumptions of linear regression analysis, and point out pitfalls in using the regression and correlation techniques.

11.1 Introduction: bivariate relationships

In many environmental and agricultural research problems you may be faced with a situation where you are interested in the relationship that exists between two different random variables X and Y. Such a relationship is known as a *bivariate* relationship. For example, an animal scientist may be interested in the relationship between the amount of total digestible nutrient (TDN) and the average daily weight gain of the animal. An agricultural economist may be interested in the bivariate relationship between the appraised value of farm real estate, X, and its sale price, Y; or an agronomist may use a crop–weather model to analyze the effects of weather (X) on the yield of a crop (Y).

To determine if one variable is a predictor of another variable, we use the bivariate modeling technique. The simplest model for relating a variable Y to a single variable X is a straight line. This is referred to as a *linear* relationship. *Simple linear regression* is used as a technique to judge whether a relationship exists between Y and X. Furthermore, the technique is used to estimate the mean value of Y, and to predict a future value of Y for a given value of X.

In simple regression analysis, we are interested in describing the pattern of the functional nature of the relationship that exists between two variables. This is accomplished by estimating an equation called the regression equation. The variable to be estimated in the regression equation is called the *dependent variable* and is plotted on the vertical (or Y) axis. The variable used as the predictor of Y, which exerts influence in explaining the variation in the dependent variable, is called the *independent variable*. This variable is plotted on the horizontal (or X) axis.

The linear relationship between the two variables Y and X is expressed by the general equation for a straight line as:

$$\hat{Y} = a + bX \qquad\qquad [11\text{--}1]$$

where $\hat{Y}$ = estimated value of dependent variable
  a = regression constant, or Y intercept
  b = regression coefficient, or slope of regression line
  X = given value of independent variable

In correlation analysis, on the other hand, we are simply interested in the *magnitude* or *closeness* of the relationship between two variables.

Regression and correlation analyses work together. One asks if there is any relationship between two variables, and the other seeks to provide an answer to how close this relationship is.

In the following sections of this chapter, you are introduced to the techniques for estimating a regression equation, the standard error, and the coefficients of determination and correlation. The concepts developed in this chapter can be applied to more than two variables, as discussed in the next chapter on *multiple regression*.

11.2 Regression analysis

As pointed out in the introduction, simple linear regression analysis is concerned with the relationship that exists between two variables. Researchers use prior knowledge and past research as a basis for selecting the independent variables that are helpful in predicting the values of the dependent variable.

After we have determined that there is a logical relationship between two variables, we can portray the relationship between the variables through a *scatter diagram*. A scatter diagram is a graph of the plotted points, each of which represents an observed pair of values of the dependent and independent variables. The scatter diagram serves two purposes: (1) it provides a visual presentation of the relationship between two variables, and (2) it aids in choosing the appropriate type of model for estimation.

In Example 11.1 a set of data is presented that is used to illustrate how a scatter diagram is helpful in determining the presence or lack of linear relationship between dependent and independent variables.

Example 11.1

An animal scientist interested in the milk yield of a lactating ewe measures the milk yield (kilograms per day) by weighing the lamb before and after suckling. The following observations are obtained at different time intervals. The scatter diagram is constructed as follows.

Day	Yield	Day	Yield
10	1.78	42	1.50
14	1.66	46	1.48
18	1.62	50	1.43
22	1.59	54	1.40
26	1.55	58	1.37
30	1.60	62	1.35
34	1.58	66	1.32
38	1.54		

Solution

By following the standard convention of plotting the dependent variable along the Y axis and the independent variable along the X axis, we have the milk yield plotted along the Y axis and the day intervals along the X axis. Figure 11.1 shows the scatter diagram for this problem.

The scatter diagram is also used to determine whether there is a *linear* or a *curvilinear* relationship between variables. If a straight line can be used to describe the relationship between variables X and Y, a linear relationship exists. If the observed points in the scatter diagram fall along a curved line, a *curvilinear relationship* exists between variables.

Figure 11.2 illustrates a number of different scatter diagrams depicting different relationships between variables. You notice that scatter diagrams *a* and *b* illustrate a positive and negative linear relationship between two variables, respectively. Diagrams *c* and *d* show the positive and negative curvilinear relationship between variables X and Y.

Another curvilinear relationship is illustrated in *e* where X and Y rise at first, and then as X increases, Y decreases. Such a relationship is observed in agricultural economics and business, for instance, where a farmer's earned income tends to rise with the age of the farmer and then declines after the farmer retires.

Figure 11.2 (*f*) shows no relationship between the variables.

Linear regression equation.   The mathematical equation of a line such as the one in the scatter diagram in Figure 11.1 that describes the relationship between two variables is called the *regression* or *estimating*

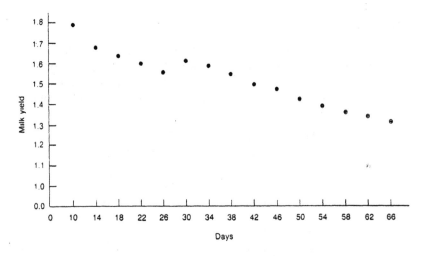

Figure 11.1   A scatter diagram of milk yield of a lactating ewe at 4-d intervals.

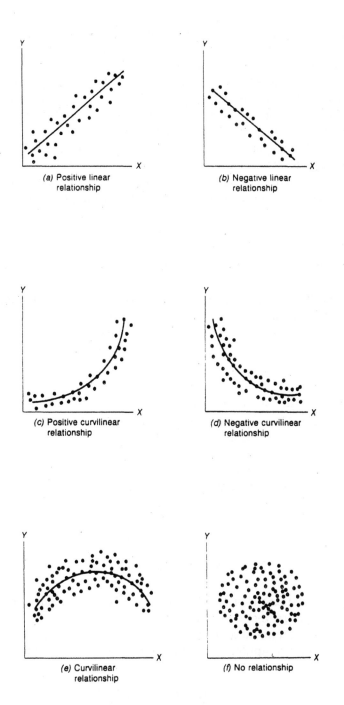

Figure 11.2 Examples of linear and curvilinear relationships found for scatter diagrams.

equation. The regression equation has its origins in the pioneering work of Sir Frances Galton (1877), who fitted lines to scatter diagrams of data on the heights of fathers and sons. Galton found that the heights of children of tall parents tended to regress toward the average height of the population. Galton referred to his equation as the regression equation.

The regression equation is determined by the use of a mathematical method referred to as the *least squares*. This method simply minimizes the sum of the squares of the vertical deviations about the line. Thus, the least squares method is a *best fit* in the sense that the $\sum (Y - \hat{Y})^2 = 0$ is less than it would be for any other possible straight line. Additionally, the least squares regression line has the following property:

$$\sum (Y - \hat{Y}) = 0 \qquad [11\text{--}2]$$

This characteristic makes the total of positive and negative deviations equal to zero.

You should note that the linear regression equation, Equation 11–1, is just an estimate of the relationship between the two variables in the population given in the following Equation 11–3

$$\mu_{y.x} = A + BX \qquad [11\text{--}3]$$

where $\mu_{y.x}$ = mean of Y variable for a given X value

A and B = population parameters that must be estimated from sample data

The regression equation can be calculated by two methods. The first involves solving simultaneously two equations called the *normal equations.* They are

$$\sum Y = na + b\sum X \qquad [11\text{--}4]$$

$$\sum XY = a\sum X + b\sum X^2 \qquad [11\text{--}5]$$

We use Equations 11–4 and 11–5 to solve for *a* and *b*, and obtain the estimating or regression equation.

The second method of arriving at a least squares regression equation is using a computationally more convenient equation given below:

$$b = \frac{n(\sum XY) - (\sum X)(\sum Y)}{n(\sum X^2) - (\sum X)^2} \qquad [11\text{--}6]$$

$$a = \frac{\Sigma Y}{n} - b\left(\frac{\Sigma X}{n}\right) = \overline{Y} - b\overline{X} \qquad [11\text{–}7]$$

The following example illustrates the computation of the regression equation, using either the normal equations or the shortcut formula given in Equations 11–6 and 11–7.

Example 11.2

An agronomist is interested in the relationship between maize yield, Y, and the amount of fertilizer applied to maize, X. To determine whether there is a relationship, the agronomist divides a field into ten plots of equal size in different localities in Ohio and applies different amounts of fertilizer to each plot. The yield data (in bushels) recorded for the different amounts of fertilizer (in pounds) applied follow:

Yield Y	Fertilizer X
50	5
57	10
60	12
62	18
63	25
65	30
68	36
70	40
69	45
66	48

Solution

To compute the regression equation, the data in Table 11.1 are used. We substitute the appropriate values from Table 11.1 into Equations 11–4 and 11–5 as follows:

$$\Sigma Y = na + b\Sigma X$$
$$\Sigma XY = a\,\Sigma X + b\Sigma X^2$$

$$630 = 10a + 269b$$
$$17{,}702 = 269a + 9.343b$$

To solve the preceding two equations for either of the unknowns *a* and *b*, we must eliminate one of the unknown coefficients. For example,

Table 11.1   Computation of Intermediate Values Needed
for Calculating the Regression Equation

Yield Y	Fertilizer X	Y^2	XY	X^2
50	5	2,500	250	25
57	10	3,249	570	100
60	12	3,600	720	144
62	18	3,844	1,116	324
63	25	3,969	1,575	625
65	30	4,225	1,950	900
68	36	4,624	2,448	1,296
70	40	4,900	2,800	1,600
69	45	4,761	3,105	2,025
66	48	4,356	3,168	2,304
630	269	40,028	17,702	9,343

to eliminate the unknown coefficient a in these two equations, we multiply the first equation by 26.9. By doing so, the value of the a coefficient in the first equation is now equal to the value of a in the second equation. We then subtract the second from the first to obtain the value of b as shown:

$$16,947 = 269a + 7,236.1b$$
$$\underline{-17,702 = -269a - 9,343b}$$
$$-755 = -2,106.9b$$
$$b = 0.358$$

The value of b can be substituted in either Equation 11–4 or 11–5 to solve for the value of a.

$$630 = 10a + 269(0.358)$$
$$630 = 10a + 96.302$$
$$533.69 = 10a$$
$$a = 53.369$$

The least squares regression equation is

$$\hat{Y} = 53.369 + 0.358X$$

The same answer would be obtained by using the shortcut formula as shown as follows:

$$b = \frac{n(\Sigma XY) - (\Sigma X)(\Sigma Y)}{n(\Sigma X^2) - (\Sigma X)^2}$$

$$= \frac{10(17,702) - (269)(630)}{10(9,343) - (269)^2}$$

$$= \frac{7,550}{21,069}$$

$$= 0.358$$

We now substitute the value of b into Equation 11–7 to obtain the intercept of the line, or a, as follows:

$$a = \frac{630}{10} - 0.358 \left(\frac{269}{10} \right)$$

$$= 63.0 - 0.358(26.9)$$

$$= 63.0 - 9.630$$

$$= 53.369$$

Hence, the least squares regression line is

$$\hat{Y} = 53.369 + 0.358X$$

In this estimated equation, $a = 53.369$ and is an estimate of the Y intercept. The value of a means that the yield of maize from a plot of land with no fertilizer added is equal to 53.369 bushels. The b value indicates that the slope of the line is positive. This means that as fertilizer usage increases, so does the yield. The value of b implies that for each additional pound of fertilizer applied, the yield increases by 0.358 bushels within the range of values observed.

The preceding regression equation can also be used to estimate values of the dependent variable for given values of the independent variable. For example, if the agronomist wishes to estimate the yield from 45 lb of fertilizer, the estimated yield is

$$\hat{Y} = 53.369 + 0.358 (45)$$

$$= 53.369 + 15.036$$

$$= 69.479 \text{ bushels of maize}$$

You must keep in mind that the sample regression equation should not be used for prediction outside the range of values of the independent variable given in a sample.

To graph the regression line, we need two points. Because we have determined only one ($X = 45$, $Y = 69.479$), we need one other point for graphing the regression line. The second point ($X = 10$, $Y = 56.949$) is shown, along with the original data in Figure 11.3. The estimated yield values of 56.949 and 69.479 should be treated as average values. This means that in the future, the average yield will vary from sample to sample because of the fertility of the soil, the temperature, the amount of water used during the growing season, and a host of other factors.

In the next section, we examine a measure that helps us determine whether the estimate made from the regression equation is dependable.

11.3 Standard error of estimate

The regression equation is primarily used for estimation of the dependent variable, given values of the independent variable. Once we have estimated a regression equation, it is important to determine whether the estimate is dependable or not.

Dependability is measured by the closeness of the relationship between the variables. If in a scatter diagram the points are scattered

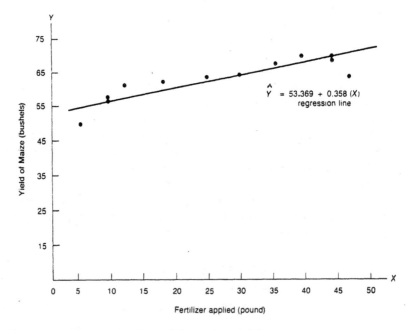

Figure 11.3 Regression line of the maize yield.

close to the regression line, a close relationship exists between the variables. If, on the other hand, there is a great deal of dispersion between the points and the regression line, the estimate made from the regression equation is less reliable.

The *standard error of estimate* is used as a measure of scatter or dispersion of the points about the regression line, just as one uses the standard deviation to measure the deviation of the individual observations about the mean of those values. The smaller the standard error of estimate, the closer the estimate is likely to be to the ultimate value of the dependent variable. In the extreme case where every point falls on the regression line, the vertical deviations are all 0; that is, $S_{y.x} = 0$. In such a situation, the regression line provides perfect predictions. On the other hand, when the scatter is highly dispersed, making the vertical deviation large ($S_{y.x}$ is large), the predictions of Y made from the regression line are subject to sampling error.

The standard error of estimate ($S_{y.x}$) is computed by solving the following equation:

$$S_{y.x} = \sqrt{\frac{\Sigma(Y - \hat{Y})^2}{n - 2}} \qquad [11\text{--}8]$$

where Y = dependent variable
$\hat{Y}$ = estimated value of dependent variable
n = sample size

The $n - 2$ value in the denominator represents the number of degrees of freedom (*d.f.*) around the fitted regression line. Generally, the denominator is $n - k$ where k represents the number of constants in the regression equation. In the case of a simple linear regression, we lose 2 *d.f.* when *a* and *b* are used as estimates of the constants in the population regression line. Notice that Equation 11–8 requires a value of Y for each value of $\hat{Y}$. We must therefore compute the difference between each $\hat{Y}$ and the observed value of Y as shown in Table 11.2.

The standard error of estimate for Example 11.2 is calculated as follows:

$$S_{y.x} = \sqrt{\frac{\Sigma(Y - \hat{Y})^2}{n - 2}}$$

$$= \sqrt{\frac{67.448}{8}}$$

$$= 2.90 \text{ bushels of maize}$$

Table 11.2   **Computation of the Intermediate Values
Needed for Calculating the Standard Error of Estimate**

Yield Y	Fertilizer X	$\hat{Y}$	$(Y - \hat{Y})$	$(Y - \hat{Y})^2$
50	5	55.159	−5.159	26.615
57	10	56.949	0.051	0.003
60	12	57.665	2.335	5.452
62	18	59.813	2.187	4.783
63	25	62.319	0.681	0.464
65	30	64.109	0.891	0.794
68	36	66.257	1.743	3.038
70	40	67.689	2.311	5.341
69	45	69.479	−0.479	0.229
66	48	70.553	−4.553	20.729
630	269	630.000	0.0	67.488

The preceding computational method requires a great deal of arithmetic, especially when large numbers of observations are involved. To minimize cumbersome arithmetic, the following shortcut formula is used in computing the standard error of estimate:

$$S_{y.x} = \sqrt{\frac{\Sigma Y^2 - a(\Sigma Y) - b(\Sigma XY)}{n - 2}} \qquad [11\text{--}9]$$

All the values needed to compute the standard error of estimate are available from Table 11.1 and the previously obtained values of a and b.

$$S_{y.x} = \sqrt{\frac{40,028 - 53.369(630) - 0.358(17,702)}{8}}$$

$$= \sqrt{\frac{68.214}{8}}$$

= 2.92 bushels of maize

The answers from the two approaches are similar, as expected. The minute difference in the two values of $S_{y.x}$ results from rounding.

Because the standard error of estimate is theoretically similar to the standard deviation, there is also similarity in the interpretation of the standard error of estimate. If the scatter about the regression line is normally distributed, and we have a large sample, approximately 68% of the points in the scatter diagram fall within one standard error of estimate above and below the regression line; 95.4% of the points fall within two standard errors of estimate above and below the regression line, and virtually all points above and below the regression line fall within three standard errors of estimate, as shown in Figure 11.4.

The standard error of estimate is used to construct confidence limits. We now turn to the discussion of confidence interval estimate.

Confidence interval estimate. We use the regression equation to estimate the value of Y given a value of X. An estimate of Y_1 is obtained by simply inserting a value for X_1 into the regression equation $Y_1 = a + bX_1$. The estimate of Y_1 is nothing more than a *point estimate*. To attach some confidence to this point estimate, we use the standard error of estimate to compute an interval estimate where a probability value may

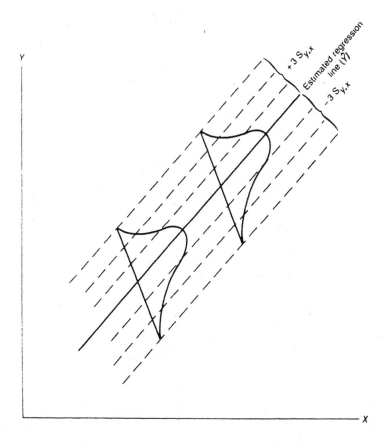

Figure 11.4 Illustration of the standard error of estimate about the estimated regression line.

be assigned to it. Thus, an interval estimate for an individual Y, denoted here as Y_i, using a small sample is computed by the following equation:

$$Y_i = \hat{Y} \pm t(S_{y.x}) \sqrt{1 + \frac{1}{n} + \frac{(X_o - \overline{X})^2}{\Sigma(X - \overline{X})^2}} \qquad [11\text{--}10]$$

We use Equation 11–10 to obtain a confidence interval for a given value of X, say X_o. A computationally more convenient form of the Equation 11–10 follows, where we wish to predict the yield of maize when 35 lb of fertilizer is applied.

$$Y_i = \hat{Y} \pm t(S_{y.x}) \sqrt{1 + \frac{1}{n} + \frac{(X_o - \overline{X})^2}{\Sigma X^2 - \frac{(\Sigma X)^2}{n}}} \qquad [11\text{--}11]$$

Substituting 35 for X in the sample regression equation gives:

$$\hat{Y} = 53.369 + 0.358\,(35) = 65.89$$

To construct a 95% prediction interval, we use Equation 11–11 and the data from Table 11.1. The critical value of t for the 95% level of confidence is given as 2.306 from Appendix E. The computation of the prediction interval is as follows:

$$Y_i = 65.89 \pm 2.306\,(2.90) \sqrt{1 + \frac{1}{10} + \frac{(35 - 26.9)^2}{9{,}343 - \frac{(269)^2}{10}}}$$

$$Y_i = 65.89 \pm 7.09$$

or

$$58.80 < Y_i < 72.98 \text{ bushels}$$

Hence, the prediction interval is from 58.80 to 72.98 bushels. To make a probabilistic interpretation of this number, we would say that if we draw repeated samples, perform a regression analysis, and construct prediction intervals for the agronomist who uses 35 lb of fertilizer, 95% of the interval include the yield. Another way of interpreting the number is that we are 95% confident that the single prediction interval constructed includes the true yield.

The prediction interval for an individual value of Y when using large samples is determined by the following expression:

$$\hat{Y} \pm z\, S_{y.x} \qquad\qquad [11\text{--}12]$$

11.4   Correlation analysis

In regression analysis we emphasize estimation of an equation that describes the relationship between two variables. In this section, we are interested in those measures that verify the degree of closeness or association between two variables, and the strength of the relationship between them. We examine two correlation measures: the *coefficient of determination* and the *coefficient of correlation*.

Before using these correlation measures, it is important to have an understanding of the assumptions of the two-variable correlation models. In correlation analysis, we make the assumption that both X and Y are random variables; furthermore, they are normally distributed. Also, the standard deviations of the Ys are equal for all values of X, and vice versa.

Coefficient of determination.   As a measure of closeness between variables, the coefficient of determination (r^2) can provide an answer to how well the least-squares line fits the observed data. The relative variation of the Y values around the regression line and the corresponding variation around the mean of the Y variable can be used to explain the correlation that may exist between X and Y. Conceptually, Figure 11.5 illustrates three different deviations — namely, the *total deviation*, the *explained deviation*, and the *unexplained deviation* — that exist between a single point Y and the mean and the regression line.

The vertical distance between the regression line and the $\overline{Y}$ line is the explained deviation, and the vertical distance of the observed Y from the regression line is the unexplained deviation. The unexplained deviation represents that portion of total deviation that was not explained by the regression line. The distance between Y and $\overline{Y}$ is called the total deviation. Stated differently, the total deviation is the sum of the explained and the unexplained deviation, or total deviation = explained deviation + unexplained deviation:

$$Y - \overline{Y} = (\hat{Y} - \overline{Y}) + (Y - \hat{Y}) \qquad\qquad [11\text{--}13]$$

To transform the preceding equation into a measure of variability, we simply square each of the deviations in Equation 11–13 and sum

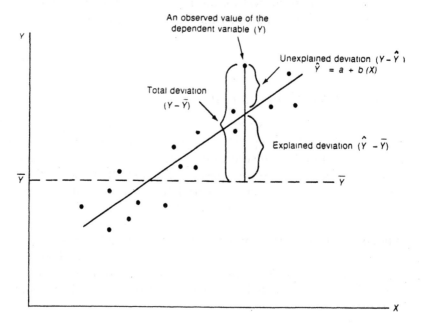

Figure 11.5 Total deviation, explained deviation, and unexplained deviation for one observed value of Y.

for all observations to obtain the squared deviations, or total sum of squares = explained sum of squares + unexplained sum of squares:

$$\Sigma(Y - \overline{Y})^2 = \Sigma(\hat{Y} - \overline{Y})^2 + \Sigma(Y - \hat{Y})^2 \qquad [11\text{--}14]$$

The term on the left-hand side of Equation 11–14 is now the *total sum of squares*, which measures the dispersion of the observed value of the Y about their mean $\overline{Y}$. Similarly, on the right-hand side of Equation 11–14 we have the explained and unexplained sum of squares, respectively.

Given the preceding relationships, we can define the sample coefficient of determination as:

$$r^2 = 1 - \frac{\text{unexplained sum of squares}}{\text{total sum of squares}}$$

or

$$r^2 = \frac{\text{explained sum of squares}}{\text{total sum of squares}}$$

or

$$r^2 = \frac{\Sigma(\hat{Y} - \overline{Y})^2}{\Sigma(Y - \overline{Y})^2} \qquad [11\text{-}15]$$

To compute r^2 using Equation 11–15, we need the value of the explained and total sum of squares, as computed in Table 11.3. Substituting the value of the explained and the total sum of squares into Equation 11–15 we get:

$$r^2 = \frac{269.97}{338} = 0.798$$

As is apparent, computation of r^2 using Equation 11–15 is tedious, particularly with a large sample. To remedy the situation, we use a shortcut formula that utilizes the estimated regression coefficients and the intermediate values used to compute the regression equation. Thus, the shortcut formula for computing the sample coefficient of determination is:

$$r^2 = \frac{a \Sigma Y + b \Sigma XY - n\overline{Y}^2}{\Sigma Y^2 - n\overline{Y}^2} \qquad [11\text{-}16]$$

Table 11.3   Calculation of the Sum of Squares when $\overline{Y}$= 63

Yield Y	Fertilizer X	$\hat{Y}$	$(Y - \overline{Y})^2$	$(\hat{Y} - \overline{Y})^2$
50	5	55.159	169	61.48
57	10	56.949	36	36.61
60	12	57.665	9	28.46
62	18	59.813	1	10.15
63	25	62.319	0	0.46
65	30	64.109	4	1.22
68	36	66.257	25	10.60
70	40	67.689	49	21.98
69	45	69.479	36	41.97
66	48	70.553	9	57.04
630	269	630.000	338	269.97

To calculate the r^2 for the agronomist in Example 11.2, we have:

$$r^2 = \frac{53.369\,(630) + .358\,(17,702) - 10\,(63)^2}{40,028 - 10\,(63)^2}$$

$$= \frac{33,622.47 + 6,337.32 - 39,690}{40,028 - 39,690}$$

$$= \frac{269.79}{338}$$

$$= 0.798$$

The calculated $r^2 = 0.798$ signifies that 79.8% of the total variation in yield of maize Y can be explained by the relationship between the yield and the amount of fertilizer X applied. Because the greatest possible value that r can have is 1, the calculated r in this example implies a strong linear relationship between X and Y. Similar to the sample coefficient of determination, the population coefficient of determination ρ^2 (the Greek letter rho) is equal to the ratio of the explained sum of squares to the total sum of squares.

Sample correlation coefficient without regression analysis. Another parameter that measures the strength of the linear relationship between two variables X and Y is the coefficient of correlation. The sample coefficient of correlation (r) is defined as the square root of the coefficient of determination. The population correlation coefficient measures the strength of the relationship between two variables in the population. Thus, the sample and population correlation coefficients are

$$r = \sqrt{r^2} \qquad\qquad\qquad [11\text{--}17]$$

$$\rho = \sqrt{\rho^2} \qquad\qquad\qquad [11\text{--}18]$$

The correlation coefficient can be any value between -1 and $+1$ inclusive. When r equals -1, there is a perfect inverse linear correlation between the variables of interest. When r equals 1, there is a perfect direct linear correlation between X and Y. When r equals 0, the variables X and Y are not linearly correlated. Figure 11.6 shows different scatter diagrams representing various simple correlation coefficients.

The algebraic sign of r is the same as that of b in the regression equation. Thus, if the regression coefficient b is positive, then r also has a positive value. Similarly, if the regression coefficient is negative, we expect r to have a negative value.

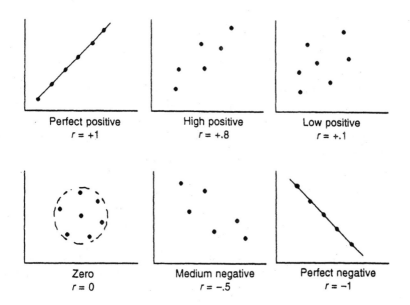

Figure 11.6   Various scatter diagrams with different correlation coefficients.

The sample coefficient of correlation for Example 11.2 is

$$r = \sqrt{0.798}$$

$$= 0.89$$

Note that the sign of this sample correlation coefficient is positive, as was the *b* coefficient.

A word of caution is needed with respect to the interpretation of the correlation coefficient. In the case of the coefficient of determination r^2, we could interpret that parameter as a proportion or as a percentage. However, when the square root of a percentage is taken, as is the case with the correlation coefficient, the specific meaning attached to it becomes obscure. Therefore, we can merely conclude that the closer the *r* value is to 1, the better the correlation is between X and Y. Because r^2 is a decimal value, its square root *r* is a larger number. This may give a false impression that a high degree of correlation exists between X and Y. For example, consider a situation where $r = 0.70$ that indicates a relatively high degree of association. However, because $r^2 = 0.49$, the reduction in total variation is only 49%, or less than half. Despite the common use of the correlation coefficient, it is best to use the coefficient of determination to explain the degree of association between X and Y when regression analysis is employed.

The correlation coefficient that is computed without performing regression analysis does provide more meaningful results. In the following section, you are shown how to compute the correlation coefficient without performing a regression analysis.

There are many instances where a researcher is not interested in making a prediction, but is interested in whether there is a relationship between X and Y. For example, an animal scientist may be interested in whether increasing the concentration of a drug's dosage reduces the symptoms. In such cases the following formula, referred to as the *Pearson sample correlation coefficient*, can be used:

$$r = \frac{\sum XY - n\overline{X}\overline{Y}}{\sqrt{\left(\sum X^2 - n\overline{X}^2\right)\left(\sum Y^2 - n\overline{Y}^2\right)}} \qquad [11\text{--}19]$$

We may use the intermediate calculations found earlier in Example 11.2 to compute the sample correlation coefficient shown as follows:

$$n = 10 \qquad \overline{X} = 26.9 \qquad \overline{Y} = 63$$

$$\sum XY = 17,702 \qquad \sum X^2 = 9,343 \qquad \sum Y^2 = 40,028$$

$$r = \frac{17,702 - 10(26.9)(63)}{\sqrt{[(9,343 - 10(26.9)^2][40,028 - 10(63)^2]}}$$

$$r = \frac{17,702 - 16,947}{\sqrt{712,132.2}}$$

$$= \frac{755}{843.881}$$

$$= 0.89$$

The correlation coefficient 0.89 computed with Equation 11–19 is the same as the value we found earlier by taking the positive square root of r^2.

Inferences regarding regression and correlation coefficients.   So far, our discussions have centered around the computation and interpretation of the regression and correlation coefficients. How confident we can be that these coefficients do not contain sampling error and that they correspond to the population parameters is the topic of this section. Hypothesis testing, or confidence interval estimation, is often used to determine whether the sample data provide sufficient evidence to indicate that the estimated regression coefficient differs from zero. If we can reject the null hypothesis that b is not equal to zero, we can conclude that X and Y are linearly related.

To illustrate the hypothesis-testing procedure for a regression coefficient, let us use the data from Example 11.2. The null and alternative hypotheses regarding the regression coefficient may be stated as:

$$H_o = \beta = 0$$
$$H_1 = \beta \neq 0$$

To test this hypothesis, the agronomist wishes to use a .05 level of significance. The procedure involves a two-tailed test in which the test statistic is

$$t = \frac{b - \beta}{s_b} \qquad\qquad [11\text{--}20]$$

where s_b is the estimated standard error of the regression coefficient and is computed as follows:

$$s_b = \frac{S_{y.x}}{\sqrt{\Sigma X^2 - n\overline{X}^2}} \qquad\qquad [11\text{--}21]$$

We substitute the appropriate values into Equations 11–21 and 11–20 to perform the test:

$$s_b = \frac{2.90}{\sqrt{9343 - 10(26.9)^2}}$$

$$= 0.06$$

The calculated test statistic is

$$t = \frac{.358}{0.06}$$

$$= 5.966$$

Given an $\alpha = .05$ and 8 *d.f.* $(10 - 2 = 8)$, the critical value of t from Appendix E is 2.306. Because the computed t exceeds the critical value of 2.306, the null hypothesis is rejected at the .05 level of significance, and we conclude that the slope of the regression line is not 0.

Similar to the test of significance of b, the slope of the regression equation, we can perform a test for the significance of a linear relationship between X and Y. In this test, we are basically interested in knowing whether there is correlation in the population from which the sample was selected. We may state the null and alternative hypotheses as:

$$H_o = \rho = 0$$

$$H_1 = \rho \neq 0$$

The test statistic for samples of small size is

$$t = \frac{r\sqrt{n-2}}{\sqrt{1-r^2}} \qquad [11\text{–}22]$$

Again, using a .05 level of significance and $n - 2$ d.f., the critical t from Appendix E is 2.306. The decision rule states that if the computed t falls within ±2.306, we should accept the null hypothesis; otherwise, reject it.

The test statistic for the current problem is

$$t = \frac{0.89\sqrt{10-2}}{\sqrt{1-(0.89)^2}}$$

$$= \frac{0.89\,(2.828)}{\sqrt{1-0.792}}$$

$$= 5.52$$

Because 5.52 exceeds the critical t value of 2.306, the null hypothesis is rejected: and we conclude that X and Y are linearly related.

For large samples, the test of significance of correlation can be performed using the following equation:

$$z = \frac{r}{\dfrac{1}{\sqrt{n-1}}} \qquad [11\text{–}23]$$

F Test: An Illustration.   Instead of using the t distribution to test a coefficient of correlation for significance, we may use an analysis of variance or the F ratio. In computing the r^2 and the $S_{y.x}$, we partition the total sum of squares into explained and unexplained sums of squares. To perform the F test, we first set up the variance table, determine the degrees of freedom, and then compute F as a test of our hypothesis. The null and alternative hypotheses are

$$H_o = \rho^2 = 0$$
$$H_1 = \rho^2 \neq 0$$

We will test the hypothesis at the .05 level of significance. Table 11.4 shows the computation of the F ratio. The numerical values of the explained and total sums of squares used in Table 11.4 were presented earlier in Table 11.3. Notice that the total degrees of freedom are always $n - 1$. Therefore, for the present problem, the total is 9 *d.f.* The degrees of freedom associated with the explained variation are always equal to the number of independent variables used to explain variations in the dependent variable ($n - k$, where k refers to the number of independent variables). Thus, for the present problem we have only 1 *d.f.* for the explained variation. The degrees of freedom associated with the unexplained variation are found simply by subtracting the degrees of freedom of the explained sum of squares from the degrees of freedom of the total sum of squares. Hence, the unexplained sum of squares is equal to 8. Computing the F ratio, as has been shown before, requires the variance estimate, which is found by dividing the explained and unexplained sums of squares by their respective degrees of freedom. The variance estimates are shown in column four of Table 11.4.

The F test is the ratio between the variance explained by the regression and the variance that is not explained by the regression. The F test is computed as follows:

$$F = \frac{\text{explained variance estimate}}{\text{unexplained variance estimate}} \qquad [11\text{--}24]$$

$$= \frac{269.97}{8.50}$$

$$= 31.76$$

The computed F ratio is compared with the critical value of F given in Appendix F. The critical F for the .05 level of significance and 1 and 8 *d.f.* is 5.32. Our tested F = 31.76 is much greater than 5.32. Therefore, we reject the null hypothesis of no correlation, and conclude that the relationship between amount of fertilizer applied and yield is significant.

Table 11.4   **Variance Table for Testing Significance of Correlation by F Ratio**

(1) Source of variation	(2) Degrees of freedom	(3) Sum of squares	(4) Variance estimate	(5) F
Explained	1	269.97	269.97	31.76
Unexplained	8	68.03	8.50	
Total	9	338.00		

11.5   An application using computer packages

In the previous sections we compute the regression coefficient, the standard error of estimate, and the coefficient of determination and correlation. These computations are time consuming, especially when large samples are involved. In this section we discuss how a *canned* software package can be used to minimize the computational burden when performing a complete regression analysis. There is a wide variety of sophisticated statistical packages that are available to students and researchers alike. The most widely used packages are the Statistical Package for the Social Sciences (SPSS), SAS, MINITAB, and StatView.

The procedure for inputting the data for all these packages is simple, as outlined in the manuals accompanying them. Figure 11.7 shows a computer printout of the MINITAB program that performs a simple linear regression analysis using the data for Example 11.2. It is important that we have a basic knowledge of how to interpret the output from a computer program.

The first item printed is the regression equation, which for our Example 11.2 is

$$Y = 53.4 + 0.358X$$

Again, Y is the dependent variable and X refers to the independent variable.

The second set of information provided on a computer printout deals with the value of a and b coefficients. You also notice that the standard deviation for a and b is given in this section. Additionally, the t ratio, which is obtained by dividing each coefficient with its respective standard deviations, is provided here. The t ratio is compared with the table value to test the hypothesis that $\beta = 0$.

The next line of the printout shows the standard error of estimate that is followed by the coefficient of determination. Note that there are two values given for the coefficient of determination. We discuss the *adjusted coefficient of determination* in the next chapter. The last part of the printout provides an analysis of variance (ANOVA) table. The

The Regression Equation is
Y = 53.4 + 0.358 X

Column	Coefficient	St. Dev. of Coef.	T-Ratio = Coef/S.D.
	53.360	1.934	27.60
X	0.35835	0.06326	5.66

S = 2.904

R-Squared = 80.0 Percent
R-Squared = 77.6 Percent, Adjusted for D.F.

Analysis of Variance

Due to	DF	SS	MS = SS/DF
Regression	1	270.55	270.55
Residual	8	67.45	8.43
Total	9	338.88	

Row	C1	Y C2	Pred. Y Value	St. Dev. Pred. Y	Residual	St. Res.
1	5.0	50.000	55.152	1.662	−5.152	−2.16R

R denotes an obs. with a large st. res.

Durbin-Watson Statistic = 0.89

Figure 11.7   MINITAB computer printout for the yield of maize data.

ANOVA table is used to perform the F test, as was done in Section 11.4. The Durbin–Watson statistic is discussed in Chapter 12.

11.6   Optional topic: curvilinear regression analysis

In our discussions so far, we have only considered the simplest form of a relationship between two variables, namely, the linear relationship. Although the simple linear regression may be considered a powerful tool in analyzing the relationship that may exist between two variables, the assumption of linearity has serious limitations. There are many instances when the straight line does not provide an adequate explanation of the relationship between two variables. When theory suggests that the underlying bivariate relationship is nonlinear, or a visual check of the scatter diagram indicates a curvilinear relationship, it is best to perform a nonlinear regression analysis. For example, the crop yield may increase with added application of fertilizer up to a point, beyond which it decreases. This type of a relationship suggests a curvilinear (in this case a parabola) model.

How we decide on which curve to use depends on the natural relationship that exists between the variables and our own knowledge and experience of the relationship. In dealing with environmental and agricultural data, there may be instances when the relation between two variables is so complex that we are unable to use a simple equation for such a relationship. Under such conditions it is best to find an equation that provides a good fit to the data, without making any claims that the equation expresses any natural relation.

There are many nonlinear models that can be used with agricultural and environmental data. The following are some examples of the non-linear equations that can easily be transformed to their linear counterparts and analyzed as linear equations.

1. $$Y = \alpha e^{\beta X} \qquad [11\text{–}25]$$

2. $$Y = \alpha \beta^X \qquad [11\text{–}26]$$

3. $$\frac{1}{Y} = \alpha + \beta X \qquad [11\text{–}27]$$

4. $$Y = \alpha + \frac{\beta}{X} \qquad [11\text{–}28]$$

5. $$Y = \left(\alpha + \frac{\beta}{X} \right)^{-1} \qquad [11\text{–}29]$$

For illustrative purposes we have selected two of the nonlinear models that are extensively used with agricultural and economic data to perform the analysis.

Exponential growth model. In its simplest form, the model is used for the decay or growth of some variable with time. The general equation for this curve is

$$Y = \alpha e^{\beta X} \qquad [11\text{–}25]$$

In this equation, X (time) appears as an exponent, and the coefficient b describes the rate of growth or decay, and $e \cong 2.718$ is *Euler's constant*, which appears in the formula for the normal curve and is also the base for natural logs. The advantage of using this base is that $\beta \cong$ the growth rate.

The assumptions of the model are that the rate of decay is proportional to the current value of Y, and the error term is multiplicative instead of additive, because it is reasonable to assume that large errors are associated with large values of the dependent variable Y. Thus, the statistical model is

$$Y = \alpha e^{\beta X} \cdot u \qquad [11\text{--}30]$$

The nonlinear models can be transformed to a linear form by taking the logarithm of the equation. For example, by transforming Equation 11–30 into its logarithm we get:

$$\log Y = \log \alpha + \beta X + \log u \qquad [11\text{--}31]$$

where
$$Y' \equiv \log Y$$
$$\alpha' \equiv \log \alpha$$
$$e \equiv \log u$$

Then, we can rewrite Equation 11–31 in the standard linear form as:

$$Y' = \alpha' + \beta X + e \qquad [11\text{--}32]$$

Exponential models have been extensively used in laboratory experiments where the studies of insects treated with chemicals are conducted. In agricultural field experiments that study the decay of chemicals in the soil or in animals, the exponential model is used. Other situations in which the exponential models have been useful are growth studies where the response variable either increases with time, t, or as a result of an increasing level of a stimulus variable, X. Figure 11.8 shows the simple exponential model with growth and decay curves.

When considering different forms of writing a particular relationship, it must be kept in mind that in fitting the model to the data, the choice of correct error structure assumption is critically important.

Example 11.3 illustrates the use of an exponential model and how it can be fitted.

Example 11.3

Since the early 1900s, there have been major resettlements of the population in many parts of California. Population data for the period 1910 to 1990 show the growth in a southern California community. Figure 11.9 shows the growth in population with a continual, though varying, percentage increase. This type of growth is similar to compound interest or unrestrained biological growth. Therefore, the appropriate model to use in analyzing the data is the exponential model.

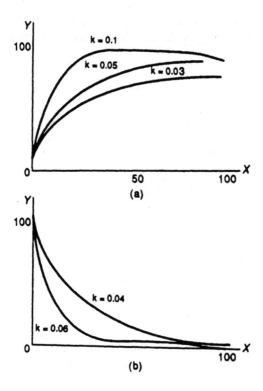

Figure 11.8   Simple exponential models: (a) gradual approach of yield to an upper limit (b) decay curve with time.

Year	Population
1910	1,520
1920	2,800
1930	3,200
1940	18,600
1950	19,800
1960	32,560
1970	65,490
1980	150,450
1990	245,900

Solution

Step one — Transform the population data into the natural logs and compute the totals as shown in Table 11.5.

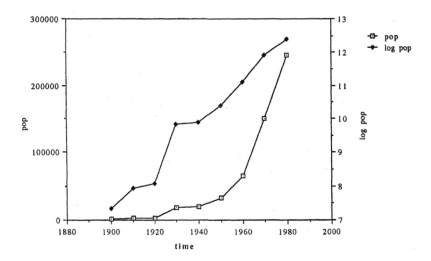

Figure 11.9 Population growth, 1910 to 1990, and its exponential fit.

Table 11.5 Population of a Southern California Community, 1910 to 1990

Year	Time (X)	Population (Y)	Log Y
1910	0	1,520	7.326
1920	1	2,806	7.940
1930	2	3,210	8.074
1940	3	18,630	9.833
1950	4	19,800	9.893
1960	5	32,560	10.391
1970	6	65,495	11.090
1980	7	150,457	11.921
1990	8	245,904	12.413
Total	36		$\Sigma y' =$ 88.881
Sum of squares	204		$\Sigma y'^2 =$ 903.627
Sum of X log Y			$\Sigma xy' =$ 394.405

Step two — Compute the regression coefficients as shown:

$$\Sigma x^2 = \Sigma X^2 - \frac{(\Sigma X)^2}{n} \qquad [11\text{--}33]$$

$$= 204 - \frac{(36)^2}{9} = 60$$

$$\Sigma y'^2 = \Sigma Y'^2 - \frac{(\Sigma Y')^2}{n} \qquad \text{[11–34]}$$

$$= 903.627 - \frac{(88.881)^2}{9} = 25.867$$

$$\Sigma xy' = \Sigma XY' - \frac{(\Sigma X \Sigma Y')}{n} \qquad \text{[11–35]}$$

$$= 394.405 - \frac{36(88.881)}{9} = 38.881$$

$$b' = \frac{\Sigma xy'}{\Sigma x^2} \qquad \text{[11–36]}$$

$$= \frac{38.881}{60} = .648$$

$$a' = \frac{\Sigma Y'}{n} - b' \frac{\Sigma X}{n} \qquad \text{[11–37]}$$

$$= \frac{88.881}{9} - .648 \left(\frac{36}{9}\right) = 7.284$$

The resultant regression equation is

$$\hat{Y}' = 7.284 + .648x$$

Step three — Transform the preceding equation back into the exponential model, as follows:

$$\log \hat{Y} = 7.284 + .648x$$

Taking antilogs (exponential)

$$\hat{Y} = e^{7.284} e^{.648x}$$

That is,

$$\hat{Y} = 1457 e^{.648x}$$

For convenience we have left time (x) in deviation form. Thus, we can interpret the coefficient $\hat{\alpha} = 1457$ as the estimate of the population in 1900 (when $x = 0$). The coefficient $\hat{\beta} = .648 = 64.80\%$ is the appropriate population growth rate every 10 years.

Step four — Compute the coefficient of determination as follows:

$$r^2 = \frac{\left(\sum xy'\right)^2}{\sum x^2 \sum y'^2} \qquad [11\text{--}38]$$

$$= \frac{(38.881)^2}{(60)(25.867)}$$

$$= 0.97$$

From the scatter diagram in Figure 11.9 and the analysis performed, it appears that the exponential curve fits the data to past population growth better than any straight line. However, it is important to keep in mind that using it for any short-term prediction of the population is unwarranted. The concern mostly stems from the fact that in this simple growth model, the error u is likely to be serially correlated and thus has to be accounted for in any prediction.

Chapter summary

Environmental and agricultural scientists often investigate the relationship between variables of interest. In this chapter you were shown how regression analysis can be applied to develop an equation showing the relationship between two variables. Methods of estimation of the regression equation were given. Once we obtain the regression coefficient, we can perform an hypothesis test to determine whether there is a statistical relationship between two variables X and Y. Keep in mind that rejecting the null hypothesis of no relationship does not guarantee that the independent variable is a useful predictor of the dependent variable. The *coefficient of determination* (r^2) and the *standard error of estimate* ($S_{y.x}$) confirm how well the regression line fits the data. The coefficient of determination explains the amount of variation in Y as explained by X. The standard error of estimate measures the dispersion about an average line, called the regression line, which is only an expression of the average relationship between the variables.

The second concept discussed in this chapter was *correlation analysis*. Correlation analysis describes the extent of the linear relationship between variables. The correlation coefficient is used as a measure of the extent of the relationship between variables. The correlation coefficients take on values that range from +1 to −1. The closer the correlation coefficient is to 1, the greater is the degree of the relationship. The sign of the coefficient indicates the nature of the relationship between variables. A positive sign shows that the independent and dependent variables are directly related. A negative sign indicates that the variables are inversely related.

As a precaution it should be emphasized that regression and correlation analysis only indicate how and to what extent the variables of interest are associated with each other. The strong relationship between variables should not be interpreted as a cause and effect relationship. Additionally, the sample regression equation should not be used to predict or estimate outside the range of values of the independent variable given in a sample.

Finally, *nonlinear regression analysis* is explained, where care must be taken before fitting the data to a linear equation. Theoretical considerations as well as a scatter diagram of the data are suggested as the basis for determining whether a linear or a curvilinear relationship exists between the variables.

Case Study

Determining the Relationship Between Global Fertilizer Use and Grain Production

World population is increasing at a rapid pace, and with it the demand for food is rising dramatically. Over the next 30 years, global population is projected to grow by nearly two thirds, from the current 5.5 billion to 8.5 billion (World Resources Institute, 1995). To cope with the increasing food demand, many countries around the world have adopted different strategies to increase food production. Such measures as bringing additional land into production and adopting high-input agriculture are the most prevalent. Given that arable land is limited in many parts of the world, countries have opted to use the high-input agriculture to increase food production.

According to the U.S. Department of Agriculture (1995), the 1994 world grain harvest was at 1,747 million tons, which was up by only 2.9% from the previous year's production of 1,697 million tons. Variations are observed in the output of major grains (wheat, rice, and corn) from year to year, because of a number of factors. Climatic changes, resource limitations, use of agricultural lands for industrialization, and government policy have all played a role.

Wheat production dropped sharply in 1994 as compared to 1990, mainly resulting from a drop in harvest in the former U.S.S.R. Severe drought in Australia also caused wheat harvest to drop from 17 million tons in 1993 to 8 million tons in 1994 (U.S. Department of Agriculture, 1995). Rice production, 90% of which takes place in Asia, has been remarkably stable. The data for 1990 and 1994 show an output of 350 and 353 million tons, respectively. This stable production trend is because the shrinkage in rice land area (of nearly 2% since 1990) was offset by a yield rise (of roughly 2%) as a result of increased use of fertilizers and other inputs (U.S. Department of Agriculture, 1995). Corn production rebounded strongly in 1994. The global corn yield of nearly 4 ton/ha is well above that of both wheat and rice (U.S. Department of Agriculture, 1995). The increase in 1994 was mainly a result of increased production in the U.S. where heavy losses had occurred in 1993 from severe flooding.

Overall, grain production around the world in the 1990s has been static. This is not surprising in view of the fact that both land planted to grain and use of fertilizer have fallen since 1990 (Isherwood and Soh, 1994). World fertilizer use was 121 million tons in 1994, down from 126 million tons. This was the fifth consecutive annual decline since the historic peak of 146 million tons in 1989.

Several factors have contributed to the decline in fertilizer use after decades of finding that increased fertilizer use led to higher yields. In many countries available crop varieties began to approach the agronomic limits of the response to fertilizer (Brown, 1995). As farmers began to realize that the relationship between increased fertilizer and higher yields was no longer holding, they stopped using more. They also reduced use where fertilizer use at the margin had become unprofitable (Isherwood and Soh, 1994). In addition, the decline in the use of fertilizer results from cutbacks in agricultural subsidies. Brown (1995) notes that in the last several years a number of major food-producing countries, such as the former U.S.S.R., India, and China have reduced or eliminated fertilizer subsidies.

As the continued use of fertilizers is debated, data such as those given in the following table can be used to determine the relationship that exists between fertilizer use and grain production.

By using these data:

1. Draw a scatter diagram to determine if there is a linear relationship between fertilizer use and grain production.
2. Perform a regression analysis.
3. Is there a significant correlation between the two variables? Use $\alpha = .01$.
4. Use the estimated regression line to identify the range of values at 95% confidence interval.
5. Use the estimated regression line to estimate grain production when the amount of fertilizer applied is 130 million tons.

World Fertilizer Use and Grain Production, 1970 to 1994

Year	Grain production (million tons)	Fertilizer use (million tons)
1970	1096	66
1971	1194	69
1972	1156	73
1973	1272	79
1974	1220	85
1975	1250	82
1976	1363	90
1977	1337	95
1978	1467	100
1979	1428	111
1980	1447	112
1981	1499	117
1982	1550	115
1983	1486	115
1984	1649	126
1985	1664	131
1986	1683	129
1987	1612	132
1988	1564	140
1989	1685	146
1990	1780	143
1991	1696	138
1992	1776	134
1993	1697	126
1994 (preliminary)	1747	121

Adapted from U.S. Department of Agriculture (USDA), *World Grain Database* (unpublished printout), 1991, Washington, D.C.; USDA, *Production, Supply, and Demand View*, (electronic database), Washington, D.C., November 1994; USDA, *World Agricultural Supply and Demand Estimates*, Washington, D.C., January 1995. Data for the fertilizer use is from FAO, *Fertilizer Yearbook*, Rome, various years.

References

Brown, L., Fertilizer use continues dropping, in Brown, L., Lenssen, N., and Kane, H., *Vital Signs 1995*, W. W. Norton, New York, 1995, 40.

Isherwood, K. F. and Soh, K. G., The agricultural situation and fertilizer demand, paper presented at the 62nd Annual Conf., International Fertilizer Industry Association, Istanbul, May 9, 1994.

World Resources Institute, *World Resources: 1994–1995*, Oxford University Press, New York, 1995.

Review questions

1. A crop scientist is interested in the dependence of the yield (milligrams per hectare) of winter wheat on the number of kernels per spike developed during the growth stage. The following results are obtained.

Grain yield (mg/ha)	Kernels per spike no.
4.03	18
5.70	22
5.90	23
7.60	28
6.20	26
6.30	24
4.50	19
5.80	21
4.60	18
6.35	24
5.60	22
4.86	20
4.10	18
4.85	19
6.40	26

   a. Develop a scatter diagram. Is there a linear relationship between the two variables? Explain.

   b. Compute a least squares regression equation, and fit it to the scatter diagram.

   c. Compute the standard error of estimate.

   d. Compute the coefficient of determination and explain its meaning.

   e. Is there a significant relationship between the two variables? Use $\alpha = .05$.

2. An environmental scientist in a recent study on the influence of temperature on germination of perennial warm-season forage grasses observes the following number of seeds to germinate per day.

Temperature (°C)	Germinated Seed (no./day)
15	10
17	15
19	18
20	19
22	21
23	20
25	22
26	24
30	28
35	34

a. Develop the least squares regression line.
b. Is there a significant correlation between the two variables? Use $\alpha = .01$.
c. Use the estimated regression line and develop a 95% confidence interval estimate of the number of germinated seeds per day when the temperature is 24°C.

3. Animal scientists have developed a genetic evaluation program for backfat and age at 90 kg for swine. A researcher is interested in determining if there is any relationship between the backfat (mm) and the number of days it takes the animal to reach 90 kg of weight. The following observations are recorded by the researcher.

Backfat	Days to 90 kg	Backfat	Days to 90 kg
14.0	160	15.0	162
15.1	165	13.9	165
15.4	163	14.6	164
14.5	164	12.9	165
12.9	167	13.5	167
13.8	159	14.1	166
14.2	168	15.0	165
15.0	162	15.6	167
14.8	167	14.8	163
14.6	163	14.9	165

a. Draw a scatter diagram.
b. Compute the least squares regression equation.

    c. Compute the standard error of estimate.

    d. Compute the coefficient of determination and correlation.

    e. Test the significance of the coefficient of correlation. Use a .05 level of significance.

4. A study (Joseph et al., 1985) on the effect of seeding rates on yield of soft red winter wheat grown on Suffolk soil provides the following observations.

Seeds per meter (no.)	Weight per kernel (mg)
186	33.3
372	30.8
558	31.9
744	29.6
1116	30.1
186	34.7
732	30.5
558	30.6

Adapted from Joseph, K. D. S. et al. *Agron. J.*, 77(2), March-April, 213, 1985.

    a. Compute the least squares regression equation. Use a computer.

    b. Interpret the meaning of the standard error of estimate.

    c. Interpret the meaning of the coefficient of determination and correlation.

    d. By testing at $\alpha = .01$, does $\beta = 0$?

References and suggested reading

Bonney, G. E., Logistic regression for dependent binary observations, *Biometrics*, 43, 951, 1987.

Breslow, N. E., Tests of hypotheses in overdispersed poison regression and other quasi-likelihood models, *J. Am. Stat. Assoc.*, 85, 565, 1990.

Cohen, J. and Cohen, P., *Applied Multiple Regression/Correlational Analysis for Behavioral Sciences*, Lawrence Erlbaum Associates, Hillsdale, NJ, 1975.

Daniel, W. W. and Terrell, J. C., *Business Statistics: Basic Concepts and Methodology*, 4th ed, Houghton Mifflin, Boston, 1986, chap. 10.

Eilers, P. H. C., Regression in action: estimating pollution roses from daily averages, *Environmetrics*, 2, 25, 1991.

Gill, J. D., *Design and Analysis of Experiments*, Vol. 1, Iowa State University Press, Ames, IA, 1978.

Hamburg, M., *Statistical Analysis for Decision Making*, 3rd ed., Harcourt Brace Jovanovich, New York, 1983, chap. 9.

Hoshmand, A. R., *Experimental Research Design and Analysis: A Practical Approach for Agricultural and Natural Sciences,* CRC Press, Boca Raton, FL, 1994, chap. 9.

Joseph, K. D. S. et al., Row spacing and seeding rate effects on yield and yield components of soft red winter wheat, *Agron. J.,* 77, 1985.

Kelejian, H. and Oates, W., *Introduction to Econometrics,* Harper & Row, New York, 1974.

Netter, J. and Wasserman, W., *Applied Linear Statistical Models,* Richard D. Irwin, Homewood, IL, 1974.

Olson, C. L. and Picconi, M. J., *Statistics for Business Decision Making,* Scott, Foresman, Glenview, IL, 1983.

Yajima, Y., On estimation of a regression model with long-memory stationary errors, *Ann. Stat.,* 16, 791, 1988.

chapter twelve

Multiple Regression

Learning objectives

The basic learning objective of this chapter is to expand your understanding of regression and correlation analysis by discussing the techniques of multiple and partial regression and correlation. You will specifically learn to:

- Define the multiple regression line and explain how it is computed.
- Define the multiple standard error of estimate.
- Explain the multiple coefficient of determination and correlation
- Describe the partial coefficient of determination.
- Explain the role of computer packages in performing multiple-regression.

- Use the t and F distributions to test for significance of relationships in multiple regression.
- Describe the assumptions of multiple linear regression — serial or autocorrelation, heteroscedasticity, and multicollinearity.

12.1 Introduction

In the previous chapter, it was shown how regression and correlation analysis are used for purposes of prediction and planning. The techniques and concepts presented were used as a tool in analyzing the relationship that may exist between two variables. A single independent variable was used to estimate the value of the dependent variable. In this chapter, you learn about the concepts of regression and correlation where two or more independent variables are used to estimate the dependent variable. As in the simple linear regression, we have only one dependent variable, but several independent variables. This improves our ability not only to estimate the dependent variable, but also to explain more fully its variations.

First, because *multiple regression and correlation* is simply an extension of the simple regression and correlation, we show how to derive the multiple regression equation using two or more independent variables. Second, attention is given to calculating the standard error of estimate and related measures. Finally, the computation of multiple coefficient of determination and correlation are explained.

The advantage of multiple regression over simple regression analysis is in enhancing our ability to use more available information in estimating the dependent variable. To describe the relationship between a single variable Y and several variables X, we may write the multiple regression equation as:

$$Y = a + b_1X_1 + b_2X_2 + \dots + b_kX_k + \varepsilon \qquad [12\text{--}1]$$

where
$$\begin{aligned} Y &= \text{dependent variable} \\ X_1, \dots, X_k &= \text{independent variables} \\ e &= \text{error term, which is a random variable with a} \\ & \quad\ \text{mean of zero and a standard deviation of } \sigma. \end{aligned}$$

The numerical constants a, and b_1 to b_k must be determined from the data, and are referred to as the *partial regression coefficients*. The underlying assumptions of the multiple regression model are:

1. The explanatory variables $(X_1, \dots, X_k)$ may be either random or nonrandom (fixed) variables

2. The value of Y selected for one value of X is probabilistically independent.
3. The random error has a normal distribution with mean equal to zero and variance equal to σ^2.

These assumptions imply that the mean, or expected value $E(Y)$, for a given set of values of $X_1, ..., X_k$ is equal to:

$$E(Y) = a + b_1X_1 + b_2X_2 + ... + b_kX_k \qquad [12\text{--}2]$$

The coefficient a is the Y intercept when the expected value of all independent variables is zero. Equation 12-2 is called a *linear* statistical model. The scatter diagram for a two-independent-variables case is a regression plane, as shown in Figure 12.1.

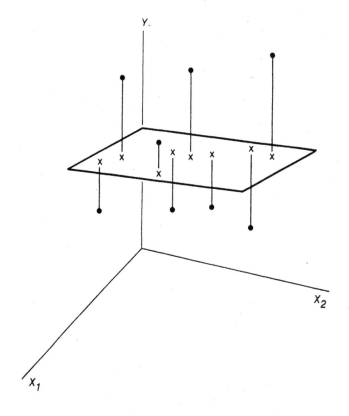

Figure 12.1   Scatter diagram for multiple regression analysis involving two independent variables.

In the next section, we gain more insight into the form of the relationship given in Equation 12–2 when we consider two independent variables in explaining the dependent variable.

12.2 *Estimating multiple regression equation: least squares method*

In the previous chapter we mentioned that if a straight line is fitted to a set of data using the least squares method, that line is the *best fit* in the sense that the sum of squared deviations is less than it would be for any other possible line. The least squares formula provides a best-fitting *plane* to the data. The *normal equations* for k variables are as follows:

$$\Sigma Y_i = na + b_1 \Sigma X_{1i} + b_2 \Sigma X_{2i} + \ldots + b_k \Sigma X_{ki}$$

$$\Sigma X_{1i} Y_i = a \, \Sigma X_{1i} + b_1 \Sigma X_{1i}^2 + b_2 \Sigma X_{1i} X_{2i} + \ldots + b_k \Sigma X_{1i} X_{ki}$$

$$\Sigma X_{2i} Y_i = a \, \Sigma X_{2i} + b_1 \Sigma X_{2i} X_{1i} + b_2 \Sigma X_{1i} \, X_{2i}^2 + \ldots + b_k \Sigma X_{2i} X_{ki}$$

$$\cdots \qquad \cdots \cdots \cdots \cdots \cdots$$

$$\Sigma X_{ki} Y_i = a \, \Sigma X_{ki} + b_1 \Sigma X_{ki} X_{1i} + b_2 \Sigma X_{2i} \, X_{ki} + \ldots + b_k X_{ki}^2 \qquad [12\text{--}3]$$

In the preceding equation, b_1, b_2, ..., b_k are estimates of β_1, β_2, ..., β_k, and the number of equations is equal to the number of parameters to be estimated. You should note that computing the coefficients, although not theoretically complex, is difficult and tedious even for k as small as four or five. We generally use a computer program (MINITAB, SPSS, or SAS) to solve for the coefficients.

To illustrate the concept and how the multiple regression coefficients are computed, let us take an example with two independent variables.

Example 12.1

In the previous chapter (Example 11.2) an agronomist takes ten observations on yield of maize as different amounts of fertilizer are applied. In this example, he wishes to consider an additional explanatory variable (the amount of rainfall). He believes that the amount of fertilizer and rainfall are better predictors of yield. Using the following information, determine for the agronomist the least squares multiple regression equation.

Yield of maize (bushels/acre) Y	Fertilizer (lb N/acre) X_1	Rainfall (in.) X_2
50	5	5
57	10	10
60	12	15
62	18	20
63	25	25
65	30	25
68	36	30
70	40	30
69	45	25
66	48	30

Solution

The first step in calculating the coefficients is to compute the intermediate values needed in the normal equations from the data given in Table 12.1.

Table 12.1 **Calculation of Coefficients for Normal Equations**

Y	X_1	X_2	X_1Y	X_1^2	X_1X_2	X_2Y	X_2^2	Y^2
50	5	5	250	25	25	250	25	2,500
57	10	10	570	100	100	570	100	3,249
60	12	15	720	144	180	900	225	3,600
62	18	20	1,116	324	360	1,240	400	3,844
63	25	25	1,575	625	625	1,575	625	3,969
65	30	25	1,950	900	750	1,625	625	4,225
68	36	30	2,448	1,296	1,080	2,040	900	4,624
70	40	30	2,800	1,600	1,200	2,100	900	4,900
69	45	25	3,105	2,025	1,125	1,725	625	4,761
66	48	30	3,168	2,304	1,440	1,980	900	4,356
630	269	215	17,702	9,343	6,885	14,005	5,325	40,028

Note: $\overline{Y} = 63$; $\overline{X}_1 = 26.9$; $\overline{X}_2 = 21.5$.

The normal equations for the two independent variables are written as follows:

$$\Sigma Y = na + b_1\Sigma X_1 + b_2\Sigma X_2 \qquad [12\text{--}4]$$

$$\Sigma X_1Y = a\,\Sigma X_1 + b_1\Sigma X_1^2 + b_2\Sigma X_1X_2 \qquad [12\text{--}5]$$

$$\Sigma X_2 Y = a \, \Sigma X_2 + b_1 \Sigma X_1 X_2 + b_2 \Sigma X_2^2 \qquad \text{[12–6]}$$

The information from Table 12.1 is substituted into the preceding normal equations to obtain:

$$630 = 10a + 269b_1 + 215b_2 \qquad \text{[12–7]}$$

$$17{,}702 = 269a + 9{,}343b_1 + 6{,}885b_2 \qquad \text{[12–8]}$$

$$14{,}005 = 215a + 6{,}885b_1 + 5{,}325b_2 \qquad \text{[12–9]}$$

To solve Equations 12–7 through 12–9 simultaneously, we first multiply Equation 12–7 by $\overline{X}_1$ or 26.9 and then subtract the result from Equation 12–8. This eliminates a and we obtain an equation that involves b_1 and b_2 only:

$$17{,}702 = 269a + 9{,}343b_1 + 6{,}885.0b_2$$
$$\underline{-16{,}947 = -269a - 7{,}236.1b_1 - 5{,}783.5b_2}$$
$$755 = 2{,}106.9b_1 + 1{,}101.5b_2 \qquad \text{[12–8A]}$$

Now multiply Equation 12–7 by $\overline{X}_2$ or 21.5 and subtract the results from Equation 12–9. The result is

$$14{,}005 = 215a + 6{,}885.0b_1 + 5{,}325.0b_2$$
$$\underline{-13{,}545 = -215a - 5{,}783.5b_1 - 4{,}622.5b_2}$$
$$460 = 1{,}101.5b_1 + 702.5b_2 \qquad \text{[12–9A]}$$

Given Equations 12–8A and 12–9A, we can now solve simultaneously for b_1 and b_2. This requires multiplying Equation 12–9A by 1,101.5/702.5 and then subtracting the result from Equation 12–8A which gives the following:

$$755.0000 = 2{,}106.9000b_1 + 1{,}101.5b_2$$
$$\underline{-721.2669 = -1{,}727.1206b_1 - 1{,}101.5b_2}$$
$$b_1 = 33.7331/379.7794$$
$$= 0.0888$$

Once we have determined the b_1 value, we substitute it in Equation 12–9A to obtain b_2 as follows:

$$460 = 1101.5 \, (0.0888) + 702.5b_2$$
$$460 = 97.8132 + 702.5b_2$$
$$460 - 97.8132 = 702.5b_2$$
$$362.187 = 702.5b_2$$
$$b_2 = 362.187/702.5$$
$$= 0.5155$$

Now we substitute the values of b_1 and b_2 into Equation 12–7 to obtain a as follows:

$$630 = 10a + 269(0.0888) + 215(0.515)$$
$$630 = 10a + 2.8872 + 110.8325$$
$$630 = 10a + 134.7197$$
$$630 - 134.71797 = 10a$$
$$a = 495.2803/10$$
$$= 49.5280$$

Thus the estimated regression equation is

$$\hat{Y} = 49.53 + 0.089X_1 + 0.515X_2$$

Another method of solving for the multiple regression equation is to reduce the column, or item, totals as shown in Table 12.1 to the corresponding deviation totals. The deviation totals are those from the respective arithmetic means as shown:

$$\Sigma y^2 = \Sigma Y^2 - \frac{(\Sigma Y)^2}{n} \qquad [12-10]$$

$$\Sigma x_1^2 = \Sigma X_1^2 - \frac{(\Sigma X_1)^2}{n} \qquad [12-11]$$

$$\Sigma x_2^2 = \Sigma X_2^2 - \frac{(\Sigma X_2)^2}{n} \qquad [12-12]$$

$$\Sigma yx_1 = \Sigma YX_1 - \frac{\Sigma Y \Sigma X_1}{n} \qquad [12-13]$$

$$\Sigma yx_2 = \Sigma YX_2 - \frac{\Sigma Y \Sigma X_2}{n} \qquad [12\text{--}14]$$

$$\Sigma x_1 x_2 = \Sigma X_1 X_2 - \frac{\Sigma X_1 \Sigma X_2}{n} \qquad [12\text{--}15]$$

Once we have computed the deviations from the mean, the following equations are used to compute a, b_1, and b_2:

$$\Sigma yx_1 = b_1 \Sigma x_1^2 - b_2 \Sigma x_1 x_2 \qquad [12\text{--}16]$$

$$\Sigma yx_2 = b_1 \Sigma x_1 x_2 - b_2 \Sigma x_2^2 \qquad [12\text{--}17]$$

By substituting the data from Table 12.1 into Equations 12–10 through 12–17, we obtain the following:

$$\Sigma y^2 = 40,028 - \frac{(630)^2}{10} = 338$$

$$\Sigma x_1^2 = 9,343 - \frac{(269)^2}{10} = 2,106.9$$

$$\Sigma x_2^2 = 5,325 - \frac{(215)^2}{10} = 702.5$$

$$\Sigma yx_1 = 17,702 - \frac{(630)(269)}{10} = 755$$

$$\Sigma yx_2 = 14,005 - \frac{(630)(215)}{10} = 460$$

$$\Sigma x_1 x_2 = 6,885 - \frac{(269)(215)}{10} = 1,101.5$$

To compute b_1 and b_2, we solve simultaneously the following two equations:

$$\Sigma yx_1 = b_1 \Sigma x_1^2 - b_2 \Sigma x_1 x_2 \qquad [12\text{--}16]$$

$$\Sigma yx_2 = b_1 \Sigma x_1 x_2 - b_2 \Sigma x_2^2 \qquad [12\text{--}17]$$

The intercept of the equation or a is found as follows:

$$a = \overline{Y} - b_1 \overline{X}_1 - b_2 \overline{X}_2 \qquad [12\text{--}18]$$

By substituting the deviation totals into Equations 12–16 and 12–17, we get:

$$755 = 2106.9b_1 + 1105.5b_2 \qquad [12\text{--}19]$$

$$460 = 1105b_1 + 702.5b_2 \qquad [12\text{--}20]$$

Equations 12–19 and 12–20 have two unknowns (namely, b_1 and b_2). To solve for one of the unknowns, say, b_2, we multiply Equation 12–20 by 1.9128 and subtract the result from Equation 12–19 to get:

$$755 = 2106.9b_1 + 1105.5b_2$$
$$\underline{-879.9 = -2106.9b_1 - 1343.7b_2}$$
$$-124.9 = -242.2b_2$$
$$b_2 = -124.9 / -242.2$$
$$b_2 = 0.515$$

By substituting the value of b_2 into Equation 12–20, we have:

$$460 = 1101.5b_1 + 702.5\,(0.515)$$
$$460 = 1101.5b_1 + 361.8$$
$$b_1 = 98.2 / 1105.5$$
$$b_1 = 0.089$$

Now we substitute the value of b_1 and b_2 and the value of the means for each variable into Equation 12–18 to get a as follows:

$$a = 63 - (0.089)(26.9) - (0.515)(21.5)$$
$$= 63 - 2.39 - 11.07$$
$$= 49.54$$

The multiple regression line is

$$\hat{Y} = 49.53 + 0.089X_1 + 0.515X_2$$

The estimated regression equation computed by either approach yields the same values for the coefficients. The slight difference in these coefficients results from rounding of decimal points.

The interpretation of the coefficients a, b_1, and b_2 is analogous to the simple linear regression. The constant a is the intercept of the regression line. However, we interpret it as the value of $\hat{Y}$ when both X_1 and X_2 are zero. Variables b_1 and b_2 are called *partial regression coefficients*; b_1 simply measures the change in $\hat{Y}$ per unit change in X_1 when X_2 is held constant. Likewise, b_2 measures the change in $\hat{Y}$ per unit change in X_2 when X_1 is held constant. Thus, we may say that the b coefficients measure the net influence of each independent variable on the estimate of the dependent variable.

Hence, in the present example the b_1 value of 0.089 indicates that for each increase of 1 lb of fertilizer, the yield increases by 0.089 bushels, regardless of the rainfall; that is, the amount of rainfall is held constant. The b_2 coefficient indicates that for each increase of 1 in. of rainfall, the yield increases by 0.515 bushels, regardless of the amount of fertilizer used.

Before we can compute the standard error of estimate and the coefficient of multiple determination, we need to partition the sum of squares for the dependent variable.

The total sum of squares (*SST*) has already been computed before as:

$$SST = \Sigma y^2 = 338$$

The explained or regression sum of squares (*SSR*) is computed as follows:

$$SSR = b_1\Sigma yx_1 + b_2\Sigma yx_2 \qquad [12\text{--}21]$$

By substituting the appropriate values into Equation 12–21, we get:

$$SSR = 0.089(755) + 0.515(460)$$
$$= 304.10$$

The unexplained or error sum of squares (*SSE*) is the difference between the *SST* and *SSR*:

$$SSE = SST - SSR \qquad [12\text{--}22]$$

Therefore, the *SSE* is

$$SSE = 338 - 304.10 = 33.90$$

We now turn our attention to computing the standard error of estimate, and the coefficient of multiple determination.

12.3 Standard error of estimate

The *standard error of estimate* measures the standard deviation of the residuals about the regression plane, and thus specifies the amount of error incurred when the least squares regression equation is used to predict values of the dependent variable. The smaller the standard error of estimate, the closer the fit of the regression equation is to the scatter of observations. The standard error of estimate is denoted by $S_{y.12}$. The subscripts indicate that we have used two independent variables to predict the dependent variable.

The standard error of estimate is computed by using the following equation:

$$S_{y.12...(k-1)} = \sqrt{\frac{SSE}{n-k}} \qquad [12\text{--}23]$$

where SSE = error sum of squares
n = number of observations
k = number of parameters

Hence, in a multiple regression analysis involving two independent variables and a dependent variable, the divisor is $n - 3$. Because we have computed the SSE we substitute its value into Equation 12–23 to compute the standard error of estimate as follows:

$$S_{y.12} = \sqrt{\frac{33.90}{7}}$$

$$= 2.20 \text{ bushels}$$

The standard error of estimate about the regression plane may be compared with the standard error of estimate of the simple regression.

In the previous chapter when only fertilizer was used to explain the variation in yield, we computed a standard error of estimate of 2.92. Inclusion of an additional variable (rainfall) to explain the variation in yield gives us a standard error of estimate of 2.20. As was mentioned before, the standard error expresses the amount of variation in the dependent variable that is left unexplained by regression analysis. Because the standard error of the regression plane is smaller than the standard error of the regression line, inclusion of this additional variable provides for a better prediction.

The standard error of estimate is interpreted like any other standard deviation. For the present example, it means that if the yield of maize is distributed normally about the multiple regression plane, approximately 68% of the yield falls within 2.20 bushels of their estimated $\hat{Y}$ value. Furthermore, 95% of yield falls within $\pm\, 2S_{y.12}$, and approximately 99.7% of the yield falls within $\pm\, 3S_{y.12}$ of the estimated $\hat{Y}$ value.

12.4 *Multiple correlation analysis*

Because we have determined the multiple regression equation and computed the standard error of estimate, we now turn to a statistical quantity that measures how well a linear model fits a set of observations. As in the case of the simple regression analysis, we use the coefficients of determination and correlation to describe the relationship between the dependent and independent variables. We first discuss the coefficient of determination and then elaborate on the coefficient of correlation. As in simple correlation, the coefficient of determination is the ratio of the explained or *SSR* to the *SST*. The *multiple coefficient of determination* denoted as $R_{y.12}^2$ is found using the equation:

$$R_{y.12}^2 = \frac{SSR}{SST} \qquad [12\text{--}24]$$

where SSR = regression sum of squares
  SST = Σy^2 = total sum of squares

By substituting the values of the *SSR* and the *SST* into Equation 12–24, we get:

$$R_{y.12}^2 = \frac{304.10}{338} = 0.90$$

$R_{y.12}^2$ is interpreted as before. This means that 90% of the variation in yield of maize is explained by the amount of fertilizer applied and the amount of rainfall in a locality. The preceding $R_{y.12}^2$ value is not adjusted for degrees of freedom (*d.f.*). Hence, we may overestimate the impact of adding another independent variable in explaining the amount of variability in the dependent variable. Thus, it is recommended that an adjusted R^2 be used in interpreting the results.

The adjusted coefficient of multiple determination is computed as follows:

$$R_a^2 = 1 - (1 - R_{y.12}^2)\,\frac{n-1}{n-k} \qquad [12\text{--}25]$$

where R_a^2 = adjusted coefficient of multiple determination
     n = number of observations
     k = total number of parameters

For the present example, the adjusted R^2 is

$$R_a^2 = 1 - (1 - .90)\frac{10 - 1}{10 - 3}$$

$$= 0.87$$

We may wish to compare the coefficient of determination of the simple regression model where one independent variable (namely, the impact of application of fertilizer) is analyzed with the coefficient of multiple determination where, in addition to fertilizer use, the impact of rainfall on yield is observed. The adjusted coefficient of determination for the simple regression is $r^2 = 0.76$, whereas the adjusted coefficient of multiple determination is 0.87. The difference of 0.11 indicates that an additional 11% of the variation in the yield of maize is explained by the amount of rainfall, beyond that already explained by the amount of fertilizer applied.

Partial correlation.   Because we include additional variables in the multiple regression equation, we should be able to measure the contribution of each variable when holding other variables in the equation constant. Partial correlation provides us with such a measure. Our primary interest in computing the partial correlation coefficient is to eliminate the influence of all other variables except the one we are interested in. For a three-variable case (Y, X_1, X_2), there are three simple correlations among these variables, written as $r_{y1.2}$, $r_{y2.1}$, and $r_{12.y}$.

To be able to compute the partial correlation coefficients, we first compute the simple correlation coefficients r_{y1}, r_{y2}, and r_{12}. These simple correlation coefficients indicate the correlation between Y and X_1, Y and X_2, and X_1 and X_2, respectively:

$$r_{y1} = \frac{\sum yx_1}{\sqrt{\sum x_1^2 \sum y^2}} \qquad\qquad [12\text{--}26]$$

$$r_{y2} = \frac{\sum yx_2}{\sqrt{\sum x_2^2 \sum y^2}} \qquad\qquad [12\text{--}27]$$

$$r_{12} = \frac{\sum x_1 x_2}{\sqrt{\sum x_1^2 \sum x_2^2}} \qquad [12\text{–}28]$$

We use the preceding simple correlation coefficients to compute the partial correlation between Y and X_1 when X_2 is held constant, Y and X_2 when X_1 is held constant, and finally X_1 and X_2 when Y is held constant as follows:

$$r_{y1.2} = \frac{\left(r_{y1} - r_{y2}r_{12}\right)}{\sqrt{\left(1 - r_{y_2}^2\right)\left(1 - r_{12}^2\right)}} \qquad [12\text{–}29]$$

$$r_{y2.1} = \frac{\left(r_{y2} - r_{y1}r_{12}\right)}{\sqrt{\left(1 - r_{y_1}^2\right)\left(1 - r_{12}^2\right)}} \qquad [12\text{–}30]$$

$$r_{12.y} = \frac{\left(r_{12} - r_{y1}r_{y2}\right)}{\sqrt{\left(1 - r_{y_1}^2\right)\left(1 - r_{y_2}^2\right)}} \qquad [12\text{–}31]$$

For the present problem, the simple correlations are found as follows:

$$r_{y1} = \frac{755}{\sqrt{(2106.9)(338)}} = 0.89$$

$$r_{y2} = \frac{460}{\sqrt{(702.5)(338)}} = 0.94$$

$$r_{12} = \frac{1101.5}{\sqrt{(2100.9)(702.5)}} = 0.91$$

Using the previous values in Equation 12–29 through 12–31, we find the partial coefficients to be

$$r_{y1.2} = \frac{[0.89 - (0.94)(0.91)]}{\sqrt{[1-(0.94)^2][1-(0.91)^2]}} = 0.24$$

$$r_{y2.1} = \frac{[0.94 - (0.89)(0.91)]}{\sqrt{[1-(0.89)^2][1-(0.91)^2]}} = 0.69$$

$$r_{12.y} = \frac{[0.91 - 0.89)(0.94)]}{\sqrt{[1-(0.89)^2][1-(0.94)^2]}} = 0.47$$

Partial coefficients of determination.   As pointed out in Chapter 11, the square of the correlation coefficient is the coefficient of determination. The partial coefficients of determination are simply the square of the partial correlation coefficients. The partial coefficients of determination for the present problem are

$$r_{y1.2}^2 = (0.24)^2 = 0.0576$$

$$r_{y2.1}^2 = (0.69)^2 = 0.4761$$

$$r_{12.y}^2 = (0.47)^2 = 0.2209$$

The partial coefficients simply show the true value of an additional independent variable. $r_{y1.2}^2 = 0.0576$ indicates that after X_2 (rainfall) has explained as much of the total variability in the dependent variable as it can, X_1 (fertilizer) explains 5.76% of the remaining variability in Y. The other partial coefficients of determination are interpreted similarly.

12.5 Inferences concerning regression and correlation coefficients

In our discussion so far, we have computed various regression measures that use sample data. For us to make inferences about the population, the assumptions regarding the standard error must be met. Just as in the case of the simple regression analysis, we are interested in the reliability of the multiple regression coefficients. A hypothesis test may be performed to determine whether there is a significant relationship among the dependent variable Y and the independent variables X_1 and X_2.

The null and alternative hypotheses may be stated as:

H_o: $\beta_1 = \beta_2 = 0$.
H_1: At least one of the two coefficients is not equal to zero.

Suppose we wish to test this hypothesis of no significant relationship between the Y, X_1, and X_2 at .05 level of significance; then if the null hypothesis is correct, we are not able to use the regression equation for prediction and estimation.

F test.   To test the preceding hypotheses, we perform an F test as done in Chapter 11. The test statistic is

$$F = \frac{MSR}{MSE}$$ [12–32]

where MSR = mean square due to regression
       MSE = mean square due to error

The sum of squares, the degrees of freedom, the corresponding mean squares, and the F ratio are summarized in Table 12.2. The critical F value for α = .05 given in Appendix F for 2 and 7 *d.f.* for the numerator and denominator, respectively, is 4.74. The decision rule states that we should accept H_0 if the computed F is <4.74, and reject H_0 if the computed F is >4.74.

Table 12.2   **ANOVA Table in the Regression Problem to Predict Yield of Maize**

Source	Degrees of freedom	Sum of squares	Mean square	F
Regression	2	304.10	152.05	31.41
Error	7	33.90	4.84	
Total	9	338.00		

Because the computed F value is 31.41, we reject the null hypothesis, and conclude that there is a significant relationship between the yield of maize and the independent variables fertilizer and rainfall. Thus, the estimated regression equation may be used for prediction and estimation within the range of values included in the sample.

Because the same procedures apply in testing the correlation between Y, X_1, and X_2, an illustration of multiple correlation is not presented.

The estimated multiple regression equation can be used to predict the average yield of maize based on the value of the independent variables.

Example 12.2

Suppose the agronomist in example 12.1 applies 40 lb of fertilizer (X_1) to a field in a locality that receives 15 in. of rainfall (X_2). What is the estimated average yield of maize?

Solution

The estimated regression equation for Example 12.1 was found to be

$$\hat{Y} = 49.53 + 0.089X_1 + 0.515X_2$$

To determine the estimated yield of maize, we now substitute the value of X_1 and X_2 into the regression equation as follows:

$$\hat{Y} = 49.53 + 0.089 \,(40) + 0.515\,(15)$$
$$\hat{Y} = 60.82 \text{ bushels of maize}$$

Given the estimated regression equation and that the agronomist applies 40 lb of fertilizer to a field that receives 15 in. of rainfall, the yield is 60.82 bushels of maize.

t test. The F test performed to determine whether there is a significant relationship between the dependent and the independent variables without specifying which variable is significant. The F test indicates that at least one of the β coefficients is not equal to zero. Researchers are usually interested in knowing whether the individual regression coefficients ($b_1, b_2, ..., b_k$) are significant. The t test allows us to perform a test of significance on each individual regression coefficient.

The null and alternative hypotheses test for each independent variable may be stated as:

$$H_o\text{: } \beta_i = 0$$
$$H_1\text{: } \beta_i \neq 0$$

where $i = 1, 2, ..., k$
The test statistic is

$$t = \frac{b_i - \beta_{io}}{S_{bi}}$$

[12–33]

where　t = t distribution with $n - k - 1$ $d.f.$
$\quad\quad\quad \beta_{io}$ = value of regression coefficient β specified by null hypothesis
$\quad\quad\quad b_i$ = sample regression coefficient
$\quad\quad\quad s_{bi}$ = standard error of regression coefficients

To perform the t test, we need an estimate of the standard error of the regression coefficients s_{bi}. For a two-variable case, as in the present example, the estimated standard error of b_1, and b_2 are found using the formulas:

$$s_{b_1} = \frac{s_{y.12}}{\sqrt{\Sigma\left(X_1 - \overline{X}_1\right)^2\left(1 - r_{12}^2\right)}} \quad\quad [12\text{--}34]$$

and

$$s_{b_2} = \frac{s_{y.12}}{\sqrt{\Sigma\left(X_2 - \overline{X}_2\right)^2\left(1 - r_{12}^2\right)}} \quad\quad [12\text{--}35]$$

Before we are able to compute the standard error of b_2, we have to calculate the deviation of sample observations from their mean. This is presented in Table 12.3.

By substituting the respective numerical values for each term into Equations 12–34 and 12–35, we find:

$$s_{b_1} = \frac{2.20}{\sqrt{2106.9[1 - (.91)^2]}}$$

$$s_{b_1} = 0.11$$

and

$$s_{b_2} = \frac{2.20}{\sqrt{702.50[1 - (.91)^2]}}$$

$$s_{b_2} = 0.20$$

Table 12.3   Calculation of the Deviation of Each Observed
Value from Their Mean

X_1	X_2	$(X_1 - \overline{X}_1)$	$(X_1 - \overline{X}_1)^2$	$(X_2 - \overline{X}_2)$	$(X_2 - \overline{X}_2)^2$
5	5	−21.9	479.61	−16.5	272.25
10	10	−16.9	285.61	−11.5	132.25
12	15	−14.9	222.01	−6.5	42.25
18	20	−8.9	79.21	−1.5	2.25
25	25	−1.9	3.61	3.5	12.25
30	25	3.1	9.61	3.5	12.25
36	30	9.1	82.81	8.5	72.25
40	30	13.1	171.61	8.5	72.25
45	25	18.1	327.61	3.5	12.25
48	30	21.1	445.21	8.5	72.25
269	215	0.0	2,106.90	0.0	702.50

Note: $\overline{X}_1 = 26.9$; $\overline{X}_2 = 21.5$.

Given the preceding information, we are now able to test the hypothesis
as follows:

$$t_1 = \frac{0.089}{0.11} = 0.80$$

$$t_2 = \frac{0.515}{0.20} = 2.57$$

By using $\alpha = .05$ and $10 - 2 - 1 = 7$ d.f. we can find the critical t
value for our hypothesis tests in Appendix E. For a two-tailed test
we have:

$$t_{.05} = 2.365$$

Because the computed t_1 value of 0.80 is less than the critical value,
we accept the null hypothesis, and conclude that b_1 does not differ
significantly from zero. On the other hand, the computed t_2 value of
2.57 is greater than the critical value of 2.365, hence the null hypothesis
is rejected, and we conclude at the 5% level of significance that b_2 differs
significantly from zero.

To summarize our findings, we conclude that rainfall has a statistically significant effect on yield of maize, but that after the rainfall
effect has been accounted for, fertilizer (X_1) does not have a statistically
significant effect.

Confidence intervals.   The multiple regression equation once again may be used to construct prediction intervals for an individual value of the dependent variable Y. For small sample observations, the following equation is used to find the prediction interval:

$$Y_i = \hat{Y} \pm t_\alpha S_{y.12} \sqrt{\frac{n+1}{n}} \qquad [12\text{--}36]$$

When the sample size is large ($n > 30$), the normal distribution replaces the t distribution, and we use the normal deviate z in Equation 12–36.

Example 12.3

Calculate the 95% prediction interval for the agronomist who used 40 lb of fertilizer (X_1), if the rainfall (X_2) was 15 in.

Solution

The sample size in this example is 10; therefore there are $10 - 2 - 1 = 7$ d.f. The critical value of $t_\alpha = t_{.025} = 2.365$. The predicted yield with 95% confidence is

$$Y_i = 60.82 \pm 2.365(2.20)\sqrt{\frac{10+1}{10}}$$

$$Y_i = 60.82 \pm 5.46$$

$$55.36 < Y_i < 66.28$$

The interpretation of this interval is that if we apply 40 lb of fertilizer to a field in a locality that receives 15 in. of rainfall, we are 95% confident that the yield of maize in the area is between 55.36 and 66.28 bushels.

12.6 Assumptions and problems in multiple linear regression

Similar to the simple regression model, a number of assumptions apply to the case of the multiple regression. These assumptions are

1. The regression model is linear and of the form

$$E(Y) = a + b_1X_1 + b_2X_2 + \ldots + b_kX_k \qquad [12\text{–}37]$$

2. The values of Y are independent of each other.
3. The values of Y are normally distributed.
4. The variance of Y values is the same for all values of $X_1, X_2 \ldots, X_k$.

Violation of these assumptions leads to a number of problems such as serial or autocorrelation, heteroscedasticity, and multicollinearity, which are explained as follows.

Serial or autocorrelation. This problem arises when the assumption of the independence of Y values is not met. That is, there is dependence between successive values. This problem is often observed when time series data are employed in the analysis. The validity of this assumption for cross-sectional data is virtually assured because of random sampling. To be sure that there is no autocorrelation, plotting the residuals against time is helpful. Figure 12.2 suggests the presence of autocorrelated terms. The figure shows a run of positive residuals followed by a run of negative residuals and the beginning of another run of positive residuals. When we observe such a pattern in the residuals, it signals the presence of auto or serial correlation.

It should be noted that in computer printouts, the points are not joined by lines or curves, but they are so depicted here to clarify the pattern. Methods for detecting serial correlation, such as the Durbin–Watson test, are often used. The Durbin–Watson statistic ordinarily tests the null hypothesis that no positive autocorrelation is present, thus implying that the residuals are random. Most computer programs including the MINITAB, SPSS, and SAS provide the

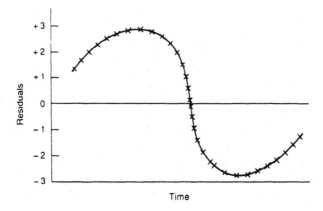

Figure 12.2 Positive and negative autocorrelation in the residuals.

Durbin–Watson statistics. The Durbin–Watson statistic, as shown in Appendix J, is defined as:

$$d = \frac{\sum\limits_{i=2}^{n}\left(u_i - u_{i-1}\right)^2}{\sum\limits_{i=1}^{n} u_i^2} \qquad [12\text{--}38]$$

where u_i = residual from the regression equation in time period i
  u_{i-1} = residual from the regression equation in time period $i - 1$, that is, the period before i

Durbin and Watson have tabulated lower and upper critical values of d statistic, d_L and d_U, such that:

1. If $d < d_L$, reject the null hypothesis of no positive autocorrelation and conclude that there is positive autocorrelation.
2. If $d > d_U$, accept the hypothesis of no positive autocorrelation and conclude that there is no positive autocorrelation.
3. If $d_L < d < d_U$, the test is inconclusive.

Serial or autocorrelation problems can be remedied by a variety of techniques. The most common approach is to work in terms of *changes* in the dependent and independent variables — referred to as *first differences* — not in terms of the original data. For example, we may use the year-to-year changes in the production of agricultural commodities instead of the original data on the production. This year-to-year change is referred to as the *first differences in production*. Other techniques such as transforming the data (see Hoshmand, 1994), adding variables, and using modified versions of the first difference transformation are also used to remedy serial correlation problems. For further discussion and more elaborate explanation of serial correlation problems, see the references at the end of this chapter.

Equal variances.   One of the assumptions of the regression model is that the error terms all have equal variances. This condition of equal variance is known as *homoscedasticity*. When this assumption is violated, the problem of *heteroscedasticity* arises. We can determine the existence of heteroscedasticity by plotting the residuals against the values of X as is shown in Figure 12.3. The pattern of residuals displayed indicates that the variance of residuals increases as the values of the independent variable increase. As an example, in studying the yield of milk with

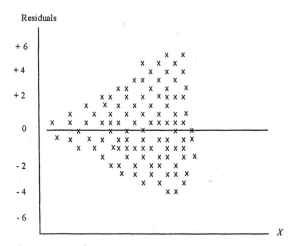

Figure 12.3 **Heteroscedasticity in residuals.**

different high-energy rations, we might find that yield rises with different high-energy rations. In such a case as this, the yield function is probably heteroscedastic. When heteroscedasticity is present, use of least squares is not the most efficient procedure for estimating coefficients of the regression model. Furthermore, the usual procedures for deriving confident intervals and tests of hypotheses for these coefficients are no longer valid.

One approach to remedy the problem of heteroscedasticity is to use the *weighed least squares* technique in estimating the regression equation. Kelejian and Oates (1974) suggest a test for heteroscedasticity and give the generalized least squares as a procedure to overcome problems arising from this error.

Multicollinearity. The problem of multicollinearity arises when two or more independent variables are highly correlated with each other. This implies that the regression model specified is unable to separate out the effect of each individual variable on the dependent variable. When multicollinearity exists between the independent variables, estimates of the parameters have larger standard errors, and the regression coefficients tend to be unreliable.

How do we know if we have a problem of multicollinearity? When a researcher observes a large coefficient of determination (R^2) accompanied by statistically insignificant estimates of the regression coefficients, the chances are that there is *imperfect multicollinearity*. When one (or more) independent variable(s) is an exact linear combination of the others, we have *perfect multicollinearity*.

Once it is determined that multicollinearity exists between the independent variables, a number of possible steps can be taken to remedy this problem:

1. Drop the correlated variable from the equation. Which independent variable to drop from the equation depends on the test of significance of the regression coefficient, and the judgment of the researcher. If the *t* test indicates that the regression coefficient of an independent variable is statistically insignificant, that variable may be dropped from the equation. Dropping a highly correlated independent variable from the equation does not affect the value of R^2 much.
2. Change the form of one or more independent variables. For example, an agricultural economist, in a demand equation for beef (Y), finds that income (X_1) and another independent variable (X_2) are highly correlated. In such a situation, dividing the income by the variable of population yields per capita income, which may result in less correlated independent variables. Other approaches are suggested in Kelejian and Oates (1974) and Hamburg (1987).

12.7 *Optional topic: curvilinear multiple regression analysis*

As discussed in Chapter 11, there are many instances where the theoretical nature of the relationship between dependent and independent variables are such that a straight line does not provide an adequate explanation between two or more variables. For example, the Cobb–Douglas production function is an example of the exponential model that is given as:

$$Q = \alpha K^\beta L^\gamma u \qquad [12\text{--}39]$$

where Q = quantity produced
   K = capital
   L = labor
   u = multiplicative error term
$\alpha, \beta,$ and γ = parameters to be estimated

As before, by taking the logs of Equation 12–39, we obtain:

$$\log Q = (\log \alpha) + \beta (\log K) + \gamma (\log L) + (\log u) \qquad [12\text{--}40]$$

which is of the standard form:

$$Y = \alpha' + \beta X + \gamma Z + e \qquad [12\text{–}41]$$

The demand function is also an exponential model shown as follows:

$$Q = \alpha P^\beta u \qquad [12\text{–}42]$$

where Q = quantity demanded
 $\quad\;\; P$ = price

Equation 12–42 can be linearized by taking its logarithms as shown:

$$\log Q = \log \alpha + \beta \log P + \log u \qquad [12\text{–}43]$$

which in standard form is

$$Y = \alpha' + \beta X + e \qquad [12\text{–}44]$$

Polynomial curve.   The polynomial is by far the most widely used equation to describe the relation between two variables. Snedecor and Cochran (1980) point out that when faced by nonlinear regression, and where one has no knowledge of a theoretical equation to use, the second-degree polynomial in many instances provides a satisfactory fit.

The general form of a polynomial equation is

$$Y = \alpha + \beta_1 X + \beta_2 X^2 + \dots + \beta_k X^k \qquad [12\text{–}45]$$

In its simplest form, the equation has the first two terms on the right-hand side of the equation, and is known as the equation for a straight line. By adding another term ($\beta_2 X^2$) to the straight-line equation, we get the second-degree or a quadratic equation, the graph of which is a parabola. As more terms are added to the equation, the degree or power of X increases. A polynomial equation with a third degree (X^3) is called a cubic; and those with a fourth and fifth degrees are referred to as the quartic and quintic, respectively. Figure 12.4 shows examples of the polynomial curves with different degrees.

To illustrate the method and some of its applications, the following example is used.

Example 12.4

To determine the relationship between yield and the protein content of soybean, an agronomist gathers the following data. The agronomist is

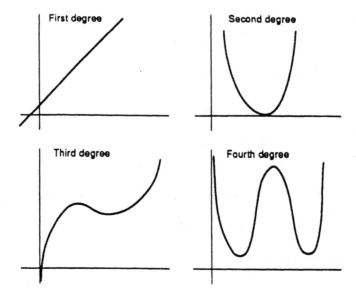

Figure 12.4   Typical shapes of polynomial curves.

particularly interested in estimating the protein content for the different yields. Furthermore, she wishes to test for any departure from linearity in the data. Thus, the regression equation is assumed to have the following functional form:

$$Y = \alpha + \beta X + \gamma X^2$$

Soybean yield (bushels/acre)	Protein (%)	Soybean yield (bushels/acre)	Protein (%)
29.50	40.80	31.20	37.60
28.60	42.10	34.50	35.10
28.10	41.30	38.10	35.80
31.20	40.30	33.20	36.90
32.40	39.60	36.10	35.80
30.20	43.40	33.60	36.80
38.30	35.30	33.00	35.20
30.40	41.50	32.30	36.60
30.00	39.70	37.30	34.40
31.30	39.20	35.50	34.00
29.80	41.90	34.30	37.20
34.50	36.05	33.90	37.60
30.20	39.00	39.20	37.80
27.00	43.10	39.15	35.60
32.00	37.30	38.40	36.25

Solution

Step one — Because it is assumed by the agronomist that the functional form of the regression equation is quadratic, and the data gathered are only on two variables (yield and protein content), we need to add another variable to the regression equation. The added variable is the square of the yield of soybean (column two of Table 12.4) and is treated like a third variable in the equation. Because of this added variable, we now have two independent variables, and thus a multiple regression equation. Because the step-by-step analysis of the multiple regression is given previously, we do not elaborate on the calculation procedures now. For this particular example, we will simply compare the estimated regression equations and their fit to the data.

Table 12.4 **Soybean Yield (X), and Protein Content (Y) from 30 Plots**

Soybean yield (bushels/acre) X	X^2	Protein (%) Y	Soybean yield (bushels/acre) X	X^2	Protein (%) Y
29.50	870.25	40.80	31.20	973.44	37.60
28.60	817.96	42.10	34.50	1190.25	35.10
28.10	789.61	41.30	38.10	1451.61	35.80
31.20	973.44	40.30	33.20	1102.24	36.90
32.40	1049.76	39.60	36.10	1303.21	35.80
30.20	912.04	43.40	33.60	1128.96	36.80
38.30	1466.89	35.30	33.00	1089.00	35.20
30.40	924.16	41.50	32.30	1043.29	36.60
30.00	900.00	39.70	37.30	1391.29	34.40
31.30	979.69	39.20	35.50	1260.25	34.00
29.80	888.04	41.90	34.30	1176.49	37.20
34.50	1190.25	36.05	33.90	1149.21	37.60
30.20	912.04	39.00	39.20	1536.64	37.80
27.00	729.00	43.10	39.15	1532.72	35.60
32.00	1024.00	37.30	38.40	1474.56	36.25

Step two — Calculate a simple regression using the yield and the protein data to determine the fit. The estimated regression equation and the coefficient of determination are given as follows:

$$\hat{Y} = 58.923 - .629X$$

$$R^2 = 0.64$$

The coefficient of determination shows that the straight line accounts for 64% of the variability in protein content. Figure 12.5 shows how the estimated simple linear equation fits the data.

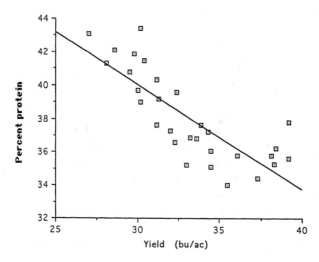

Figure 12.5   Graph of the soybean data fitted to a linear equation.

Step three — To determine whether the relationship between the yield and the protein content is quadratic, as hypothesized by the agronomist, we estimate a quadratic equation with the square of the yield data serving as the third variable. The estimated regression equation and the multiple coefficient of determination are given as follows:

$$\hat{Y} = 148.832 - 6.039X + 0.081X^2$$
$$R^2 = .77$$

Figure 12.6 shows that the quadratic equation accounts for 77% of the variability in the protein content, and is a much better fit than the straight line.

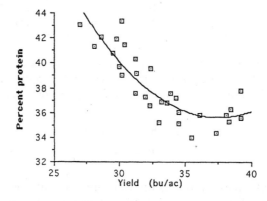

Figure 12.6   Graph of the soybean data fitted to a quadratic equation.

Step four — Compare the results from the simple and quadratic equations as summarized in Table 12.5.

Table 12.5 **Test of Significance of Departure from Linear Regression**

Source of variation	Degree of freedom	Sum of squares	Mean square	F
Deviations from linear regression	28	75.357		
Deviations from quadratic regression	27	49.217	1.823	14.34[a]
Reduction in sum of squares	1	26.140	26.140	

[a] = Highly significant.

From Table 12.5 we observe that the reduction in the sum of squares tested against the mean square remaining after curvilinear regression is highly significant, thus confirming the agronomist's hypothesis that the relationship between yield and protein content is curvilinear.

12.8 A computer application

It should be apparent by now that calculation of a multiple regression equation when only two independent variables are considered in the regression model is tedious and cumbersome. Inclusion of additional explanatory variables in the regression model makes it difficult to compute. Computer programs such as MINITAB, SPSS, and SAS all perform multiple regression analysis.

The MINITAB multiple regression program is used to perform all the needed computation for Example 12.1. Figure 12.7 illustrates a computer printout of Example 12.1. Let us examine the information contained in this printout.

1. The regression equation is given as:

$$\hat{Y} = 49.53 + 0.089X_1 + 0.516X_2$$

The preceding values of the coefficients given above are slightly different from those computed earlier. This difference is simply due to rounding error.

2. In this section of the printout the regression coefficients, the standard deviation of each coefficient, and the t ratio for each coefficient are given. The t coefficients are simply compared with the critical t values of Appendix E. Given different values, we

1.	**The Regression Equation is** $C1 = 49.5 + 0.089\,C2 + 0.516\,C3$

2.	Column	Coefficient	St. Dev. of Coef.	T-Ratio = Coef/S.D.
		49.527	2.061	24.03
	C2	0.0888	0.1127	0.79
	C3	0.5155	0.1953	2.64

3.	$S = 2.197$

4.	R-Squared = 90.0 Percent R-Squared = 87.1 Percent, Adjusted for D.F.

5.	**Analysis of Variance**			
	Due to	DF	SS	MS = SS/DF
	Regression	2	304.21	152.10
	Residual	7	33.79	4.83
	Total	9	338.00	

Further Analysis of Variance
SS Explained By Each Variable When Entered in the Order Given

Due to	DF	SS
Regression	2	304.21
C2	1	270.55
C3	1	33.65

6.	**Durbin-Watson Statistic = 1.85**

MTB >

Figure 12.7   A computer printout of the regression equation, the coefficient of determination, and the ANOVA table.

are able to determine whether the coefficients are statistically significant or not.

3. The standard error for the equation is given as $S = 21.79$. Interpretation and use of the standard error of estimate is the same as given in Section 12.3.
4. The coefficient of determination (R^2) and the adjusted coefficient of determination are given in this part of the printout.
5. In this section of the computer printout, an analysis of variance (ANOVA) table is given. The ANOVA table provides information on the source of variations, the degrees of freedom ($d.f.$), sum of squares (SS), and the mean square (MS). The ratio of the regression mean square and the residual mean square give us the F value. The t test can be performed as before to determine whether there is a relationship between the dependent and independent variables.
6. The Durbin–Watson statistic is given in this part of the computer printout. This statistic is compared with the critical value given in Appendix J to determine the presence of autocorrelated terms. The Durbin–Watson statistic is mentioned in Section 12.6.

Chapter summary

In this chapter we elaborated on the concepts of the multiple linear regression model. Notice that *multiple regression–correlation analysis* is simply an extension of the simple linear regression model. In multiple regression we consider the relationship between a dependent variable and two or more independent variables.

The technique of least squares was used in estimating the multiple regression equation. The geometric interpretation of the model involves *plane* instead of straight lines.

The purpose of multiple regression equation is to predict and estimate the value of the dependent variable. The strength of the prediction depends on the multiple standard error of estimate.

The strength of the relationship between the dependent and the independent variables is measured by the coefficient of multiple correlation (R). The range of values that R can take is from zero to 1. An R value of zero indicates no correlation between the dependent and independent variables. On the other hand, an R value of 1 implies perfect correlation.

The assumptions of the multiple regression model are similar to those of the simple regression case. When the assumptions are not met, there are problems of serial or *autocorrelation, heteroscedasticity,* and *multicollinearity.* Serial correlation exists when it is observed that the successive data points of the dependent variable Y are dependent on each other. A widely used method of detecting the presence of auto-correlation involves the Durbin–Watson statistic. When the assumption that the error terms have equal variances is violated, the problem of heteroscedasticity arises. Techniques such as the generalized least squares are used to correct for this problem. Finally, we may encounter the problem of multicollinearity. This problem arises when two or more independent variables are highly correlated with each other. There are a number of steps that can be taken to solve the problem. The simplest step is to delete one or more of the correlated variables. The researcher's judgment and test of significance for the net regression coefficients can determine which variable to drop. Remember that in general if one of two highly correlated independent variables is dropped, the R^2 value does not change greatly. Additional material on these topics can be found in the references at the end of this chapter.

Case Study

Global Warming as a Function of Environmental Pollution

Climatologists around the world are monitoring the data on the average daily temperature that has been rising during the last century. According to the 1992 report of the Intergovernmental Panel on Climate Change (IPCC) the global temperature today is 0.3 to 0.6°C above what it was in 1880. The steady rise in temperature over the years is a major concern for environmentalists. Some believe that global warming may have severe consequences for the world community. Higher temperatures can cause melting of the polar ice resulting in sea levels rising by a projected 6 cm per decade (Warrick and Oerlemans, 1990). Additionally, local climate patterns can be altered such that precipitation, soil moisture, and the length of seasons are seriously affected. These changes can destroy many ecosystems, threatening thousands of species (U.S. Congress, 1993).

Various factors are linked to the temperature changes around the world. Among them are volcanic eruptions, dust storms, environmental pollution (carbon emission from fossil fuel, and emission of other particulates), and deforestation. The eruption of Mount Pinatubo in the Philippines, the largest volcanic eruption in the century, injected millions of tons of dust into the upper atmosphere, blocking enough sunlight to depress temperatures in the lower atmosphere by about half a degree Celsius (Kerr, 1993).

Most climatologists believe that the warming trend is the result of the buildup of greenhouse gases in the air, particularly carbon dioxide (CO_2). Historical data show that global temperature consistently reflects the amount of carbon dioxide in the air (Barnola et al., 1994). In the following table, data show that the amount of carbon from fossil fuel released into the air has been rising since 1970. Other greenhouse gases including chlorofluorocarbons (CFCs), methane, and nitrous oxide also are contributing to global warming. Their impact, while significant, is not as great as carbon dioxide. It is reported that the warming impact of the climb in carbon dioxide concentrations alone is at least 60% higher than those of other gases combined (IPCC, 1994). There appears to be a clear relationship between the rise in global temperature and the use of fossil fuels.

Unfortunately, the economies of many nations are tied with the use of fossil fuels. Fossil fuels are used in the manufacturing, agriculture, and transportation sectors of countries around the world. The fastest rise in emission is occurring in the industrializing countries of Asia and Latin America. Although the developing countries emit only 0.5 ton of carbon per person as compared to 3.0 tons per person in the developed countries, they already account for a third of the global total; and their contribution to pollution is growing fast enough to double every 14 years (Marland et al., 1994; Roodman, 1995).

Although the emission of CFCs is decreasing because of regulations, the link between CFCs, bromine, and other ozone-depleting substances and the destruction of stratospheric ozone layer is unquestionable. It is believed that ozone depletion causes increases in biologically damaging ultraviolet radiation to reach the earth's surface and causes human health and ecological problems (Ryan, 1995). There continues to be a need to monitor the level of CFC in the air.

The following table shows the average global temperature and the presence of carbon emission and CFC over the past quarter of a century.

To determine if there is a relationship between the three variables:

1. Perform a multiple regression analysis.
2. Test the significance of b_1 and b_2 at $\alpha = .05$.
3. Determine if the overall model is useful. Test using $\alpha = .01$ level.

Global Average Temperature, Carbon Emission, and CFC Production, 1970 to 1994

Year	Temperature (°C)	Carbon emissions (million tons)	CFC production (thousand tons)
1970	15.04	4006	640
1971	14.89	4151	690
1972	14.93	4314	790
1973	15.19	4546	900
1974	14.93	4553	970
1975	14.95	4527	860
1976	14.79	4786	920
1977	15.16	4920	880
1978	15.09	4960	880
1979	15.14	5239	850
1980	15.28	5172	880
1981	15.39	5000	890
1982	15.07	4960	870
1983	15.29	4947	950
1984	15.11	5109	1050
1985	15.11	5282	1090
1986	15.16	5464	1130
1987	15.32	5584	1250
1988	15.35	5801	1260
1989	15.25	5912	1150
1990	15.47	5941	820
1991	15.41	6026	720
1992	15.13	5910	630
1993	15.20	5893	520
1994 (preliminary)	15.32	5925	295

Adapted from Roodman, D. M., Global temperature rises again, in Brown. L. et al., *Vital Signs 1995*, W. W. Norton, New York, 1995, 65; Data for CFC production from Ryan, M., 1995. CFC Production Plummeting, in Brown, L. et al., *Vital Signs 1995*, W. W. Norton, New York, 1995, 63.

References

Barnola, J. M. et al., Historical CO_2 record from the Vostok ice core, in Boden, T. A. et al., Eds., *Trends '93: A Compendium of Data on Global Change*, Oak Ridge National Laboratory, Oak Ridge, TN, 1994.

Intergovernmental Panel on Climate Change (IPCC), *Climate Change 1992: The IPCC Supplementary Report*, Cambridge University Press, Cambridge, 1992.

Intergovernmental Panel on Climate Change (IPCC), Radiative forcing of climate change, The 1994 Report of the Scientific Assessment Working Group of IPCC, Summary for Policymakers, World Meteorological Organization, Geneva, 1994.

Kerr, R. A., Pinatubo global cooling on target, *Science*, January 29, 1993.

Marland, G., Andres, R. J., and Boden, T. A., Global Regional, and National CO_2 Emissions, in Boden, T. A. et al., Eds., *Trends '93: A Compendium of Data on Global Change*, Oak Ridge National Laboratory, Oak Ridge, TN, 1994.

Office of Technology Assessment, U.S. Congress, Preparing for an Uncertain Climate, Vol. 1, U.S. Government Printing Office, Washington, D.C., 1993.

Roodman, D. M., Carbon Emissions Resume Rise, in Brown, L. et al., *Vital Signs 1995*, W. W. Norton, New York, 66, 1995.

Ryan, M., CFC Production Plummeting, in Brown, L. et al., *Vital Signs 1995*, W. W. Norton, New York, 62, 1995.

Warrick. R. A. and Oerlemans, H., Sea level rise in Intergovernmental Panel on Climate Change, *The IPCC Scientific Assessment*, Cambridge University Press, Cambridge, 1990.

Review questions

1. The coefficient of multiple determination is equal to the _____ divided by the total sum of squares.

2. _____ refers to a condition in multiple regression analysis when there is a high level of correlation between independent variables.

3. The multiple regression equation is used for _____ and _____.

4. The _____ is used to determine the significance of a regression equation.

5. An animal researcher is interested in the relationship that exists between urea nitrogen and the energy balance and the phosphorous balance. The researcher randomly selects 15 cows fed rations that contain high and low levels of energy and phosphorus. The data from the experiment are given as follows.

Y Urea N (mg/100 ml)	X_1 Energy Balance (k cal/d)	X_2 P Balance (g/d)
9.8	8.2	22.4
10.1	8.3	26.2
10.2	8.6	25.4
11.1	10.6	28.2
12.4	7.6	−2.5
10.6	7.9	22.1
10.9	11.9	19.2
11.2	−5.6	8.6

(continued)

Y Urea N (mg/100 ml)	X_1 Energy Balance (k cal/d)	X_2 P Balance (g/d)
11.4	7.8	18.5
11.9	8.2	-9.7
12.3	9.8	-8.6
11.6	11.3	-10.2
11.1	10.1	12.4
10.8	- 5.2	7.7
10.9	5.8	9.8

a. Determine the estimated regression equation.
b. Calculate the standard error of estimate.
c. Test the significance of b_1 and b_2 at $\alpha = .05$.
d. Compute R^2 and interpret its meaning.

6. An agricultural engineer wishes to determine the relationship that exists between the monthly electricity usage in a greenhouse and the size of the greenhouse. He believes that the relationship is best exemplified by the following model:

$$Y = a + b_1X_1 + b_2X_1^2 + \varepsilon$$

Given the following data:

Monthly electrical usage (kW/h)	Greenhouse size (ft²)
2200	3000
2350	3200
2630	3220
2700	3600
3100	3640
3350	3700
3420	3710
3600	3800
3720	3900
3800	4200

a. Estimate the multiple regression equation. (Use a computer to solve this problem.)
b. Investigate and determine whether the overall model is useful. Test using $\alpha = .01$.
c. Find R^2 for the fitted model. Interpret your results.

7. The agricultural engineer in the previous example also hypothesizes that there is a strong relationship between the monthly electrical usage in the greenhouse and the size of the greenhouse, as well as the average monthly temperature. Data gathered are shown as follows.

Monthly electrical usage (kW/h)	Greenhouse size (ft²)	Average monthly temperature (°F)
2200	3000	70
2350	3200	72
2630	3220	80
2700	3600	85
3100	3640	86
3350	3700	82
3420	3710	74
3600	3800	71
3720	3900	68
3800	4200	62

a. Determine the estimated regression equation. (Use a computer to solve this problem.)
b. State the meaning of the partial regression coefficients b_1 and b_2.
c. Suppose that we have a greenhouse of size 3500 ft², and that the average monthly temperature is 75°F. Determine the prediction interval for the monthly electrical usage at 95% level of confidence.
d. Test the overall significance of the regression model at $\alpha = .01$. (Hint: F test).

8. An ornamental horticulturist is interested in the impact of the amount of nitrogen fertilization and exposure to light on the thickness of the leaves of a new variety of photoplant. After 2 weeks of experimentation, the following observations of leaf thickness, amount of light in the greenhouse, and fertilization are recorded for 20 plants of the same age grown in the same environment.

Leaf thickness (mm)	Nitrogen fertilization (mg/pot)	Hours of sunlight
.21	.40	2.0
.68	.35	2.5
.77	.25	3.0

(continued)

Leaf thickness (mm)	Nitrogen fertilization (mg/pot)	Hours of sunlight
.62	.75	4.5
.45	.65	4.0
.50	.45	4.5
.59	.35	4.0
.75	.40	2.0
.65	.35	2.5
.60	.50	6.0
.54	.60	1.5
.76	.70	2.0
.60	.20	6.0
.62	.35	5.5
.67	.50	5.0
.68	.15	5.0
.65	.20	6.0
.68	.25	2.5
.64	.15	4.5
.59	.45	5.0

a. Fit a multiple regression equation to the data.
b. Interpret the slope coefficient (b_i values) specifically in terms of the variables and units in this problem.
c. Calculate the coefficient of multiple determination and multiple correlation. Which is easier to interpret?
d. Calculate the standard error of estimate.
e. Using a 5% level of significance, perform an appropriate test of the overall significance of the regression equation.

9. An equine researcher postulates that there is a nonlinear relationship between the life span of Arabian horses and the gestation period. The researcher believes that the following model best exemplifies the relationship:

$$Y = a + b_1 X_1 + b_2 X_1^2 + \varepsilon$$

Life span (years)	Gestation period (d)
15.6	220
18.1	240
20.0	275
22.1	290
23.2	380
20.5	340

Life span (years)	Gestation period (d)
22.1	290
21.4	390
19.8	300
18.5	279
20.5	320
22.5	310
20.0	400
18.5	290
15.6	320

a. Estimate the multiple regression equation. (Use a computer to solve this problem.)
b. Interpret the meaning of the coefficient of multiple determination.
c. Test the significance of each at $\alpha = .05$.
d. Determine the p value.

References and suggested reading

Barnett, V. and Turkman, K. F., Eds., *Statistics for the Environment*, John Wiley & sons, New York, 1993, chap. 16.

Cohen, J. and Cohen, P., *Applied Multiple Regression/Correlation Analysis for Behavioral Sciences*, Lawrence Erlbaum Associates, Hillsdale, NJ, 1975.

Cothern, C. R. and Ross, N. P., Eds., *Environmental Statistics, Assessment, and Forecasting*, CRC Press, Boca Raton, FL, 1994.

Draper, N. R. and Smith, H., *Applied Regression Analysis*, 2nd ed., John Wiley & Sons, New York, 1981.

Galton, F., *Memories of My Life*, E. P. Dutton, New York, 1908.

Hamburg, M., *Statistical Analysis for Decision Making*, 3rd ed., Harcourt Brace Jovanovich, New York, 1987, chap. 10.

Hoshmand, A. R., *Experimental Research Design and Analysis: A Practical Approach for Agricultural and Natural Sciences*, Boca Raton, FL, CRC Press, 1994, chap. 9.

Kelejian, H. H. and Oates, W. E., *Introduction to Econometrics*, Harper & Row, New York, 1974.

Mead, R. and Curnow, R. N., *Statistical Methods in Agriculture and Experimental Biology*, Chapman & Hall, New York, 1983.

Netter, J. and Kutner, M. H., *Applied Linear Regression Models*, Irwin, Homewood, IL, 1983.

Snedecor, G. W. and Cochran, W. G., *Statistical Methods*, Iowa State University Press, Ames, IA, 1980.

Yajima, Y., On estimation of a regression model with long-memory stationary errors, *Ann. Stat.*, 16, 791, 1988.

PART VI

Analysis of Change over Time

chapter thirteen

Time Series Analysis

Learning objectives

The overall learning objective of this chapter is to master the methods of measuring data over time. This requires an understanding of the components of a time series. Specifically, you should learn to:

- Define the classical time series model of trend, seasonal, cycle, and irregular.
- Describe the methods of computing and projecting a linear trend.
- Fit a nonlinear trend line.
- Measure rates of growth and rates of acceleration.
- Forecast time series using trend projections.
- Compute and use a seasonal index by the method of ratio-to-moving average.
- Deseasonalize the original data from seasonal variations.
- Do cycle-irregular analysis by the residual method.

13.1   Introduction

In this chapter we study methods of analyzing data over time. Researchers and managers alike are interested in planning for future changes. A *time series* analysis provides the basis for such planning. Time series is nothing more than observed successive values of a variable or variables over regular intervals of time. Examples of a time series include the annual production of wheat in the U.S. from 1940 to the present, monthly price of beef, and quarterly production of tractors. The review of historical data over time provides the decision maker with a better understanding of what has happened in the past, and how estimates of future values may be obtained.

Time series analysis is mostly used in the economic and business aspects of agriculture. The aim in this chapter is to provide a sufficient technical base for performing a time series analysis in any discipline within agriculture including environmental sciences. However, the emphasis of the chapter is for those specifically interested in agricultural economics and business. The use of time series analysis for environmental issues is also highlighted.

Time series analysis is basically descriptive in nature and does not permit us to make any probability statement concerning future events. If used properly, however, the analysis provides good forecasts.

Classical time series model.   To have a better understanding of what is meant by a time series model, we should know the basic components of time series data, each of which can influence our forecast of future outcomes. The four components are

1. Secular trend (T)
2. Seasonal variation (S)
3. Cyclical variation (C)
4. Random or irregular variation (I)

The time series model generally used is a multiplicative model that shows the relationship between each component and the original data of a time series (Y) as:

$$Y = T \times S \times C \times I \qquad\qquad [13\text{--}1]$$

In the following sections of this chapter, we learn the techniques of computing and identifying each component of a time series. The trend (T) is computed directly from the original data using different approaches. The seasonal variations (S) in a time series are represented by a seasonal index that is also computed from the original data by eliminating the other components. The cycle (C) and irregular (I)

movements in a time series are difficult to separate because both affect the series in similar ways. Because of this fact, the time series model may be written as:

$$Y = T \times S \times C I \qquad [13\text{--}2]$$

Once we have computed the trend and seasonal, the cycle and irregular components are found by dividing the original data by them using the following formula:

$$\frac{Y}{T \times S} = \frac{\cancel{T} \times \cancel{S} \times CI}{\cancel{T} \times \cancel{S}} = CI \qquad [13\text{--}3]$$

Before applying the time series model in detail, let us examine the nature of each component.

Secular trend.   *Secular trend* is a long-term growth movement of a time series. The trend may be upward, downward, or steady. It can be pictured as a smooth linear or nonlinear graph. Figure 13.1 shows the trend of fertilizer nutrients used per acre in the U.S. between 1955 and 1982.

Seasonal variation.   *Seasonal variation* refers to repetitive fluctuations that occur within a period of one year. The repetitive patterns of seasonal variation are best exemplified by the weather. Agricultural commodities are subject to seasonal variations. These seasonal variations, although completed within a year, are repeated year after year.

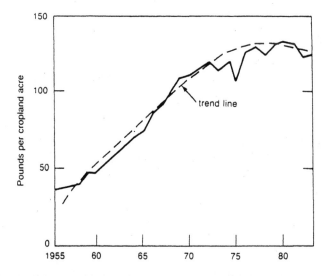

Figure 13.1   Secular trend of the fertilizer nutrient used per acre.

Cyclical variation.   Another component of time series is *cyclical variation*. Cyclical variations are wave-like movements that are observable over extended periods of time. Depending on the type of cycle, the length varies from 3 to 15 years. The *business cycle* in the American economy, for instance, usually lasts about eight years (Mason, 1982). Cyclical variations are less predictable and may have different causes.

Random or irregular variation.   *Random variation* refers to variation in a time series that is not accounted for by trend, seasonal, or cyclical variations. Because of its unsystematic nature, the random or irregular variation is erratic with no discernible pattern. Such factors as strikes, wars, and heat waves contribute to the irregular or random variation of a time series.

Figure 13.2 presents examples of trend (a), seasonal (b), cycle (c), and irregular (d) components of a time series. The next section addresses each component of time series and methods for measuring each.

13.2   Secular trend analysis

One of the purposes of time series analysis is to give the researcher and manager alike a method for measuring change over time. If the change over time is constant, a straight line may be an appropriate way of describing the trend. The linear trend is expressed as:

$$Y_t = a + bx \qquad\qquad [13\text{--}4]$$

where Y_t = trend value of Y variable for a given value of x (subscript t) necessary to distinguish trend values from data values)

   a = Y intercept or estimated value of Y when x is equal to zero

   b = slope of line or average change in Y per unit change in time

   x = unit of time

There are other forms of trend lines such as exponential, as shown in Equation 13–5, or quadratic, as shown in Equation 13–6:

$$Y_t = e^{a+b_t} \qquad\qquad [13\text{--}5]$$

$$Y_t = a + b_t + c_t^2 \qquad\qquad [13\text{--}6]$$

The basic pattern of the trend line is determined by the coefficients a, b, and c. In general, we use annual data to fit a trend line, thus

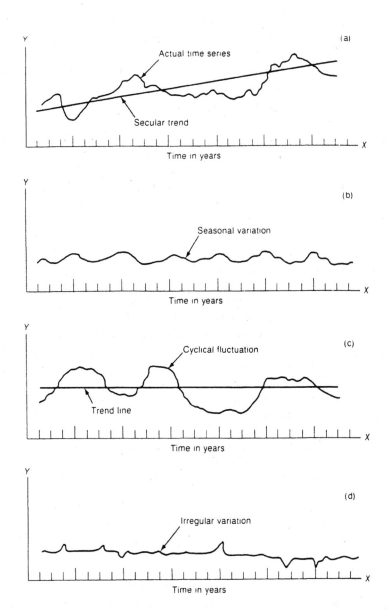

FIGURE 13.2   The components of a time series: (a) secular trend, (b) seasonal variation, (c) cyclical fluctuation, and (d) irregular variation.

eliminating the seasonal variations that may be present in the data. The choice of which equation to use in fitting a trend line is based on theoretical considerations and historical patterns the data may have followed. Therefore, it is the responsibility of the researcher to choose the best equation that fits the data.

Before we discuss the various methods of fitting a trend line, it is essential that we understand how we can transform the time variable that is measured in weeks, months, and years into a *coded time*. By using the coded time, the computation is simplified because we do not have to deal with the square of large numbers such as 1995, 1996, and so on. Furthermore, it simplifies the equations that are used in computing the intercept and the slope of the trend line, as shown later.

To transform the traditional measures of time into a coded time we simply find the mean time and then subtract the value from each of the sample times. This is shown in Table 13.1 for both the *odd* and *even* number of years in a data set. To transform an odd number of years in a data set (1994, 1995, and 1996) to coded time, we find the mean year first, and then subtract each year in the data set from it. In this case the mean year is 1995 ($x = 0$). The corresponding code times are –1, 0, and 1. In this instance –1 represents the first year (1994 to 1995), and 1 represents the last year (1995 to 1996). Column three of the table shows the coded time for odd number of years. When we encounter an even number of years in the data set, the mean time period, at which $x = 0$, falls midway between the two central years. For example, if our data set included the years 1993, 1994, 1995, and 1996, then the two central years, 1994 and 1995, would deviate from this origin by –1/2 and +1/2, respectively. To avoid the use of fractions in our computations, we multiply each time element by 2. Thus the coded times will be –3, –1, 0, +1, and +3. The coded time for the even number of years is shown in column seven of the table.

Table 13.1   Transforming time into coded time

Odd number of years[a]			Even number of years[b]			
X	$X - \overline{X}$	Coded time (x)	X	$X - \overline{X}$	$(X - \overline{X}) \times 2$	Coded time (x)
(1)	(2)	(3)	(4)	(5)	(6)	(7)
1990	1990–1993	– 3	1990	1990–1992.5 = -2.5×2 =		– 5
1991	1991–1993	– 2	1991	1991–1992.5 = -1.5×2 =		– 3
1992	1992–1993	– 1	1992	1992–1992.5 = -0.5×2 =		– 1
1993	1993–1993	0	1993	1993–1992.5 = 0.5×2 =		1
1994	1994–1993	1	1994	1994–1992.5 = 1.5×2 =		3
1995	1995–1993	2	1995	1995–1992.5 = 2.5×2 =		5
1996	1996–1993	3				

[a] $\overline{X} = \dfrac{\Sigma X}{n} = \dfrac{13,951}{7} = 1993$, $\overline{X}$ (the mean year) $= 0$.

[b] $\overline{X} = \dfrac{\Sigma X}{n} = \dfrac{11,955}{6} = 1992.5$; $\overline{X}$ (the mean year) $= 0$.

Next we turn to the various techniques in fitting a trend line.

There are several approaches to measuring the secular trend for a linear equation such as Equation 13–4. As in Chapter 11, we have to compute a value for *a* and *b* in Equation 13–4. We may apply:

1. The *freehand* or graphic method of fitting the line to the data
2. The *method of semiaverage*
3. The *least squares* method discussed in the previous two chapters

Although the results of fitting a trend line by any of the preceding methods may be satisfactory, the first two approaches have certain limitations that preclude them from being used often. Most researchers find that the method of least squares not only is easy to use, but also mathematically fits the line to the data.

Freehand method. The freehand method simply approximates a linear trend equation. The trend line is approximated by drawing a straight line through the middle of the data. Table 13.2 shows the total farm output in the U.S. in current dollars between 1984 and 1993. Figure 13.3 shows the freehand method trend line for the given data.

Table 13.2 **U.S. Farm Output
Between 1984 and 1993**

Year	Total farm output ($ billion)
1984	160.0
1985	152.7
1986	144.0
1987	152.1
1988	158.2
1990	186.4
1991	181.7
1992	188.6
1993	184.5

From U.S. Department of Commerce, *Statistical Abstract of the United States — 1995. The National Data Book*, Washington, D.C., 1995, 676, Table no. 1109.

To determine the linear trend line by this method, we first assign the beginning year (1984) as the origin or 0 year, and code each successive year with 1, 2, 3, ..., and finally 1993 as year 9. Because 1984 is the origin or year 0 in this illustration, the intercept of the line or *a* is $160 billion dollars.

$ billion

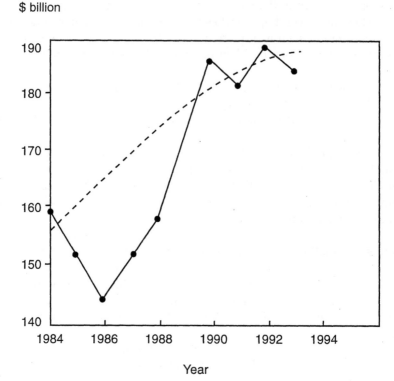

Figure 13.3   Trend line by the method of *freehand*.

Second, we draw a straight line from the point of intercept through the middle of the data. The straight line drawn indicates the path of average increase in the total farm output in the U.S. We observe that the total farm output increased from $160 billion in 1984 to $184.5 billion in 1993. This represents an increase of $24.5 billion in 9 years, or an average of $2.72 billion per year. This value of $2.72 represents the slope of the line or value of b in our Equation 13–4.

The straight line equation for the preceding illustration is

$$Y_t = 160 + 2.72x$$

where x = 0 in 1984
x = 1 year
Y = billion dollars

We may use the trend line to forecast per capita personal disposable income for U.S. farmers in the year 2000. Because 2000 is our coded year 16, the forecast for total farm output, under the assumption that production remains the same, is

$$Y_t = 160 + 2.72 \ (16)$$

$$Y_{2000} = 203.5 \text{ dollars}$$

Semiaverage method. The trend line fitted by the freehand method, as observed, is based on the judgment of the person drawing the line. This subjective approach as expected may yield unsatisfactory results. The semiaverage method provides a simple but objective method in fitting a trend line. To fit a trend line through the method of semiaverage, the following steps are followed:

Step one — Prior to fitting the data to a trend line, graph the data to determine whether a straight line or any other forms mentioned earlier best describe the data.

Step two — Divide the data into two equal periods, and compute an average for each period.

Step three — Given the two averages, we now have two points on a straight line. The slope of the line is computed by taking the difference between the averages in the second and the first period, and dividing them by half of the total number of years in the observation.

Let us use an example to illustrate the fitting of a trend line by this method.

Example 13.1

The following table gives the total cash income between 1984 and 1993 for farmers in the U.S. Fit a straight line trend line by the method of semiaverage.

Year	Millions of dollars
1984	156.1
1985	157.9
1986	152.9
1987	165.2
1988	172.9
1989	179.8
1990	186.8
1991	184.9
1992	188.2
1993	197.2

Adapted from *Statistical Abstract of the United States — 1995: The National Data Book*, U.S. Department of Commerce, Washington, D.C., 1995, 676, Table no. 1110.

Solution

We follow these three steps in fitting a straight trend line by the semi-average method.

Step one — Figure 13.4 shows the graph of the total cash income between 1984 and 1993.

$ million

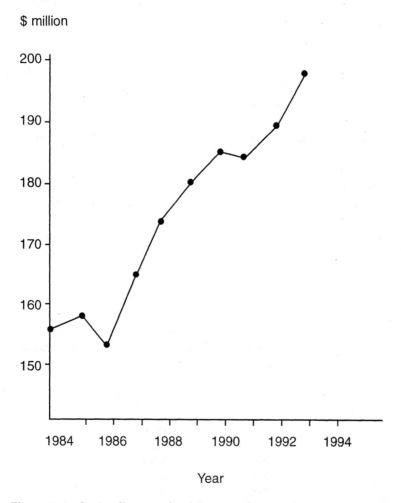

Figure 13.4 Scatter diagram of cash income of farmers between 1984 and 1993.

Step two — The data are divided into two equal periods, and the average for each period is computed as shown in Table 13.3. The average annual cash income for the first period of 5 years is $161 million; and for the second period, $187.38 million. These two points serve as the basis for drawing a trend line. The slope of the line is found as:

$$b = \frac{187.38 - 161.00}{5} = 5.28$$

The annual increase in cash income is therefore $5.28 million. The *a* value for the equation is the value of Y_t when *x* equals zero. For the present problem, $a = 161$. The semiaverage trend equation is

$$Y_t = 161.0 + 5.28x$$

$$x = 0 \text{ in } 1986$$

$$1x = 1 \text{ year}$$

$$Y = \text{millions of dollar}$$

Table 13.3 Fitting a Straight Line by the Method of Semiaverage to Cash Income of Farmers, 1984 to 1993

Year	Income	Semitotal	Semiaverage	Coded time
1984	156.1			-2
1985	157.9			-1
1986	152.9	805.0	161.00	0
1987	165.2			1
1988	172.9			2
1989	179.8			3
1990	186.8			4
1991	184.9	936.9	187.38	5
1992	188.2			6
1993	197.2			7

It is important to remember that because both the origin and the unit of time (*x*) are variable, they are a necessary part of the trend equation.

The trend values for any year can be obtained by substituting the appropriate *x* value into the trend equation. For instance, the *x* value for 1993 is 7, so the trend for that year is

$$Y_t = 161.0 + 5.28 \, (7)$$

$$Y_t = \$197.96 \text{ million}$$

If we assume that the past pattern is likely to persist into the future, we may use the trend equation to project future cash income of the farmers. This is done by simply substituting in the *x* value a future date into the trend equation. For the year 2000, the projected cash income of the farmers is

$$Y_t = 161.0 + 5.28\ (14)$$

$$Y_t = \$234.92 \text{ million}$$

Least squares method. Earlier it was pointed out that this method of fitting a straight line is not only simple, but provides the best fit. It should be mentioned that the assumptions of the least squares models of the simple and multiple regression are not met when time series data are used. For further explanation of the violation of the least squares assumptions when time series data are used, see Hamburg (1987).

The equations for fitting a least squares trend line are

$$a = \frac{\Sigma Y}{n} \qquad\qquad [13\text{--}7]$$

$$b = \frac{\Sigma xY}{\Sigma x^2} \qquad\qquad [13\text{--}8]$$

Equations 13–7 and 13–8 are similar to those used in the simple regression model where we estimated the values of a and b. However, in the present situation, the value of X is substituted with a coded time scale denoted by lower case x, where $\Sigma x = 0$. This sum must equal zero whether we fit an odd or an even number of observations.

To fit a trend line to an even number of points as in Example 13.1, we first find the middle of the time period or origin and assign a value of zero to it. Then we count in consecutive odd integers. This is illustrated in Table 13.4, where the origin is halfway between 1988 and 1989.

The values computed in Table 13.4 are now substituted into Equations 13–7 and 13–8 to obtain the following:

$$a = \frac{1741.9}{10} = 174.19$$

$$b = \frac{813.7}{330} = 2.47$$

Thus, the least squares trend equation is

$$Y_t = 174.19 + 2.47x$$

$$x = 0 \text{ in } 1988\ 1/2$$

$$1x = 1/2 \text{ year}$$

$$Y = \text{millions of dollars}$$

Table 13.4 **Fitting a Straight Line by the Method of Least Squares to Cash Income of Farmers, 1984 to 1993**

Year	Income Y	Coded time x	xY	x^2
1984	156.1	− 9	−1404.9	81
1985	157.9	− 7	−1105.3	49
1986	152.9	− 5	− 764.5	25
1987	165.2	− 3	− 495.6	9
1988	172.9	− 1	− 172.9	1
1989	179.8	1	179.8	1
1990	186.8	3	560.4	9
1991	184.9	5	924.5	25
1992	188.2	7	1317.4	49
1993	197.2	9	1774.8	81
Total	1,741.9	0	813.7	330

We may once again use the trend equation to determine the trend value for, say, 1993. By substituting the value for x, we have

$$Y_t = 174.19 + 2.47 \ (9)$$

$$Y_t = \$196.42 \text{ million}$$

The least squares line has an annual trend increase of \$4.94 million. Because in the present problem $1x$ equals one-half year, we multiply 2.47 by 2 to obtain the value for a year.

To fit a trend line to an odd number of points, we first find the origin that is the midpoint of the number of observations and assign a value of zero to it. Second, we count backward and forward in consecutive integers. Example 13.2 illustrates fitting a trend line to an odd number of points.

Example 13.2

An agricultural economist observes that consumption expenditure on food has been rising. Fit a trend line to the following data using the least squares method.

Year	Food expenditure (million dollars)	Year	Food expenditure (million dollars)
1987	375.5	1991	465.1
1988	398.8	1992	474.5
1989	419.4	1993	491.3
1990	449.8		

Adapted from *Statistical Abstract of the United States — 1995: The National Data Book*, U.S. Department of Commerce, Washington, D.C., 1995, 681, Table no. 1118.

Solution

Given that we have data on an odd number of years, the midpoint is 1990. This year is the year of the origin to which we assign a value of zero. We count in consecutive integers forward and backward so that the $\Sigma x = 0$. Table 13.5 shows the intermediate calculations needed to compute a and b of Equations 13–4 and 13–5.

Table 13.5   **Fitting a Straight Line by the Method of Least Squares to Food Expenditure, 1987 to 1993**

Year X	Food expenditure (million dollars) Y	Coded time x	xY	x²
1987	375.5	– 3	– 1126.5	9
1988	398.8	– 2	– 797.6	4
1989	419.4	– 1	– 419.4	1
1990	449.8	0	0	0
1991	465.1	1	465.1	1
1992	474.5	2	949.0	4
1993	491.3	3	1473.9	9
Total	3074.4	0	+ 544.5	28

$$a = \frac{3074.4}{7} = 439.2$$

$$b = \frac{544.5}{28} = 19.45$$

Therefore, the least squares trend line is

$$Y_t = 439.2 + 19.45x$$

$$x = 0 \text{ in } 1990$$

$$1x = 1 \text{ year}$$

$$Y = \text{millions of dollars}$$

Trend projection.   The least squares trend equation computed in the previous section may be used for projecting food expenditures in the future. The basic assumption for making a projection is that the past pattern of expenditure persists into the future. Suppose we wish to project food expenditures in the year 1998. To be able to make such a projection, the time scale displayed in column three of Table 13.5 must be extended to 1998. The value of x for 1993 (the last year for which we have an observation) is 3. The value of x for 1998 is 8. The projected consumption of food in the year 1998 is

$$Y_{1998} = 439.2 + 19.45\ (8)$$

$$= 594.8 \text{ million dollars}$$

Nonlinear trend. In the preceding analysis we assume that Y_t increases and decreases at a constant rate. In many agricultural and environmental science experiments we observe that the time series does not follow a constant rate of increase or decrease, but instead follows an increasing or a decreasing pattern. To fit a trend to such time series data, we frequently use a polynomial function such as a second-degree parabola. A second-degree parabola provides a good historical description of increase or decrease per time period.

The trend equation for a second-degree parabola is written as:

$$\hat{Y}_t = a + b_t + c_t^2 \qquad\qquad [13\text{--}9]$$

where $\hat{Y}_t$ = estimated trend value
a, b, c = constants to be determined
x = deviations from middle time period or coded time

The least squares equations used to solve for a, b, and c are

$$\Sigma Y = na + c\Sigma x^2 \qquad\qquad [13\text{--}10]$$

$$\Sigma x^2 Y = a\Sigma x^2 + c\Sigma x^4 \qquad\qquad [13\text{--}11]$$

$$b = \frac{\Sigma xY}{\Sigma x^2} \qquad\qquad [13\text{--}12]$$

Equations 13–10 and 13–11 are solved simultaneously for a and c, whereas b is determined by using Equation 13–12. The following example illustrates the least squares method of fitting a trend line to a second-degree parabola.

Example 13.3

With increased industrialization around the world, carbon emissions, especially from fossil fuels, are rising. The deleterious effects of carbon emissions are numerous, and environmental scientists and policy makers are concerned with this trend. The following data show world carbon emissions from fossil fuels between 1982 and 1994. Fit a trend line using the least squares method to determine whether the trend for carbon emission is rising or falling.

World Carbon Emissions from Fossil Fuel
Burning 1982 to 1994

Year	Million tons of carbon
1982	4960
1983	4947
1984	5109
1985	5282
1986	5464
1987	5584
1988	5801
1989	5912
1990	5941
1991	6026
1992	5910
1993	5893
1994 (preliminary)	5925

Adapted from Roodman, D. M., Carbon emissions resume rise, in Brown, L. et al., *Vital Signs 1995*, W. W. Norton, New York, 1995, 67.

Solution

There are 13 years of observations on carbon emissions. Because of the odd number of years, $x = 0$ in the middle year, 1988. Table 13.6 is the work table for the present example. The relevant information from

Table 13.6   **World Carbon Emissions from Fossil Fuel Burning 1982 to 1994**

X (Year)	Y (Million tons)	x	xY	x^2Y	x^2	x^4
1982	4,960	−6	−29,760	178,560	36	1,296
1983	4,947	−5	−24,735	123,675	25	625
1984	5,109	−4	−20,436	81,744	16	256
1985	5,282	−3	−15,846	47,538	9	81
1986	5,464	−2	−10,928	21,856	4	16
1987	5,584	−1	−5,584	5,584	1	1
1988	5,801	0	0	0	0	0
1989	5,912	1	5,912	5,921	1	1
1990	5,941	2	11,882	23,764	4	16
1991	6,026	3	18,078	54,234	9	81
1992	5,910	4	23,640	94,560	16	256
1993	5,893	5	29,465	147,325	25	625
1994	5,925	6	35,550	213,300	36	1,296
Total	72,754	0	17,238	998,061	182	4,550

Table 13.6 is substituted into Equations 13–10 and 13–11 to solve simultaneously for *a* and *c* as follows:

$$72{,}754 = 13a + 182c \qquad [13\text{–}10\text{A}]$$

$$998{,}061 = 182a + 4{,}550c \qquad [13\text{–}11\text{A}]$$

To be able to solve for either of the two unknown coefficients (*a* or *c*) we must eliminate one of them. For example, to eliminate *c*, we first multiply Equation 13–10A by 25. This makes the *c* coefficients in both equations equal to each other. We then subtract Equation 13–11A from the results to obtain:

$$
\begin{array}{l}
1{,}818{,}850.0 = 325.0a + 4{,}550c \\
\underline{-998{,}052.0 = -182.0a - 4{,}550c} \\
820{,}798.0 = 143.0a
\end{array}
$$

$$a = 5{,}739.8$$

We now substitute the value of *a* into Equation 13–10A to determine the value for *c* as follows:

$$72{,}754 = 13\,(5{,}739.8) + 182c$$
$$72{,}754 = 74{,}617.4 + 182c$$
$$-1{,}864 = 182c$$
$$c = -10.24$$

By using Equation 13–12, we solve for *b* and obtain:

$$b = \frac{17{,}238}{182} = 94.71$$

Therefore, our trend equation is

$$\hat{Y}_t = 5379.8 + 94.71x - 10.24x^2$$
$$x = 0 \text{ in } 1988$$
$$1x = 1 \text{ year}$$
$$Y = \text{million tons}$$

An interpretation that may be given to the constants *a*, *b*, and *c* is that *a* is the intercept of the line when $x = 0$. This means that the trend for

1988 is 5379.8 million tons. The constant b is the slope of the parabola at the time we specified as the origin.

For the present problem, the slope of the parabola, given that 1988 served as the origin, is 94.71. This implies that for each year, the average increase in carbon emission is 94.71 million tons. Constant c is the amount of acceleration (if the constant has a positive sign) or deceleration (if the constant has a negative sign) in the curve. Constant c also indicates the amount by which the slope changes per time period. In this example the decrease in the slope per time period is 10.24 million tons per year.

Logarithmic trend.   Trend lines may also be fitted to the percentage rates of change by using a logarithmic straight line. Such trend lines describe a time series to be increasing or decreasing at a constant rate. Agricultural growth and development in the less developed countries often show such growth patterns. The logarithmic trend line equation is

$$\log Y_t = \log a + x \log b \qquad\qquad [13\text{--}13]$$

To fit a least squares trend line to Equation 13–13, we determine the constants $\log a$ and $\log b$ using the formulas:

$$\log a = \frac{\Sigma \log Y}{n} \qquad\qquad [13\text{--}14]$$

$$\log b = \frac{\Sigma (x \log Y)}{\Sigma x^2} \qquad\qquad [13\text{--}15]$$

Equations 13–14 and 13–15 hold true as long as $\Sigma x = 0$. Notice that the procedure for obtaining a and b is similar to that in our previous examples, except that in the present situation we replace Y with log of Y (to base ten). To determine Y for any time period, we take the antilog of $\log Y_t$. If log of Y is plotted against x, the graph is a straight line. A logarithmic straight line shows how a series is increasing at a constant rate. To determine the numerical value of that rate, we use the antilog of $\log b$. This value is the relative change (R) per unit of x. We can find the rate of change (r) by subtracting 1 from R. An example of how we may fit a logarithmic trend line and determine the rate of change from year to year is given next.

Example 13.4

Global exports expanded to an estimated $3.92 trillion in 1994 (Gardner, 1995). Much of this increase is a by-product of economic recovery in industrial nations, the export-led trade strategies of the developing countries, and regional trade agreements such as the North American Free Trade Agreement. The following table shows world exports between 1981 and 1994.

Year	Exports (billion, 1990 dollars)
1981	2230
1982	2154
1983	2207
1984	2400
1985	2475
1986	2492
1987	2643
1988	2920
1989	3137
1990	3337
1991	3487
1992	3658
1993	3839
1994 (preliminary)	3924

Adapted from Gardner, G., World trade climbing, in Brown, L. et al., *Vital Signs 1995*, W. W. Norton, New York, 1995, 75.

Solution

Table 13.7 shows the calculations necessary for fitting the trend line. The first step in fitting the line is to transform the Y values into log Y values as shown in column three. Next, the coded time x is shown in column four. Column five is labeled as x log Y, which is the product of the previous two columns. Finally, in the last column we present the square of the time values. By substituting the values from Table 13.7 into Equations 13–14 and 13–15, we obtain:

$$\log a = \frac{\Sigma \log Y}{n} = \frac{48.386}{14} = 3.456$$

$$\log b = \frac{\Sigma (x \log Y)}{\Sigma x^2} = \frac{10.052}{910} = 0.0110$$

Thus, the estimated trend line equation is

$$\log \hat{Y} = 3.456 + 0.0110x$$

$$x = 0 \text{ in } 1987 \ \tfrac{1}{2}$$

$$1x = \tfrac{1}{2} \text{ year}$$

Table 13.7   Computations for Least-Squares of a Logarithmic
Trend Line

Year (1)	World exports (billion 1990 dollars) (2)	log Y (3)	x (4)	x log Y (5)	x^2 (6)
1981	2230	3.348	−13	−43.524	169
1982	2154	3.333	−11	−36.663	121
1983	2207	3.344	−9	−30.096	81
1984	2400	3.380	−7	−23.660	49
1985	2475	3.394	−5	−16.970	25
1986	2492	3.397	−3	−10.191	9
1987	2643	3.422	−1	−3.422	1
1988	2920	3.465	1	3.465	1
1989	3137	3.497	3	10.491	9
1990	3337	3.523	5	17.615	25
1991	3487	3.542	7	24.794	49
1992	3658	3.563	9	32.067	81
1993	3839	3.584	11	39.424	121
1994	3924	3.594	13	46.722	169
Total		48.386	0	10.052	910

To check the goodness-of-fit of the line and to solve for values of Y_t, we must solve for log Y, by substituting appropriate values of x into the estimated trend equation and then taking the antilog of log Y_t.

For example, let us substitute the x value for the year 1983 and 1992. This provides us with two points on the trend line. By drawing a straight line connecting these two points, we determine graphically whether the fit is good. By substituting the x value for 1983 (−9) and 1992 (9) into the equation, we obtain:

$$\log Y_t = 3.456 + 0.0110(-9)$$

$$\log Y_t = 3.357$$

and

$$Y_t = \text{antilog (log } Y_t) = \text{antilog } 3.357$$

$$Y_t = 2275.0 \text{ for 1983}$$

$$\log Y_t = 3.456 + 0.0110(9)$$

$$\log Y_t = 3.555$$

and

$$Y_t = \text{antilog (log } Y_t) = \text{antilog } 3.555$$
$$= 3589.2 \text{ for } 1992$$

To determine the rate of change from year to year, we take the antilog of log b, which is

$$R = \text{antilog } 0.0110 = 1.026$$

Because the rate of change (r) was defined as $R - 1$, r equals:

$$r = 1.025 - 1 = 0.025$$

or

$$r = 2.5\% \text{ per half year}$$

To determine the rate of change for a year in this problem, we have:

$$R = \text{antilog } [2(\log b)]$$
$$R = \text{antilog } 0.022 = 1.05$$

Thus:

$$r = R - 1 = 1.05 - 1$$
$$r = 0.05$$

or

$$r = 5\% \text{ per year}$$

This implies that world exports increased at a constant rate of 5% per year between 1981 and 1994.

13.3 *Seasonal variation*

In the introduction to this chapter it was mentioned that a time series is affected by its four components, one of which is the seasonal variation. Seasonal variation is defined as a predictable and repetitive movement observed around a trend line within a period of one year or less.

This means that to be able to analyze seasonal variations we must have data that are reported weekly, monthly, quarterly, etc. The most important seasonal variation in agriculture results from weather. For example, production of certain commodities in the winter months is lower than at any other time of the year.

There are several reasons for measuring seasonal variations:

1. When analyzing the data from a time series, it is important to know how much of the variation in the data results from seasonal factors.
2. We may use seasonal variation patterns in making projections or forecast of a short-term nature.
3. By eliminating the seasonal variation from a time series, we may discover the *cyclical* pattern of the time series.

Many agricultural statistics reported by the U.S. Department of Agriculture are seasonally adjusted. This means that the variation from seasonal factors has been removed from the data.

To measure seasonal variation we must construct an *index of seasonal variation*. The method of *ratio to moving average* is the most popular approach to computing a seasonal index. The following steps are used in constructing an index and then using the index to deseasonalize the data.

Step one — Compute a moving total of the original data. To compute a 12-month moving total, we sum the value of the first 12 months and place it on the opposite of the seventh month in the next column. The second 12-month total is computed by dropping the first month of the first 12-month totals and adding the 13th observation of the data. We continue with the same procedure until all the data have been accounted for. If we are computing a quarterly moving total, the value of the first four quarters are summed and placed opposite the third quarter in a column.

Step two — Compute a moving average either for the monthly or quarterly data. This is accomplished by dividing the sum of each of the 12-month totals by 12. In the case of quarterly data, the moving average is obtained by dividing the sum of each four quarters by 4.

Step three — Compute the specific seasonals. This is simply the ratio of original data to the moving average, and is expressed as a percentage.

Step four — Compute a seasonal index. To accomplish this task, we have to first arrange the specific seasonals computed in step three in a table by months or quarters for each year. We then average the ratios by month or quarters in an attempt to remove any nonseasonal variations (irregulars and cycles). Any number of averaging techniques may be used. Most often we use what is known as a *modified mean*.

A modified mean is computed by taking the mean of a set of numbers after extreme values have been eliminated. The modified mean may be the average of the middle three ratios for months or quarters.

Once we have computed the mean for months or quarters, they are then summed. A seasonal index, in theory, should total 1200 for monthly observations and 400 for quarterly observations. Because our computations will most likely result in a value other than 1200 (for monthly data) or 400 (for quarterly data), we have to adjust our results by dividing the computed total into 1200 or 400 (monthly or quarterly data, respectively) to obtain a correction factor. Each modified mean is then multiplied by the correction factor to get the *seasonal index*.

The following example illustrates the computation of a seasonal index using quarterly data.

Example 13.5

Data given next contain quarterly information on tomato production in California between 1989 and 1995. Compute a quarterly seasonal index.

Year and quarter	Tomato production (million tons)	Year and quarter	Tomato production (million tons)
1989 III	12.5	1992 III	15.0
IV	12.7	IV	15.4
1990 I	12.9	1993 I	15.4
II	13.4	II	15.9
III	13.5	III	15.1
IV	13.2	IV	15.4
1991 I	12.5	1994 I	15.6
II	12.5	II	16.0
III	14.0	III	15.1
IV	14.2	IV	13.4
1992 I	15.0	1995 I	12.1
II	14.6	II	12.5

Solution

Table 13.8 illustrates the first three steps necessary to compute a seasonal index.

Table 13.8 Computation of Ratio of Original Data to Moving Average

Year and quarter (1)	Tomato production (in million tons) (2)	Moving four-quarter total (3)	Moving average (4)	Ratio of original data to moving average (%) (5)
1989 III	12.5			
IV	12.7			
1990 I	12.9	51.5	12.9	100.0
II	13.4	52.5	13.1	102.1
III	13.5	53.0	13.3	101.5
IV	13.2	52.6	13.2	100.0
1991 I	12.5	51.7	12.9	96.9
II	12.5	52.2	13.1	95.4
III	14.0	53.2	13.3	105.3
IV	14.2	55.7	13.9	102.2
1992 I	15.0	57.8	14.5	103.4
II	14.6	58.8	14.7	99.3
III	15.0	60.0	15.0	100.0
IV	15.4	60.4	15.1	102.0
1993 I	15.4	61.7	15.4	100.0
II	15.9	61.8	15.5	102.6
III	15.1	61.8	15.5	97.4
IV	15.4	62.0	15.5	99.4
1994 I	15.6	62.1	15.5	100.6
II	16.0	62.1	15.5	103.2
III	15.1	60.1	15.0	100.7
IV	13.4	56.6	14.2	94.4
1995 I	12.1	53.1		
II	12.5			

The next step is to average the ratios by quarters. This is done to average out any remaining nonseasonal variations. Table 13.9 shows how the ratios are rearranged into quarters. To compute the modified mean, we eliminate the extreme values in each quarter. For example, to get a mean of 100.2 in the first column of Table 13.9, we eliminate the extreme values of 96.9 and 103.4. The modified mean is simply the sum of the middle three divided by 3. The same procedure is followed for computing the mean for each of the second quarters, third quarters, and fourth quarters.

Finally, to compute a seasonal index we sum the modified means as computed in Table 13.9. This sum is 402.7. Because 402.7 is greater than 400, we have to make an adjustment to force the four modified means to total 400. This is done by dividing 400 by 402.7 to get the correction factor as follows:

$$\text{Correction factor} = \frac{400}{402.7} = .9933$$

Table 13.9   Ratios of Original Data to Moving
Average by Quarters

Year	Quarters			
	I	II	III	IV
1990	100.0	102.1	101.5	100.0
1991	96.9	95.4	105.3	102.2
1992	103.4	99.3	100.0	102.0
1993	100.0	102.6	97.4	99.4
1994	100.6	103.2	100.7	94.4
Mean	100.2	101.3	100.7	100.5

Next, the modified means are multiplied by the correction factor of .9933 to obtain the seasonal index as shown in Table 13.10.

Table 13.10   Computing a Seasonal Index

Quarter	Mean middle three ratios	Seasonal index
I	100.2	99.5
II	101.3	100.6
III	100.7	100.0
IV	100.5	99.8

After we have computed a seasonal index, we can deseasonalize the original data by first dividing it by its respective seasonal index and then multiplying it by 100 as shown in Table 13.11.

13.4   *Identifying cycles and irregular variations*

Measuring cycles-irregulars directly from the data has been difficult and not very satisfactory. The irregular nature of cycles makes it impossible to find an average cycle. The best approach in measuring cycle–irregular is to determine the cyclical–irregular fluctuations through an indirect method called the *residual* approach.

In the introduction section it was indicated that the classical time series model is multiplicative in nature. This means that the relationship between the components of a time series and the original data may be stated as follows:

$$Y = T \times S \times CI \qquad [13–2]$$

Given this model, we can determine the variation resulting from cycles–irregulars using the residual approach. In this approach the cycle–irregular variations of a time series are found by removing the

Table 13.11   **Deseasonalized Tomato Production Data, 1989 to 1995**

Year and quarter (1)	Tomato production (in million tons) (2)	Seasonal index (3)	Deseasonalized tomato production [col 2 ÷ col 3] × 100
1989 III	12.5	100.0	12.5
IV	12.7	99.8	12.7
1990 I	12.9	99.5	13.0
II	13.4	100.6	13.3
III	13.5	100.0	13.5
IV	13.2	99.8	13.2
1991 I	12.5	99.5	12.6
II	12.5	100.6	12.4
III	14.0	100.0	14.0
IV	14.2	99.8	14.2
1992 I	15.0	99.5	15.1
II	14.6	100.6	14.5
III	15.0	100.0	15.0
IV	15.4	99.8	15.4
1993 I	15.4	99.5	15.5
II	15.9	100.6	15.8
III	15.1	100.0	15.1
IV	15.4	99.8	15.4
1994 I	15.6	99.5	15.7
II	16.0	100.6	15.9
III	15.1	100.0	15.1
IV	13.4	99.8	13.4
1995 I	12.1	99.5	12.2
II	12.5	100.6	12.4

effect of trend and seasonal from the original data using the formula given earlier:

$$\frac{Y}{T \times S} = \frac{\cancel{T} \times \cancel{S} \times CI}{\cancel{T} \times \cancel{S}} = CI \qquad [13\text{--}3]$$

Equation 13–3 holds true when we have a model that specifies monthly or quarterly data. In monthly data, the seasonal factor is present, whereas in annual data they do not appear, as all seasons are represented. Thus if we use a time series model composed of annual data, we can find the cycle–irregular fluctuations using the formula:

$$\frac{Y}{T} = \frac{\cancel{T} \times CI}{\cancel{T}} = CI \qquad [13\text{--}16]$$

Equation 13–16 indicates that dividing the values in the original series by the corresponding values yields a measure of the cycle–irregular. Because cycle–irregular fluctuations are measured as a *percent of trend*, we multiply them by 100. When multiplied by 100, the measure is called a *cyclical irregular*. An illustration of the procedure follows.

Example 13.6

From the following data, a citrus marketing cooperative wants to measure the variations in its members' citrus production over a 13-year period. Determine the cyclical variation from the data.

X Year	Y 1000 box
1983	6.5
1984	7.2
1985	7.6
1986	8.4
1987	8.5
1988	8.0
1989	8.6
1990	8.9
1991	9.5
1992	10.2
1993	10.6
1994	10.8
1995	11.0

Solution

To be able to determine the cyclical fluctuations using the residual approach, we first compute the trend using the least squares method, Table 13.12 shows the estimated trend values for each year.

The estimated trend line is graphed in Figure 13.5. Observe that the actual values move above and below the estimated trend line. This reflects the cyclical fluctuation in the data.

To determine the cyclical variation, we have to compute the percent of trend. The computation of the percent of trend is shown in Table 13.13. Column four of Table 13.13 shows the percent of trend for each of the years for which we have an observation. The actual amount of citrus delivered to the cooperative varies around the estimated trend (91.9 to 106.3). These cyclical variations may have resulted from such factors as rainfall and frost conditions in the producing area. As mentioned earlier, because of the unpredictable nature of rainfall, or frost, or other factors that contribute to cyclical variations, we are unable to forecast any specific patterns of variation using the residual method.

Table 13.12　Citrus Received by the
Cooperative Between 1973 to 1985, and
the Estimated Trend

Year X	Boxes of citrus (in 1000) Y	Trend Y_t
1983	6.5	6.7
1984	7.2	7.1
1985	7.6	7.5
1986	8.4	7.9
1987	8.5	8.3
1988	8.0	8.7
1989	8.6	8.9
1990	8.9	8.7
1991	9.5	9.5
1992	10.2	9.9
1993	10.6	10.3
1994	10.8	10.7
1995	11.0	11.1

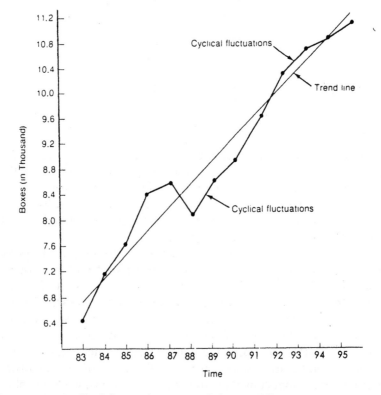

Figure 13.5　**Cyclical fluctuations around the trend line.**

Table 13.13   Calculation of Percent Trend

Year X	Boxes of citrus (in 1,000) Y	Trend Y_t	Percent of trend $(Y/Y_t \times 100)$
1983	6.5	6.7	97.0
1984	7.2	7.1	101.4
1985	7.6	7.5	101.3
1986	8.4	7.9	106.3
1987	8.5	8.3	102.4
1988	8.0	8.7	91.9
1989	8.6	8.9	96.6
1990	8.9	8.7	102.3
1991	9.5	9.5	100.0
1992	10.2	9.9	103.0
1993	10.6	10.3	102.9
1994	10.8	10.7	100.9
1995	11.0	11.1	99.1

To illustrate the interpretation of a specific year's value, let us consider the 1983 percent of trend or cyclical relative of 97.0%. This value indicates that the number of boxes of citrus delivered to the cooperative is 97.0% of trend, or 3% below trend in 1983 because of cyclical and irregular factors. Similarly, the 1986 percent of trend value of 106.3 indicates that the producers' delivery of citrus to the cooperative is 6.3% above trend because of cyclical and irregular factors in the data.

Frequently, the cyclical variations as measured by the percent of trend are graphed so that the cyclical pattern may be more readily observed. Figure 13.6 illustrates how this process eliminates the trend line and isolates the cyclical component of the time series. You should remember that the procedures discussed in determining the cyclical component only apply to past cyclical variation, and cannot be used to predict future cyclical variations. Methods that are applied in predicting the future cyclical variation are beyond the scope of this text.

It should also be kept in mind that movements of a time series may be separated from the cyclical movement by constructing a moving average of three or more periods to smooth out the irregularities of the data. For the present analysis, we need not concern ourselves with such refined analysis. For more information on how to eliminate irregular movement from a time series, see Enns (1985).

Chapter summary

The purpose of this chapter is to introduce the techniques involved in making forecasts. Managers and researchers use *time series analysis* as

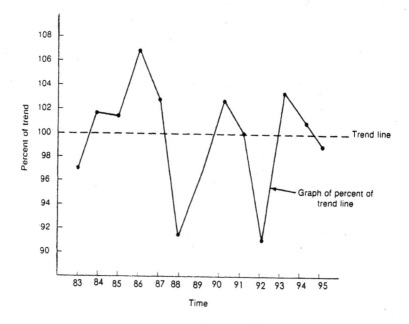

Figure 13.6   Cycle chart of citrus delivered to the marketing cooperative, 1983 to 1995.

a tool in forecasting and understanding past and current patterns of change.

You were introduced to the four components of a time series, namely, the *trend, seasonal, cycle,* and *irregular or random* variations. By isolating these components and measuring their effects, it is possible to forecast future values of the time series.

We use secular trend or movement that is typical of a long period to describe historical patterns in a series and project future patterns of change. Three separate approaches, namely, the *freehand*, the *semiaverage*, and the *least squares* were discussed in computing a trend. The method of least squares is preferred over the other two approaches because it provides the best fit. In discussing the method of fitting a trend line, it was pointed out that not all trend lines are linear. To fit a nonlinear trend line, we use a polynomial function. The most common polynomial function used in agricultural experiments is the second degree parabola. This function provides a good description of an increase or decrease per time period. Logarithmic trend lines are used when we are interested in fitting *percentage* rate of change over time. The logarithmic trend lines describe a time series to be increasing or decreasing at a constant rate.

The second component of a time series is the seasonal variation. Seasonal variations are mostly repetitive movements or fluctuations with a duration of not more than a year. Seasonal movements in agriculture result from climatic and weather factors. The seasonal variation of a time series is measured through a seasonal index called the *ratio to moving average*. After the index is computed, it is then used to deseasonalize the original data. This is accomplished by dividing the original data by its respective seasonal index and multiplying the result by 100.

Cyclical variation is the third component of a classical time series. It refers to wavelike movements that occur over a period of several years. Because of their varied causes, cycles are analyzed indirectly through a method called the *residual*. The residual method requires computing the trend only if annual data are used, and then dividing the original data by the trend to determine the cyclical variations. If, on the other hand, monthly data are used, we must compute the trend as well as the seasonal variation before we divide the original data by these two values.

Irregular movements are variations in a time series that do not fit into any one of the three former categories. Separating the effect of irregular movements from a cyclical movement has proved to be difficult. However, constructing a moving average of three or more periods may be used to smooth out irregular movements.

Case Study

Projecting Global Demand for Feed Grain

The consumption of grain is driven primarily by population growth, whereas that for feed grain is largely driven by increasing income around the world. Economic development and rising wages and expectations have led to an increasing number of people consuming more meat and animal-derived products. As the demand for meat and other animal-derived products rises, so does the demand for feed grain.

Global per capita meat consumption varies among countries. For example, in India it is 2 kg/year, whereas Egypt, the Philippines, and Turkey consume 15 to 20 kg/year. Mexicans, Japanese, and Brazilians are reported to consume approximately 40 to 50 kg per year. Major meat consumers are found in France, Australia, and the U.S. where per capita consumption is reported to be around 90 to 110 kg/year (U.S. Department of Agriculture 1986; U.N. Food and Agriculture Organization, 1988).

Such high consumption patterns have negative effects on the ecological base of many of these countries. Livestock production practices of the past integrated food crop production with animal production, thus creating an ecologically balanced system. Often food crops such as wheat, corn, and potatoes were rotated with fodder crops such as hay, clover, and alfalfa that were fed to the livestock. Manure from the livestock was used to fertilize the soil. Unfortunately, the current demand for meat cannot be met with traditional practices. Large portions of food crops must now be used as animal feed. This means increased use of fertilizers and pesticides that affect the ecosystem.

The world's use of feed grain in 1994 is estimated at 665 million tons out of a total of 1.76 billion tons. This represents 38% of the total (U.S. Department of Agriculture, 1995). China, one of the world's largest consumers of feed grain, increased her share from 7% in 1978 when economic reforms were launched, to 22% in 1994 (U.S. Department of Agriculture, 1995).Over the long term, the demand for livestock products is projected to rise, pulling the demand for feed grain upward (U.S. Department of Agriculture, 1994). How much the demand will rise remains to be seen.

The following table shows the trend of feed grain use from 1980 to 1994. By using the data given:

1. Determine the trend line for the feed grain.
2. Would you use a linear or a nonlinear trend line, and why?
3. Interpret the meaning of each of the coefficients you have computed.

Year	Grain used for feed (million tons)
1980	560
1981	574
1982	592
1983	583
1984	610
1985	620
1986	652
1987	661
1988	627
1989	649
1990	666
1991	657
1992	652
1993	652
1994 (preliminary)	665

Adapted from U.S. Department of Agriculture, *Production, Supply, and Demand View*, Electronic Database, Washington, D.C., November, 1994.

References

U.N. Food and Agriculture Organization (FAO), *World Agriculture: Toward 2000,* Rome, 1988.

U.S. Department of Agriculture, *World Agricultural Supply and Demand Estimates,* Washington, D.C., January, 1995.

U. S. Department of Agriculture, *Feed Situation and Outlook Yearbook,* Washington, D.C., October, 1994.

U.S. Department of Agriculture, *Crop Yields and Crop Change to the Year 2000,* Washington, D.C., 1986.

Review questions

1. Agricultural exports have played a significant role in the balance of trade of the U.S. The following data on agricultural exports are from 1970 to 1980.

Year	Agricultural exports (million dollars)
1970	6,958
1971	7,955
1972	8,242
1973	14,984
1974	21,698
1975	21,854
1976	22,760
1977	23,974
1978	27,290
1979	31,975
1980	40,480

From U.S. Department of Agriculture, *FATUS*, February 1979; and U.S. Department of Agriculture, *Outlook for U.S. Agricultural Exports*, November 1980.

   a. Graph the data.

   b. Fit a linear trend by the method of least squares to (1) the original data and (2) the logarithms of the data. Is there a good fit?

   c. Calculate projected trend values for 1988, using the least squares method.

2. One of the most rapidly rising farm expenses is the interest paid by farmers on borrowed money. Interest paid data collected by the U.S. Department of Agriculture are given next.

Year	Interest paid (million dollars)
1971	3,372
1972	3,700
1973	4,474
1974	5,509
1975	5,973
1976	6,703
1977	7,951
1978	9,531
1979	12,150
1980	15,140
1981	18,967

From Economic Indicators of the
Farm Sector, Farm Sector Review,
1982, USDA/ERS/ECIFS 2–1,
May 1983, 104.

a. Plot the data on a sheet of arithmetic graph paper.
b. Fit a logarithmic straight line trend to the series by the method of least squares.
c. Compute the annual rate of growth in the trend of interest paid by the farmers.

3. Sales of small tractors have been rising in recent years. The information on sales during the past 10 years indicates the regional sales in an agricultural state as follows:

Year	Sales (units)	Year	Sales (units)
1987	350	1992	290
1988	360	1993	285
1989	310	1994	300
1990	320	1995	305
1991	330	1996	315

a. Plot the time series and comment on the appropriateness of a linear trend.
b. Fit a straight line trend by the method of semiaverage.
c. Fit a straight line trend by the method of least squares.
d. What can you say about the trend?
e. Forecast the annual sales of small tractors for 1998.

4. The following data pertain to the number of cartons of lettuce sold by the Green Leaf Farm:

Year	Quarter	Cartons of lettuce (1000)	Year	Quarter	Cartons of lettuce (1000)
1986	I	12	1991	I	35
	II	16		II	37
	III	20		III	38
	IV	18		IV	36
1987	I	13	1992	I	34
	II	14		II	33
	III	18		III	31
	IV	22		IV	34
1988	I	23	1993	I	33
	II	25		II	35
	III	24		III	38
	IV	27		IV	39
1989	I	25	1994	I	38
	II	27		II	39
	III	29		III	42
	IV	31		IV	43
1990	I	28	1995	I	43
	II	33		II	44
	III	36		III	46
	IV	37		IV	48

a. By using the ratio to moving average method, determine the seasonal indices for each of the four quarters.
b. Adjust the quarterly sales between 1986 and 1995 for seasonal variations.

5. The following table gives the total cash income between 1960 and 1981 for farmers in the U.S. Fit a straight line trend by the method of semiaverage.

Year	Millions of dollars	Year	Millions of dollars
1960	34.3	1971	52.7
1961	35.2	1972	61.1
1962	36.5	1973	86.9
1963	37.5	1974	92.4
1964	37.3	1975	88.9
1965	39.4	1976	95.4
1966	43.4	1977	96.2
1967	42.8	1978	112.7
1968	44.2	1979	132.7
1969	48.2	1980	139.1
1970	50.5	1981	141.5

Adapted from Economic Indicators of the Farm Sector-Income and Balance Sheet Statistics, 1981, USDA/ERS, August 1982, 13, Table 3.

6. An agricultural economist observes that consumption expenditure on food has been rising. Fit a trend line to the following data using the least squares method.

Year	Food expenditure (million dollars)
1976	200.5
1977	217.4
1978	240.9
1979	272.3
1980	299.9
1981	329.1
1982	349.7

From Economic Indicator of the Farm Sector, Farm Sector Review, 1982, USDA/ERS, May 1983, 55, Table 26.

7. With increased price instability in the market for agricultural commodities, the number of acres under production in the U.S. varies. The following data show the number of total acres harvested between 1968 and 1980. Fit a trend line using the least squares method.

Acres of Harvested Cropland in the
U.S., 1968 to 1980

Year	Million acres
1968	296
1969	286
1970	289
1971	300
1972	289
1973	316
1974	322
1975	330
1976	331
1977	338
1978	331
1979	343
1980	346

From Economic Indicator of the Farm Sector: Production and Efficiency Statistics, 1980, USDA/ERS/Bulletin 679, January 1982, 20, Table 13.

8. Greater industrialization of the agricultural sector means displacement of labor from farms. An increasing number of farmers are supplementing their incomes with jobs from outside the agricultural sector. The following data show the amount of income from off-farm sources. Fit a logarithmic trend line.

Year	Off-farm income (million $)	Year	Off-farm income (million $)
1960	8,482	1971	19,110
1961	9,163	1972	21,265
1962	9,904	1973	24,714
1963	11,020	1974	28,135
1964	11,637	1975	23,905
1965	12,727	1976	26,426
1966	13,882	1977	25,623
1967	14,495	1978	28,721
1968	15,466	1979	33,782
1969	16,612	1980	36,569
1970	17,617	1981	39,329

From Economic Indicator of the Farm Sector: Income and Balance Sheet Statistics, 1981, USDA/ERS/ECIFS1-1, August 1982, 12, Table 2.

References and suggested reading

Anderson, T. W., *The Statistical Analysis of Time Series*, John Wiley & Sons, New York, 1971.

Armstrong, J. S., *Long-Range Forecasting*, John Wiley & Sons, New York, 1978.

Barnett, V. and Turkman, K. F., Eds., *Statistics for the Environment*, John Wiley & Sons, New York, 1993.

Bowerman, B. L. and O'Connell, R. T., *Time Series and Forecasting*, Duxbury, North Situate, MA, 1979.

Box, G. E. P. and Jenkins, G. M., *Time Series Analysis Forecasting and Control*, revised ed., Holden-Day, San Francisco, 1976.

Enns, P. G., *Business Statistics: Methods and Applications*, Richard D. Irwin, Homewood, IL, 1985, chap. 16.

Gardner, G., World trade climbing, in *Vital Signs 1995*, Brown, L. et al., W. W. Norton, New York, 1995, 74.

Hamburg, M., *Statistical Analysis for Decision Making*, 3rd ed., Harcourt Brace Jovanovich, New York, 1987, chap. 11.

Kitagawa, G., A nonlinear smoothing method for time series analysis, *Stat. Sinica*, 1, 371, 1990.

Mason, R. D., *Statistical Techniques in Business and Economics*. 5th ed., Richard D. Irwin, Homewood, IL, 1982, chap. 7.

Skaliski, J. R., A design for long-term and trends, *J. Environ. Manage.,* 30, 139, 1990.

Smith, R. L., Extreme value analysis of environmental time series: an application to trend detection in ground-level ozone, *Stat. Sci.,* 4, 367, 1989.

U.S. Department of Agriculture, Economic Indicators of the Farm Sector: Farm Sector Review, Washington, D.C., USDA/ERS/ECIFS 2-1, May 1983, 104.

U.S. Department of Agriculture, Economic Indicators of the Farm Sector-Income and Balance Sheet Statistics, 1981, USDA/ERS/ELIFSI -1, August 1982, Economic Indicators of the Farm Sector. Production and Efficiency Statistics, USDA/ERS/ Bulletin 679, January 1980, 20, Table 13.

U.S. Department of Agriculture, FATUS, February 1979 Outlook for U.S. Agricultural Exports, USDA/ERS, November 1980.

Young, P. C. and Minchin, P. E. H., Envirometric time-series analysis: modeling natural systems from experimental time-series data, *Int. J. Biol. Macromol.,* 13, 190, 1991.

chapter fourteen

Index Numbers

Learning objectives

After mastering this chapter and working through the review questions, you should be able to:

- Explain the use of index numbers.
- Construct a simple and composite index.
- Construct index numbers using the aggregative and the average of relatives methods.
- Discuss the problems associated with the index numbers.

14.1 Introduction

For several years in the early and mid-1970s, the price of certain agricultural products such as wheat, soybeans, feedstuffs, and beef rose rapidly (U.S. Department of Agriculture, 1983, p. 21). As a result the income of producers of such commodities also increased. Some states reported a rise in income of as much as 30% over a four-year period for individual producers. Given this statistic, and the fact that those years also represented an inflationary period, can we say that the increase in income has improved the purchasing power of agricultural producers? To answer such a question, we have to know whether the general price level has increased, decreased, or remained the same. *Index number* is such a measure that provides relative comparison between groups of related items or measured change in magnitude over periods of time.

Index numbers may be classified as either a *simple index* or a *composite index*. The simple index expresses the relationship between two numbers with one of the numbers used as a base. In this basic form the index number is referred to as a *percentage relative*. We use an index to show the percent change from one time period to another. For example, a time series of price of beef may be used to construct a price index.

The composite index number combines information from a number of related time series into a single series. For example, the consumer price index (CPI) takes into account the price of a number of related items. This composite index measures the general movement of prices of all those commodities included in the price index.

Indices may be used to measure movement of prices, quantities, or values. The price index measures the change in prices received by producers and consumers. The CPI and the wholesale price index are examples of such an index. The quantity index measures change in production. The Federal Reserve Board's Index of Industrial Production is an example of a quantity index. The value index that has two components, namely, the quantity and price, measures the change in total value. The change in value may result from change from either the quantity or the price.

There are certain advantages to the use of index numbers. First, they allow us to combine time series that otherwise cannot be added because they are not stated in comparable units such as tons, dollars, cubic feet, bales, etc. For example, bushels of wheat and bales of hay cannot properly be added. If, however, wheat production is 120% of the previous year's production and hay production is 110%, it is appropriate to average these two percentages, and report that the average production of these two commodities is 115% of the previous year.

Second, the index number simplifies data presentation. For example, a single composite index such as the CPI that periodically measures the cost of hundreds of items, may be used to show overall change in an economic variable. If the CPI for year 1995 was 186, and the base period was 1990, the index indicates the relative change in prices at two different time periods. The index of 186 means that it would take $86 more to purchase the same quantity of items that could have been purchased for $100 in 1990. Had we not used an index number to determine the purchasing power of the consumers, we would have had to deal with prices of hundreds of items collected throughout the country. Thus, an index aids in communication.

Third, an index number may be used to measure seasonal variation. In Chapter 13, we computed a seasonal index to remove seasonal movements. In Table 13.10 a seasonal index of 99.5 for the first quarter of the year indicates that tomato production is 0.5% below an average quarter.

14.2 Basic construction techniques

As is mentioned earlier, index numbers may be classified as either simple or composite. For example, to construct a simple index for beef prices given a time series of prices on beef, we express the relationship between any two prices as a percentage by dividing each figure by the price in the base period. These percentages are referred to as the *price relatives*. A simple price index is computed using the following formula:

$$I = \frac{P_n}{P_o} \cdot 100 \qquad \qquad [14\text{--}1]$$

where P_n = price in given period
P_o = price in base period

It is convenient to use the calendar years as subscripts. For instance, P_{1997} indicates the price of a commodity in the year 1997. Example 14.1 illustrates the basic construction of a simple index number.

Example 14.1

Given the information on the average retail price per pound of ground chuck 100% beef for the years 1988 to 1994, construct a simple price index using 1988 as a base. Interpret the meaning of the price index.

Year	Price of beef per pound
1988	1.79
1989	1.88
1990	2.02
1991	1.93
1992	1.91
1993	1.91
1994	1.84

Adapted from U.S. Department of Commerce, *Statistical Abstract of the United States — 1995*, 1995, 507, Table no. 777.

Solution

As pointed out earlier an index number is simply a percentage relative that shows the relationship between two values. Thus, to construct an index, we use 1988 as a base and divide each year's price by the price in 1988 to obtain the relatives. We then multiply the relatives by 100 to convert each relative to a percentage relative as shown in Table 14.1 column four. Column four is the simple index number that shows the relationship between yearly beef prices. In interpreting the index value you observe that the 1988 index, which is the arbitrary base period, has an index number value of 100. For 1989, the index number of 105 implies that the price of beef is 105% of the beef price in 1988. Similar interpretations can be made of index numbers for other years. For instance, the 1992 index number indicates that the price of beef in 1992 is 106.7% of the 1988 price. Alternatively, we may say that the price of beef in 1992 is 6.7% higher than it was in 1988.

To construct a composite index, we may use one of two methods:

1. *Aggregative price indices*
2. *Average of relatives indices*

The term *aggregate* in the first approach in computing a composite index means the sum of a series of values. The following example illustrates the construction techniques used for each method.

Table 14.1 **Simple Price Index of ground beef, 1988 to 1994**

(1) Year	(2) Price in year ÷ price in 1988	(3) Price relative	(4) Percent relative or index number
1988	$1.79 ÷ 1.79	1.000 × 100	100.0
1989	1.88 ÷ 1.79	1.050 × 100	105.0
1990	2.02 ÷ 1.79	1.128 × 100	112.8
1991	1.93 ÷ 1.79	1.078 × 100	107.8
1992	1.91 ÷ 1.79	1.067 × 100	106.7
1993	1.91 ÷ 1.79	1.067 × 100	106.7
1994	1.84 ÷ 1.79	1.027 × 100	102.7

Aggregative price indices.   A price index for several items can be constructed in a manner similar to that used in Example 14.1. A composite price index may be either *unweighted* or *weighted*. The unweighted and weighted aggregate price indices, respectively, are computed using the following formulas:

Unweighted index:

$$I = \frac{\sum P_n}{\sum P_o} \cdot 100 \qquad \text{[14–2]}$$

Weighted index:

$$I = \frac{\sum P_n Q_o}{\sum P_o Q_o} \cdot 100 \qquad \text{[14–3]}$$

where P_o = price in base period
P_n = price in nonbase period, which may be any time unit
Q_o = quantity in base period

Equation 14–3 is also called the *Laspeyres index*. The Laspeyres index uses quantity weights (Q_o) from the base period so that only the prices are allowed to change. It is desirable to use a quantity weight that is not too far in the distant past, because weights become out of date with passage of time due to changes in buying patterns.

A similar aggregate index called the *Paasche index* uses the quantities of the current period (Q_n) as weights. The Paasche index is more realistic than Laspeyres in that it allows for changes in taste and income of the consumers. However, the Laspeyres index is generally preferred over the Paasche index. The reason for this preference stems from the fact that to collect data on the quantity on a yearly basis would be expensive and time consuming. For further comments on the advantages and disadvantages of these two indices, see Lapin (1987).

Examples 14.2 and 14.3 illustrate the technique of constructing a composite price index.

Example 14.2

Given the prices on the following agricultural commodities in 1992, 1993, and 1994, construct an unweighted price index.

Commodities	Average price		
	1992	1993	1994
Eggs (dozen)	$0.93	$0.87	$0.87
Butter (pound)	1.64	1.61	1.54
Milk (1/2 gal)	1.39	1.43	1.44
Bread (1-lb loaf)	0.74	0.76	0.75

Adapted from U.S. Department of Commerce, *Statistical Abstract of the United States — 1995*, 1995, 507, Table no. 777.

Solution

Our interest is in the change in these prices from 1992 to 1994. We first sum (or aggregate) the prices per unit for each year as shown in Table 14.2. In this example, we use 1992 as the base and compute an index for any given year by dividing the sum of price for that year by the sum of the base period. The summed values given in Table 14.2 are now substituted into Equation 14–2 to calculate the index for the years 1993 and 1994 as follows:

$$I_{1993} = \frac{\Sigma P_{1993}}{\Sigma P_{1992}} \cdot 100$$

$$= \frac{4.67}{4.70} \cdot 100$$

$$= 99.36$$

and for 1994, the index is

$$I_{1994} = \frac{4.60}{4.70} \cdot 100$$

$$= 97.87$$

The indices for 1993 and 1994 of 99.36 and 97.87 may be interpreted as percentage decrease in prices of items of goods between 1992 and 1994. In terms of percentage change, it costs 0.64% less in 1993 than in 1992 to purchase the agricultural commodities indicated. Similarly, the index number of 97.87 means that in 1994 the cost of the same commodities is 2.13% less than in 1992.

The overall index shows a decline in prices. First, the unweighted index has some inherent problems associated with it. For instance, the

Table 14.2 **Calculation of the Unweighted Aggregates Index for Agricultural Commodities in 1993, 1994, and 1995**

Commodities	1992	Average price 1993	1994
Eggs (dozen)	$0.93	$0.87	$0.87
Butter (pound)	1.64	1.61	1.54
Milk (1/2 gallon)	1.39	1.43	1.44
Bread (1-lb loaf)	0.74	0.76	0.75
Total	$4.70	$4.67	$4.60

unweighted price index is affected by the units on which the prices are based. Second, it does not take into account the relative importance of a particular item in any single consumer's budget. That is, certain items are consumed relatively more than other items.

To remedy this problem, a weight may be given to each item in relation to its importance. In the next example, we illustrate how weights are attached to the aggregates so that a much more meaningful measure of price change is obtained.

Example 14.3

Suppose the following annual consumption data for a family of four, as well as the prices on the commodities given in Example 14.2, are provided to us. Construct a weighted price index.

| | Average price | | | Annual consumption, 1992 |
Commodities	1992	1993	1994	(Q_o)
Eggs (dozen)	$0.93	$0.87	$0.87	104
Butter (pound)	1.64	1.61	1.54	20
Milk (1/2 gal)	1.39	1.43	1.44	156
Bread (1-lb loaf)	0.74	0.76	0.75	200

Solution

We use the quantity consumed in the base period as weights to construct this price index. Table 14.3 illustrates the calculation method.
The index numbers given in Table 14.3 are calculated as follows:

$$I = \frac{\Sigma P_n Q_o}{\Sigma P_o Q_o} \cdot 100$$

$$I_{1994} = \frac{497.76}{494.36} \cdot 100$$

$$= 100.7$$

The 1993 index of 100.7 shows that the price of 104 dozen eggs, 20 lb of butter, 156 half gallons of milk, and 200 loaves of bread using the 1992 quantities consumed as a base is 0.7% higher than the price of the same commodities in 1992. Again, we use the constant weights to provide a measure of comparability of prices from one period to another. However, it is desirable to review, revise, and update the weights because they become out of date and unrealistic with time.

Table 14.3 **Calculation of the Weighted Aggregates Index for Agricultural Commodity Prices**

Commodities	Average Price			Annual consumption			
	1992 P_0	1993 P_1	1994 P_2	1992 (Q_0)	1992 P_0Q_0	1993 P_1Q_0	1994 P_2Q_0
Eggs (dozen)	$0.93	$0.87	$0.87	104	$96.72	$90.48	90.48
Butter (pound)	1.64	1.61	1.54	20	32.80	32.20	30.80
Milk (1/2 gal)	1.39	1.43	1.44	156	216.84	223.08	224.64
Bread (1-lb loaf)	0.74	0.76	0.75	200	148.00	152.00	150.00
Total					494.36	497.76	495.92
Aggregate index					100.00	100.7	100.3

Average of relatives indices. This is the second approach in constructing an index number. In constructing the index by this method, first calculate a price relative for each commodity by dividing its value (price) in a nonbase period by the value (price) in the base period. Then an average of these (price) relatives is calculated. This is, in effect, the construction of a simple index for each series that is to make up the composite index. As in the case of the aggregative indices, we may construct an unweighted or weighted index. The unweighted and weighted arithmetic means of relatives are calculated using the following formulas:

$$\text{Unweighted index} = \frac{\Sigma\left(\frac{P_n}{P_0} \cdot 100\right)}{n} \qquad [14\text{--}4]$$

and

$$\text{Weighted index} = \frac{\Sigma\left(\frac{P_n}{P_0} \cdot 100\right)w}{\Sigma w} \qquad [14\text{--}5]$$

where $\dfrac{P_n}{P_0}$ = the price relative

n = number of commodities and services
w = weight applied to price relative

Examples 14.4 and 14.5 illustrate the construction method for unweighted and weighted averages of relative indices.

Example 14.4

Given the following information, construct an unweighted arithmetic mean of relatives index.

Commodities	Price 1992	Price 1994
Eggs(dozen)	$0.93	$0.87
Butter (pound)	1.64	1.54
Milk (1/2 gal)	1.39	1.44
Bread (1-lb loaf)	0.74	0.75

Solution

Table 14.4 shows the calculation of the price relatives as used in the construction of the unweighted arithmetic mean of relatives.

Table 14.4   Calculation of the Unweighted Mean of Relatives Index of Food Prices for 1994 on a 1992 Basis

(1) Commodities	(2) p 1992	(3) p 1994	(4) Price relatives $P_{94}/P_{92} \cdot 100$
Eggs (dozen)	$0.93	$0.87	93.55
Butter (pound)	1.64	1.54	93.90
Milk (1/2 gal)	1.39	1.44	103.59
Bread (1-lb loaf)	0.74	0.75	101.35
Total			392.39

The unweighted arithmetic mean of relative index using Equation 14–4 is

$$\text{Index} = \frac{392.39}{4}$$

$$= 98.09$$

The advantage of using this index is its simplicity. Additionally, the index is independent of the units in which prices are expressed. That is, whether butter is measured in pounds or ounces, the index value is the same. The disadvantage of the unweighted arithmetic mean of relatives is that it does not provide for explicit weighting in terms of the importance of the commodities whose prices have changed. The importance assigned to relative price change is taken into account by using a weighted arithmetic mean of relatives using the base period as weights, as illustrated in Example 14.5.

Example 14.5

Construct a weighted arithmetic mean of relatives index using the four food items previously considered in Example 14.3.

Commodities	Average price		Annual consumption
	P_0 1992	P_1 1994	1992 (Q_0)
Eggs (dozen)	$0.93	$0.87	104
Butter (pound)	1.64	1.54	20
Milk (1/2 gal)	1.39	1.44	156
Bread (1-lb loaf)	0.74	0.75	200

Solution

In general practice, the weights used in constructing this type of index are the values of goods produced, consumed, purchased, or sold. In the present problem, the weights are values of goods consumed. For instance, the value or dollar spent on purchasing eggs for a family of four ($96.72) is the product of the price of eggs and the quantity of eggs consumed. We use the value of each item in the base period as the weight. Table 14.5 illustrates the computation of this index. Substituting the values from Table 14.5 into Equation 14–5 yields the following weighted arithmetic mean of relatives for 1994, using the 1992 as base period value weights:

$$\text{Weighted index} = \frac{\Sigma \left(\frac{P_{94}}{P_{92}} \cdot 100 \right) (P_{92}Q_{92})}{\Sigma P_{92}Q_{92}}$$

$$= \frac{49,590.34}{494.36}$$

$$= 100.31$$

This index is interpreted in the same manner as the other indices. Using P_nQ_0 and P_nQ_n as weights creates interpretational difficulties. For this reason, their usefulness is limited.

Table 14.5 **Calculation of the Weighted Arithmetic Mean of Relatives Index of Food Prices for 1994 on a 1992 Base, Using Base Period Weights**

Commodities	Price 1992 P_o	Price 1994 P_1	Annual consumption 1992 (Q_o)	Price relatives $P_1/P_o \cdot 100$	P_oQ_o	Weighted price relatives $(P_1/P_o \cdot 100)(P_oQ_o)$
Eggs (dozen)	$0.93	$0.87	104	$93.55	96.72	$9,048.16
Butter (pound)	1.64	1.54	20	93.90	32.80	3,079.92
Milk (1/2 gal)	1.39	1.44	156	103.59	216.84	22,462.46
Bread (1-lb loaf)	0.74	0.75	200	101.35	148.00	14,999.80
Total					$494.36	$49,590.34

14.3 Pitfalls and some considerations in use of index numbers

Index numbers do provide us with a measure of change when we are unable to combine statistical series that are not in comparable units. However, there are pitfalls in the use of index numbers that we must recognize.

Shifting the base. It is important to understand that the base period of the index may be changed from one base period to another. This is called *shifting* the base. Researchers often find it necessary to shift the base so that several series, when compared with each other, have the same base period.

In other circumstances, we may wish to reduce the possible bias introduced in an index through passage of time. That is, we may want to state the series in terms of a more recent period. As time passes, an index may either over- or understate the change being measured. This is particularly true of weighted indices. For instance, when we use quantity consumed of a product as a weight in constructing an index, a bias may be introduced if through time the quantity consumed varies. Such bias (the decline in the relative importance of an item in a series) may lead to inappropriate conclusions. The implications of a shifted base and the procedure for accomplishing a base shift are illustrated in the following example.

Example 14.6

The consumer price index for food in the U.S. between 1981 and 1994 is given as follows. By using 1990 as the base, shift the index.

Year	Consumer price index for food
1981	97.7
1982	99.2
1983	99.9
1984	100.9
1985	101.6
1986	88.2
1987	88.6
1988	89.3
1989	94.3
1990	102.1
1991	102.5
1992	103.0
1993	104.2
1994	104.6

From U.S. Department of Commerce, *Statistical Abstract of the United States–1995*, Washington, D.C., September, 1995, 492, Table no. 761.

Solution

The shift to a 1990 base period is accomplished by dividing each figure in the original series by 102.1, which is the original index number for 1990, and then multiplying the result by 100. Table 14.6 shows the original food price index in column two and the shifted base index in column three.

Note that the original and the shifted base index provide the same year-to-year percentage price increases. For example, in 1992 food prices increased by 0.5% over 1991. In column three the 1991 index is 100.39 and the 1992 index is 100.88; thus, using the new base, 1992 food prices are still $100.88/100.39 = 1.005$ or 100.5% of those for 1991. This procedure for shifting base is widely used because it is considered by analysts to be the most practical approach to restating the index of a series to a more recent period.

Splicing index series.   Whenever we have a substantial change in index series from period to period or if it becomes necessary to include new commodities in an index, we must *splice* the old and the new series. Splicing follows an arithmetic procedure similar to shifting the base. When we are splicing an old and a revised series, we must have an overlapping period between the old and the revised series. To illustrate the procedure for splicing two index series, let us take the following example.

Table 14.6 **Shifting the Base**

Year (1)	Food price index (1982–1984 = 100) (2)	Food price index (1990 = 100) (3)
1981	97.7	95.69
1982	99.2	97.16
1983	99.9	97.84
1984	100.9	98.82
1985	101.6	99.51
1986	88.2	86.38
1987	88.6	86.77
1988	89.3	87.46
1989	94.3	92.36
1990	102.1	100.00
1991	102.5	100.39
1992	103.0	100.88
1993	104.2	102.05
1994	104.6	102.44

Example 14.7

Given the following price indices, splice the two series using 1991 as the base.

Year	(1) Old index	(2) Revised index
1986	98.9	
1987	100.0	
1988	101.2	
1989	105.3	
1990	104.2	
1991	106.6	100.0
1992		107.3
1993		108.2
1994		107.6
1995		111.2
1996		114.0

Solution

To splice the old and the revised indices we must have an overlapping period of old and revised indices. In this problem, 1991 is the period of overlap. To splice the two series, we divide each figure in the old series by the old index figure in 1991, which is 106.6. This restates the old series on the new base of the period 1991 and following. Column three of Table 14.7 shows the spliced series. The spliced series can also be stated in terms of the old base period of 1986. This requires multiplying the revised series by 106.6.

Table 14.7   **Splicing Two Index-Number Time Series**

Year	(1) Old index (1986 = 100)	(2) Revised index (1991 = 100)	(3) Spliced index (1991 = 100)
1986	98.9		92.8
1987	100.0		93.8
1988	101.2		94.9
1989	105.3		98.9
1990	104.2		97.7
1991	106.6	100.0	100.0
1992		107.3	107.3
1993		108.2	108.2
1994		107.6	107.6
1995		111.2	111.2
1996		114.0	114.0

Qualitative changes should also be taken into account when constructing indices. Technological changes produce higher quality products at lower prices. When this happens, comparison of a price index may be difficult if care is not given to the qualitative aspects. For instance, tractor tires produced today are qualitatively different than those produced in earlier years. For instance, if a small tractor tire costs $250 today and delivers 40 to 50% more use than those built 10 to 15 years ago, it should be treated differently when included in the CPI.

Chapter summary

In this chapter, you were introduced to another descriptive measure called an *index number*. Index numbers are used to describe changes in economic and physical variables over time. They are essentially summary measures that express a relative comparison between groups of related items. That is, they state the relationship between two figures, one of which is referred to as the *base*.

Indices, for example, are used to show changes over time in production of agricultural commodities (quantity index), prices (price index), or value index. They are a useful tool in simplifying data, facilitating comparisons, and showing seasonal variations. Whether they are a quantity, price, or value index, they are constructed by either an aggregative or an average of relatives method.

Problems associated with constructing a composite index are mostly in the choice of items to be included in it. We must have a clear understanding of what type of an index we are constructing, and what specific series of items is to be included in such an index. Related to the problem of deciding which items are to be included is how much weight each item is to be given in a composite index. Additionally, researchers must consider carefully the time period to be used as a base in constructing an index. Choosing a recent time as a base has the advantage of being readily recalled, and more useful comparisons can be made. Furthermore, a time period considered normal, that is, where there are few erratic fluctuations, is more desirable for a base period. This improves comparison of values of the index because they are not be adversely affected by unduly high or low base values.

Case Study

Indexing Economic Growth with Price

Prices serve an important role in the economies of nations. Policy makers use them to guide their decisions in the allocation of resources, and consumers gauge their purchasing power with prices. As economies expand and grow we want to know whether the benefits from such expansion and growth are real or eroded by rising prices. One way to evaluate the nominal growth of an economic indicator such as gross national product (GNP) is to develop an index for prices, quantities, or value of goods and services.

Economists use the GNP or gross domestic product (GDP) of countries as indicators of economic growth. The comparisons of GNP or GDP estimates among countries based on market exchange rates must be carefully made because they are affected by the rapid variations in those rates. GDP estimates based on purchasing power parity (PPP), which reflects relative price levels in different countries, are designed to provide a more stable and realistic comparison. For example, Japan's 1991 per capita GDP based on exchange rate was $26,984 compared with $19,390 based on PPPs. Similarly, for many developing countries, PPP estimates of per capita GDP improve their relative position. For example, India jumps from $288 to $1150, Chile from $2538 to $7060, and Kenya from $399 to $1350 (World Resources Institute, 1995, p. 255).

In 1994, world economic output increased 3.1% from the previous year (International Monetary Fund, 1994). The world economy was hailed as moving out of the recession of the early 1990s. Other reports indicate phenomenal economic growth in some developing countries of Asia (such as China, Thailand, and Singapore.) In China, the economy expanded at 10% in 1995. This followed a remarkable double-digit growth in the previous four years of 13, 13, 11, and 10% (Asian Development Bank, 1994). Economic expansions in Thailand and Singapore are reported in the double digits for the years 1993 and 1994 (International Monetary Fund, 1994).

A strong (though less dramatic) economic growth is reported in some African countries. Africa as a region began to recover in 1994 with an overall expansion of 3.3%, marking the first time economic growth had exceeded that of population growth in several years (International Monetary Fund, 1994). Africa's expansion was led by strong commodity prices for nearly all the region's export products (Brown, 1995). The following table provides the data on selected commodity prices of goods exported from Africa.

Selected Commodity Prices
(in Constant 1990 U.S. Dollars)

Year	Cocoa (kg)	Coffee (kg)	Groundnut oil (mt)	Sisal (mt)
1983	3.05	4.19	1022.4	809.7
1984	3.52	4.69	1494.0	858.2
1985	3.29	4.71	1319.2	766.8
1986	2.56	5.31	703.4	635.4
1987	2.24	2.82	562.8	576.3
1988	1.66	3.18	619.3	578.1
1989	1.31	2.52	818.6	689.9
1990	1.27	1.97	963.7	715.0
1991	1.17	1.83	876.6	655.4
1992	1.03	1.32	572.3	474.4

Adapted from World Resources Institute, *World Resources 1994–1995*, Oxford University Press, New York, 1995, Table 15.4.

By using the preceding information:

1. Develop a price index for coffee.
2. Develop an aggregate price index for all the commodities.

References

Asian Development Bank, 1994. *Asian Development Bank Outlook 1994.* New York: Oxford University Press.

Brown, L., World Economy Expanding Faster, in Brown, L., Lenssen, N., and Kane, H., *Vital Signs 1995,* W. W. Norton, New York, 1995, 70.

International Monetary Fund (IMF), World Economic Outlook, October 1994. Washington, D.C., 1994.

World Resources Institute, *World Resources 1994–1995,* Oxford University Press, New York, 1995.

Review questions

1. The cost of a gallon of milk in 1995 was $2.15, whereas in 1985 a gallon of milk could be purchased for $1.75. Construct the price relative to the 1985 price, and interpret the index.
2. Surveys of a family of four in 1975 and in 1985 have indicated consumption habits as given in the following table:

Food item	1985			1995		
	Q_o	P_o	$Q_o P_o$	Q_1	P_1	$Q_1 P_1$
Bread	60 lb	$0.40/lb	$24.00	40 lb	$0.55	$22.00
Milk	120 gal	1.75/gal	210.00	140 gal	2.15	301.00
Eggs	80 doz	0.75/doz	60.00	60 doz	1.05	63.00
Cheese	55 lb	2.10/lb	115.50	80 lb	2.20	176.00

   a. Construct a price index series using the weighted aggregative method.
   b. Construct a weighted aggregative quantity index.
3. By using 1977 as a base, compute a simple index of current account deficit series for Asia between 1977 and 1986.

Year	Current account deficit (in billion U.S. dollars)	Year	Current account deficit (in billion U.S. dollars)
1977	0.9	1982	19.8
1978	8.9	1983	16.3
1979	15.2	1984	7.9
1980	21.8	1985	8.2
1981	23.4	1986	8.7

From International Monetary Fund, *World Economic Outlook*, International Monetary Fund, Washington, D.C., April 1985, 252, Table 39.

4. The U.S. Department of Agriculture reports the following food price index. Shift the price index, using 1980 as the base period.

Year	Food price index (1970 = 100)
1970	100.0
1971	98.3
1972	99.9
1973	99.3
1974	100.2
1975	101.3
1976	104.5
1977	106.4
1978	107.6
1979	108.4
1980	111.0
1981	112.3
1982	114.5
1983	116.3
1984	115.2
1985	117.4

5. Given the following two-index-number series, determine the values of the spliced index using 1990 as the base period.

Year	Old price index	Revised price index
1985	98.3	
1986	99.5	
1987	99.9	
1988	102.4	
1989	103.2	
1990	105.2	100.0
1991		104.5
1992		106.3
1993		108.4
1994		110.2
1995		111.8

6. One of the main concerns about economic development is the effect it has on land use. In particular, it has been argued that species habitat and environmentally sensitive areas should be converted to other uses. In the past, wetlands have been converted into farmland. The following table shows losses of wetlands between 1987 and 1994. Use 1987 as a base and compute a simple index of loss of wetlands in the U.S.

Year	Loss of wetlands (1000 acres)
1987	135
1988	131
1989	126
1990	121
1991	116
1992	112
1993	107
1994	102

From Soil Conservation Service, U.S. Department of Agriculture, *National Resources Inventory*, Washington, D.C., various years.

7. The index of farm production in the U.S. between 1967 and 1980 are given as follows. Use 1977 as the base and shift the index.

Year	Index of farm output
1967	100
1968	106
1969	118
1970	111
1971	131
1972	138
1973	142
1974	122
1975	139
1976	140
1977	156
1978	168
1979	182
1980	162

From U.S. Department of Agriculture, *Economic Indicators of the Farm Sector: Production and Efficiency Statistics*, ERS/Statistical Bulletin no. 679, 1980, Table 49.

8. Agricultural production has consistently outpaced population growth over the past 30 years. The pattern of per capita agricultural production has also been fairly consistent: a period of impressive growth, followed by a decline in production. The following table shows the agricultural production index for the world. Use 1988 as the base and shift the index.

Year	Agricultural production index (1979–1981 = 100)
1977	93.29
1978	97.87
1979	98.42
1980	99.08
1981	102.50
1982	106.17
1983	106.16
1984	111.60
1985	114.23
1986	116.09
1987	116.64
1988	118.73
1989	123.27
1990	126.06
1991	125.77
1992	127.88

From World Resources Institute, World Resources Database, DSC Data Services, Washington, D.C., 1994.

References and suggested reading

Crow, W. R., *Index Numbers*, Macdonald & Evans, London, 1965.

Eisele, C. F., *The Consumer Price Index: Description and Discussion*, Center Report Series no. 3, Center for Labor and Management, College of Business Administration, University of Iowa, Iowa City, 1975.

Hamburg, M., *Statistical Analysis for Decision Making*, 3rd ed., Harcourt Brace Jovanovich, New York, 1987, chap. 12.

International Monetary Fund, *World Economic Outlook*, International Monetary Fund, Washington, D.C., April 1985.

Lapin, L. L., *Statistics for Modern Business Decisions*, 4th ed, Harcourt Brace Jovanovich, New York, 1987, chap. 19.

U.S. Department of Agriculture, *Economic Indicators of the Farm Sector: Production and Efficiency Statistics*, 1980, ERS/Statistical Bulletin no. 679, 1982.

U.S. Department of Agriculture, *Economic Indicators of the Farm Sector. Farm Sector Review*, USDA/ERS/ECIFS2-1, May 1983.

U.S. Department of Labor, *BLS Handbook Methods*, Vol. 11, The Consumer Price Index, Bureau of Labor Statistics, Bulletin 2134-2, April 1984.

U.S. Department of Labor, *CPI Issues*, Bureau of Labor Statistics, Report S93, February 1980.

Appendices

Appendix A

Binomial distribution

$$P\{r\} = \frac{n!}{r!(n-r)!}\ p^r q^{n-r}$$

n	r	.05	.10	.15	.20	.25	.30	.35	.40	.45	.50
1	0	.9500	.9000	.8500	.8000	.7500	.7000	.6500	.6000	.5500	.5000
	1	.0500	.1000	.1500	.2000	.2500	.3000	.3500	.4000	.4500	.5000
2	0	.9025	.8100	.7225	.6400	.5625	.4900	.4225	.3600	.3025	.2500
	1	.0950	.1800	.2550	.3200	.3750	.4200	.4550	.4800	.4950	.5000
	2	.0025	.0100	.0225	.0400	.0625	.0900	.1225	.1600	.2025	.2500
3	0	.8574	.7290	.6141	.5120	.4219	.3430	.2746	.2160	.1664	.1250
	1	.1354	.2430	.3251	.3840	.4219	.4410	.4436	.4320	.4084	.3750
	2	.0071	.0270	.0574	.0960	.1406	.1890	.2389	.2880	.3341	.3750
	3	.0001	.0010	.0034	.0080	.0156	.0270	.0429	.0640	.0911	.1250
4	0	.8145	.6561	.5220	.4096	.3164	.2401	.1785	.1296	.0915	.0625
	1	.1715	.2916	.3685	.4096	.4219	.4116	.3845	.3456	.2995	.2500
	2	.0135	.0486	.0975	.1536	.2109	.2646	.3105	.3456	.3675	.3750
	3	.0005	.0036	.0115	.0256	.0469	.0756	.1115	.1536	.2005	.2500
	4	.0000	.0001	.0005	.0016	.0039	.0081	.0150	.0256	.0410	.0625
5	0	.7738	.5905	.4437	.3277	.2373	.1681	.1160	.0778	.0503	.0312
	1	.2036	.3280	.3915	.4096	.3955	.3602	.2314	.2592	.2059	.1562
	2	.0214	.0729	.1392	.2048	.2637	.3087	.3364	.3456	.3369	.3225
	3	.0011	.0081	.0244	.0512	.0879	.1323	.1811	.2304	.2757	.3125
	4	.0000	.0004	.0022	.0064	.0146	.0284	.0489	.0768	.1128	.1562
	5	.0000	.0000	.0001	.0003	.0010	.0024	.0053	.0102	.0185	.0312
6	0	.7351	.5314	.3771	.2621	.1780	.1176	.8754	.0467	.0277	.0156
	1	.2321	.3543	.3993	.3932	.3560	.3025	.2437	.1866	.1350	.0938
	2	.0305	.0984	.1762	.2458	.2966	.3241	.3280	.3110	.2780	.2344
	3	.0021	.0146	.0415	.0819	.1318	.1852	.2355	.2765	.3032	.3125
	4	.0001	.0012	.0055	.0154	.0330	.0595	.0951	.1382	.1861	.2344
	5	.0000	.0001	.0004	.0015	.0044	.0102	.0205	.0369	.0609	.0938
	6	.0000	.0000	.0000	.0001	.0002	.0007	.0018	.0041	.0083	.0156
7	0	.6983	.4783	.3206	.2097	.1335	.0624	.8490	.0280	.0152	.0078
	1	.2573	.3720	.3960	.3670	.3115	.2471	.1848	.1306	.0872	.0547
	2	.0406	.1240	.2097	.2753	.3115	.3177	.2985	.2613	.2140	.1641
	3	.0036	.0230	.0617	.1147	.1730	.2269	.2679	.2903	.2918	.2734
	4	.0002	.0026	.0109	.0287	.0577	.0972	.1442	.1935	.2388	.2734
	5	.0000	.0002	.0012	.0043	.0115	.0250	.0466	.0774	.1172	.1641
	6	.0000	.0000	.0001	.0004	.0013	.0036	.0084	.0172	.0320	.0547
	7	.0000	.0000	.0000	.0000	.0001	.0002	.0004	.0016	.0037	.0078
8	0	.6634	.4305	.2725	.1678	.1002	.0576	.0319	.0168	.0084	.0039
	1	.2793	.3826	.3847	.3355	.2670	.1977	.1373	.0896	.0548	.0312

n	r	.05	.10	.15	.20	.25	.30	.35	.40	.45	.50
	2	.0515	.1488	.2376	.2936	.3115	.2065	.2587	.2090	.1549	.1094
	3	.0054	.0331	.0839	.1468	.2076	.2541	.2786	.2787	.2568	.2188
	4	.0004	.0046	.0185	.0459	.0865	.1361	.1875	.2322	.2627	.2734
	5	.0000	.0004	.0026	.0092	.0231	.0467	.0808	.1239	.1719	.2188
	6	.0000	.0000	.0002	.0011	.0038	.0100	.0217	.0413	.0403	.1094
	7	.0000	.0000	.0000	.0001	.0004	.0012	.0033	.0079	.0164	.0312
	8	.0000	.0000	.0000	.0000	.0000	.0001	.0002	.0007	.0017	.0039
9	0	.6302	.3874	.2316	.1342	.0751	.0404	.0207	.0101	.0046	.0020
	1	.2985	.3874	.3679	.3020	.2253	.1556	.1004	.0605	.0339	.0176
	2	.0629	.1722	.2597	.3020	.3003	.2668	.2162	.1612	.1110	.0703
	3	.0077	.0446	.1069	.1762	.2336	.2668	.2716	.2508	.2119	.1641
	4	.0006	.0074	.0283	.0661	.1168	.1715	.2194	.2508	.2600	.2461
	5	.0000	.0008	.0050	.0165	.0389	.0735	.1181	.1672	.2128	.2461
	6	.0000	.0001	.0006	.0028	.0087	.0210	.0424	.0743	.1160	.1641
	7	.0000	.0000	.0000	.0003	.0012	.0039	.0098	.0212	.0407	.0703
	8	.0000	.0000	.0000	.0000	.0001	.0004	.0013	.0035	.0083	.0176
	9	.0000	.0000	.0000	.0000	.0000	.0000	.0001	.0003	.0008	.0020
10	0	.5987	.3487	.1969	.1074	.0563	.0282	.0135	.0060	.0025	.0010
	1	.3151	.3874	.3474	.2684	.1877	.1211	.0725	.0403	.0207	.0098
	2	.0746	.1937	.2759	.3020	.2816	.2335	.1757	.1209	.0763	.0439
	3	.0105	.0574	.1298	.2013	.2503	.2668	.2522	.2150	.1665	.1172
	4	.0010	.0112	.0401	.0881	.1460	.2001	.2377	.2508	.2384	.2051
	5	.0001	.0015	.0085	.0264	.0584	.1029	.1536	.2007	.2340	.2461
	6	.0000	.0001	.0012	.0055	.0162	.0368	.0689	.1115	.1596	.2051
	7	.0000	.0000	.0001	.0008	.0031	.0090	.0212	.0425	.0746	.1172
	8	.0000	.0000	.0000	.0001	.0004	.0014	.0043	.0106	.0229	.0439
	9	.0000	.0000	.0000	.0000	.0000	.0001	.0005	.0016	.0042	.0098
	10	.0000	.0000	.0000	.0000	.0000	.0000	.0000	.0001	.0003	.0010
11	0	.5688	.3138	.1673	.0859	.0422	.0198	.0088	.0036	.0014	.0005
	1	.3293	.3835	.3248	.2362	.1549	.0932	.0518	.0266	.0125	.0054
	2	.0867	.2131	.2866	.2953	.2581	.1998	.1395	.0887	.0513	.0269
	3	.0137	.0710	.1517	.2215	.2581	.2568	.2254	.1774	.1259	.0806
	4	.0014	.0158	.0536	.1107	.1721	.2201	.2428	.2365	.2060	.1611
	5	.0001	.0025	.0132	.0388	.0803	.1321	.1830	.2207	.2360	.2256
	6	.0000	.0003	.0023	.0097	.0268	.0566	.0985	.1471	.1931	.2256
	7	.0000	.0000	.0003	.0017	.0064	.0173	.0379	.0701	.1128	.1611
	8	.0000	.0000	.0000	.0002	.0011	.0037	.0102	.0234	.0462	.0806
	9	.0000	.0000	.0000	.0000	.0001	.0005	.0018	.0052	.0126	.0269
	10	.0000	.0000	.0000	.0000	.0000	.0000	.0002	.0007	.0021	.0054
	11	.0000	.0000	.0000	.0000	.0000	.0000	.0000	.0000	.0002	.0005
12	0	.5404	.2824	.1422	.0687	.0317	.0138	.0057	.0022	.0008	.0002
	1	.3413	.3766	.3012	.2062	.1267	.0712	.0368	.0174	.0075	.0029
	2	.0988	.2301	.2924	.2835	.2323	.1678	.1088	.0639	.0339	.0161
	3	.0173	.0852	.1720	.2362	.2581	.2397	.1954	.1410	.0923	.0537
	4	.0021	.0213	.0683	.1329	.1936	.2311	.2367	.2128	.1700	.1208
	5	.0002	.0038	.0193	.0532	.1032	.1585	.2039	.2270	.2225	.1934
	6	.0000	.0005	.0040	.0155	.0401	.0792	.1281	.1766	.2124	.2256
	7	.0000	.0000	.0006	.0033	.0115	.0291	.0591	.1009	.1489	.1934
	8	.0000	.0000	.0001	.0005	.0024	.0078	.0199	.0420	.0762	.1208
	9	.0000	.0000	.0000	.0001	.0004	.0015	.0048	.0125	.0277	.0537

n	r	.05	.10	.15	.20	.25	.30	.35	.40	.45	.50
							p				
	10	.0000	.0000	.0000	.0000	.0000	.0002	.0008	.0025	.0068	.0161
	11	.0000	.0000	.0000	.0000	.0000	.0000	.0001	.0003	.0010	.0029
	12	.0000	.0000	.0000	.0000	.0000	.0000	.0000	.0000	.0001	.0002
13	0	.5133	.2542	.1209	.0550	.0238	.0097	.0037	.0013	.0004	.0001
	1	.3512	.3672	.2774	.1787	.1029	.0540	.0259	.0113	.0045	.0016
	2	.1109	.2448	.2937	.2680	.2059	.1388	.0836	.0453	.0220	.0095
	3	.0214	.0997	.1900	.2457	.2517	.2181	.1651	.1107	.0660	.0349
	4	.0028	.0277	.0838	.1535	.2097	.2337	.2222	.1845	.1350	.0873
	5	.0003	.0055	.0266	.0691	.1258	.1803	.2154	.2214	.1989	.1571
	6	.0000	.0008	.0063	.0230	.0559	.1030	.1546	.1968	.2169	.2095
	7	.0000	.0001	.0011	.0058	.0186	.0442	.0833	.1312	.1775	.2095
	8	.0000	.0001	.0001	.0011	.0047	.0142	.0336	.0656	.1089	.1571
	9	.0000	.0000	.0000	.0001	.0009	.0034	.0101	.0243	.0495	.0873
	10	.0000	.0000	.0000	.0000	.0001	.0006	.0022	.0065	.0162	.0349
	11	.0000	.0000	.0000	.0000	.0000	.0001	.0003	.0012	.0036	.0095
	12	.0000	.0000	.0000	.0000	.0000	.0000	.0000	.0001	.0005	.0016
	13	.0000	.0000	.0000	.0000	.0000	.0000	.0000	.0000	.0000	.0001
14	0	.4877	.2288	.1028	.0440	.0178	.0068	.0024	.0008	.0002	.0001
	1	.3593	.3559	.2539	.1539	.0832	.0407	.0181	.0073	.0027	.0009
	2	.1229	.2570	.2912	.2501	.1802	.1134	.0634	.0317	.0141	.0056
	3	.0259	.1142	.2056	.2501	.2402	.1943	.1366	.0845	.0462	.0222
	4	.0037	.0349	.0998	.1720	.2202	.2290	.2022	.1549	.1040	.0611
	5	.0004	.0078	.0352	.0860	.1468	.1963	.2178	.2066	.1701	.1222
	6	.0000	.0013	.0093	.0322	.0734	.1262	.1759	.2086	.2088	.1833
	7	.0000	.0002	.0019	.0092	.0280	.0618	.1082	.1574	.1952	.2095
	8	.0000	.0000	.0003	.0020	.0082	.0232	.0510	.0918	.1398	.1833
	9	.0000	.0000	.0000	.0003	.0018	.0066	.0183	.0408	.0762	.1222
	10	.0000	.0000	.0000	.0000	.0003	.0014	.0049	.0136	.0312	.0611
	11	.0000	.0000	.0000	.0000	.0000	.0002	.0010	.0033	.0093	.0222
	12	.0000	.0000	.0000	.0000	.0000	.0000	.0001	.0005	.0019	.0056
	13	.0000	.0000	.0000	.0000	.0000	.0000	.0000	.0001	.0002	.0009
	14	.0000	.0000	.0000	.0000	.0000	.0000	.0000	.0000	.0000	.0001
15	0	.4633	.2059	.0874	.0352	.0134	.0047	.0016	.0005	.0001	.0000
	1	.3658	.3432	.2312	.1319	.0668	.0305	.0126	.0047	.0016	.0005
	2	.1348	.2669	.2856	.2309	.1559	.0916	.0476	.0219	.0090	.0034
	3	.0307	.1285	.2184	.2501	.2252	.1700	.1110	.0634	.0318	.0139
	4	.0049	.0428	.1156	.1876	.2252	.2186	.1792	.1268	.0780	.0417
	5	.0006	.0105	.0449	.1032	.1651	.2061	.2123	.1859	.1404	.0916
	6	.0000	.0019	.0132	.0430	.0917	.1472	.1906	.2066	.1914	.1527
	7	.0000	.0003	.0030	.0138	.0393	.0611	.1319	.1771	.2013	.1964
	8	.0000	.0000	.0003	.0035	.0131	.0348	.0710	.1181	.1647	.1964
	9	.0000	.0000	.0001	.0007	.0034	.0116	.0298	.0612	.1048	.1527
	10	.0000	.0000	.0000	.0001	.0007	.0030	.0096	.0245	.0515	.0916
	11	.0000	.0000	.0000	.0000	.0001	.0006	.0024	.0074	.0191	.0417
	12	.0000	.0000	.0000	.0000	.0000	.0001	.0004	.0016	.0052	.0139
	13	.0000	.0000	.0000	.0000	.0000	.0000	.0001	.0003	.0010	.0032
	14	.0000	.0000	.0000	.0000	.0000	.0000	.0000	.0000	.0001	.0005
	15	.0000	.0000	.0000	.0000	.0000	.0000	.0000	.0000	.0000	.0000
16	0	.4401	.1853	.0873	.0281	.0100	.0033	.0010	.0003	.0001	.0000
	1	.3706	.3294	.2097	.1126	.0535	.0228	.0087	.0030	.0009	.0002
	2	.1463	.2745	.2775	.2111	.1336	.0732	.0353	.0150	.0056	.0018

n	r	.05	.10	.15	.20	.25	.30	.35	.40	.45	.50
							p				
	3	.0359	.1423	.2285	.2463	.2079	.1465	.0888	.0468	.0215	.0085
	4	.0061	.0514	.1311	.2001	.2252	.2040	.1553	.1014	.0572	.0278
	5	.0008	.0137	.0555	.1201	.1802	.2099	.2008	.1623	.1123	.0667
	6	.0001	.0028	.0180	.0550	.1101	.1649	.1982	.1983	.1684	.1222
	7	.0000	.0004	.0045	.0197	.0524	.1010	.1524	.1889	.1969	.1746
	8	.0000	.0001	.0009	.0055	.0197	.0487	.0923	.1417	.1812	.1964
	9	.0000	.0000	.0001	.0012	.0058	.0185	.0442	.0840	.1318	.1746
	10	.0000	.0000	.0000	.0002	.0014	.0056	.0167	.0392	.0755	.1222
	11	.0000	.0000	.0000	.0000	.0002	.0013	.0049	.0142	.0337	.0667
	12	.0000	.0000	.0000	.0000	.0000	.0002	.0011	.0040	.0115	.0278
	13	.0000	.0000	.0000	.0000	.0000	.0000	.0002	.0008	.0029	.0085
	14	.0000	.0000	.0000	.0000	.0000	.0000	.0000	.0001	.0005	.0018
	15	.0000	.0000	.0000	.0000	.0000	.0000	.0000	.0000	.0001	.0002
	16	.0000	.0000	.0000	.0000	.0000	.0000	.0000	.0000	.0000	.0000
17	0	.4181	.1668	.0631	.0225	.0075	.0023	.0007	.0002	.0000	.0000
	1	.3741	.3150	.1893	.0957	.0426	.0169	.0060	.0019	.0005	.0001
	2	.1575	.2800	.2673	.1914	.1136	.0581	.0260	.0102	.0035	.0010
	3	.0415	.1556	.2359	.2393	.1893	.1245	.0701	.0341	.0144	.0052
	4	.0076	.0605	.1457	.2093	.2209	.1868	.1320	.0796	.0411	.0182
	5	.0010	.0175	.0668	.1361	.1914	.2081	.1849	.1379	.0875	.0472
	6	.0001	.0039	.0236	.0680	.1276	.1784	.1991	.1839	.1432	.0944
	7	.0000	.0007	.0065	.0267	.0668	.1201	.1685	.1927	.1841	.1484
	8	.0000	.0001	.0014	.0084	.0279	.0644	.1134	.1606	.1883	.1855
	9	.0000	.0000	.0003	.0021	.0093	.0276	.0611	.1070	.1540	.1855
	10	.0000	.0000	.0000	.0004	.0025	.0095	.0263	.0571	.1008	.1494
	11	.0000	.0000	.0000	.0001	.0005	.0026	.0090	.0242	.0525	.0944
	12	.0000	.0000	.0000	.0000	.0001	.0006	.0024	.0081	.0215	.0472
	13	.0000	.0000	.0000	.0000	.0000	.0001	.0005.	.0021	.0066	.0182
	14	.0000	.0000	.0000	.0000	.0000	.0000	.0001	.0004	.0016	.0052
	15	.0000	.0000	.0000	.0000	.0000	.0000	.0000	.0001	.0003	.0010
	16	.0000	.0000	.0000	.0000	.0000	.0000	.0000	.0000	.0000	.0001
	17	.0000	.0000	.0000	.0000	.0000	.0000	.0000	.0000	.0000	.0000
18	0	.3972	.1501	.0536	.0180	.0056	.0016	.0004	.0001	.0000	.0000
	1	.3763	.3002	.1704	.0811	.0338	.0126	.0042	.0012	.0003	.0001
	2	.1683	.2835	.2556	.1723	.0958	.0458	.0190	.0069	.0022	.0006
	3	.0473	.1680	.2406	.2297	.1704	.1046	.0547	.0246	.0095	.0031
	4	.0093	.0700	.1592	.2153	.2130	.1681	.1104	.0614	.0291	.0117
	5	.0014	.0218	.0787	.1507	.1988	.2017	.1664	.1146	.0666	.0327
	6	.0002	.0052	.0301	.0816	.1436	.1873	.1941	.1655	.1181	.0708
	7	.0000	.0010	.0091	.0350	.0820	.1376	.1792	.1892	.1657	.1214
	8	.0000	.0002	.0022	.0120	.0376	.0811	.1327	.1734	.1664	.1669
	9	.0000	.0000	.0004	.0033	.0139	.0386	.0794	.1284	.1694	.1855
	10	.0000	.0000	.0001	.0008	.0042	.0149	.0385	.0771	.1348	.1669
	11	.0000	.0000	.0000	.0001	.0010	.0046	.0151	.0374	.0742	.1214
	12	.0000	.0000	.0000	.0000	.0002	.0012	.0047	.0145	.0354	.0708
	13	.0000	.0000	.0000	.0000	.0000	.0002	.0012	.0045	.0134	.0327
	14	.0000	.0000	.0000	.0000	.0000	.0000	.0002	.0011	.0039	.0117

n	r	.05	.10	.15	.20	.25	.30	.35	.40	.45	.50
	15	.0000	.0000	.0000	.0000	.0000	.0000	.0000	.0002	.0009	.0031
	16	.0000	.0000	.0000	.0000	.0000	.0000	.0000	.0000	.0001	.0006
	17	.0000	.0000	.0000	.0000	.0000	.0000	.0000	.0000	.0000	.0001
	18	.0000	.0000	.0000	.0000	.0000	.0000	.0000	.0000	.0000	.0000
19	0	.3774	.1351	.0456	.0144	.0042	.0011	.0003	.0001	.0000	.0000
	1	.3774	.2852	.1529	.0685	.0268	.0093	.0029	.0008	.0002	.0000
	2	.1787	.2852	.2428	.1540	.0803	.0358	.0138	.0046	.0013	.0003
	3	.0533	.1796	.2428	.2182	.1517	.0669	.0422	.0175	.0062	.0018
	4	.0112	.0798	.1714	.2182	.2023	.1491	.0909	.0467	.0203	.0074
	5	.0018	.0266	.0907	.1636	.2023	.1916	.1468	.0933	.0497	.0222
	6	.0002	.0069	.0374	.0955	.1574	.1916	.1844	.1451	.0949	.0518
	7	.0000	.0014	.0122	.0443	.0974	.1525	.1844	.1797	.1443	.0961
	8	.0000	.0002	.0032	.0166	.0487	.0981	.1489	.1797	.1771	.1442
	9	.0000	.0000	.0007	.0051	.0198	.0514	.0080	.1464	.1771	.1762
	10	.0000	.0000	.0001	.0013	.0066	.0220	.0528	.0976	.1449	.1762
	11	.0000	.0000	.0000	.0003	.0018	.0077	.0233	.0532	.0970	.1442
	12	.0000	.0000	.0000	.0000	.0004	.0022	.0083	.0237	.0529	.0961
	13	.0000	.0000	.0000	.0000	.0001	.0005	.0024	.0085	.0233	.0518
	14	.0000	.0000	.0000	.0000	.0000	.0001	.0006	.0024	.0082	.0222
	15	.0000	.0000	.0000	.0000	.0000	.0000	.0001	.0005	.0022	.0074
	16	.0000	.0000	.0000	.0000	.0000	.0000	.0000	.0001	.0005	.0018
	17	.0000	.0000	.0000	.0000	.0000	.0000	.0000	.0000	.0001	.0003
	18	.0000	.0000	.0000	.0000	.0000	.0000	.0000	.0000	.0000	.0000
	19	.0000	.0000	.0000	.0000	.0000	.0000	.0000	.0000	.0000	.0000
20	0	.3585	.1216	.0388	.0115	.0032	.0008	.0002	.0000	.0000	.0000
	1	.3774	.2702	.1368	.0576	.0211	.0068	.0020	.0005	.0001	.0000
	2	.1887	.2852	.2293	.1369	.0669	.0278	.0100	.0031	.0008	.0002
	3	.0596	.1901	.2428	.2054	.1339	.0718	.0323	.0123	.0040	.0011
	4	.0133	.0898	.1821	.2182	.1897	.1304	.0738	.0350	.0139	.0046
	5	.0022	.0319	.1028	.1746	.2023	.1789	.1272	.0746	.0365	.0148
	6	.0003	.0089	.0454	.1091	.1686	.1916	.1712	.1244	.0746	.0370
	7	.0000	.0020	.0160	.0545	.1124	.1643	.1844	.1659	.1221	.0739
	8	.0000	.0004	.0046	.0222	.0609	.1144	.1614	.1797	.1623	.1201
	9	.0000	.0001	.0011	.0074	.0271	.0654	.1158	.1597	.1771	.1602
	10	.0000	.0000	.0002	.0020	.0099	.0308	.0686	.1171	.1593	.1762
	11	.0000	.0000	.0000	.0005	.0030	.0120	.0336	.0710	.1185	.1602
	12	.0000	.0000	.0000	.0001	.0008	.0039	.0136	.0355	.0727	.1201
	13	.0000	.0000	.0000	.0000	.0002	.0010	.0045	.0146	.0368	.0739
	14	.0000	.0000	.0000	.0000	.0000	.0002	.0012	.0049	.0150	.0370
	15	.0000	.0000	.0000	.0000	.0000	.0000	.0003	.0013	.0049	.0148
	16	.0000	.0000	.0000	.0000	.0000	.0000	.0000	.0003	.0013	.0046
	17	.0000	.0000	.0000	.0000	.0000	.0000	.0000	.0000	.0002	.0011
	18	.0000	.0000	.0000	.0000	.0000	.0000	.0000	.0000	.0000	.0002
	19	.0000	.0000	.0000	.0000	.0000	.0000	.0000	.0000	.0000	.0000
	20	.0000	.0000	.0000	.0000	.0000	.0000	.0000	.0000	.0000	.0000

Extracted from Tables of the Binomial Probability Distribution, National Bureau of Standards, U.S. Department of Commerce, Applied Mathematics Series 6, 1952.

Appendix B

Areas under the standard normal probability distribution between the mean and successive values of z

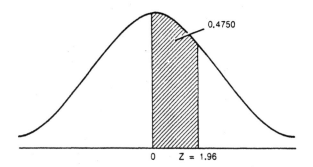

Example: If $z = 1.96$, then the area between the mean and this value of z is 0.4750

z	0.00	0.01	0.02	0.03	0.04	0.05	0.06	0.07	0.08	0.09
0.0	0.0000	0.0040	0.0080	0.0120	0.0160	0.0199	0.0239	0.0279	0.0319	0.0359
0.1	0.0398	0.0438	0.0478	0.0517	0.0557	0.0596	0.0636	0.0675	0.0714	0.0753
0.2	0.0793	0.0832	0.0871	0.0910	0.0948	0.0987	0.1026	0.1064	0.1103	0.1141
0.3	0.1179	0.1217	0.1255	0.1293	0.1331	0.1368	0.1406	0.1443	0.1480	0.1517
0.4	0.1554	0.1591	0.1628	0.1664	0.1700	0.1736	0.1772	0.1808	0.1844	0.1879
0.5	0.1915	0.1950	0.1985	0.2019	0.2054	0.2088	0.2123	0.2157	0.2190	0.2224
0.6	0.2257	0.2291	0.2324	0.2357	0.2389	0.2422	0.2454	0.2486	0.2518	0.2549
0.7	0.2580	0.2612	0.2642	0.2673	0.2704	0.2734	0.2764	0.2794	0.2823	0.2852
0.8	0.2881	0.2910	0.2939	0.2967	0.2995	0.3023	0.3051	0.3078	0.3106	0.3133
0.9	0.3159	0.3186	0.3212	0.3238	0.3264	0.3289	0.3315	0.3340	0.3365	0.3389
1.0	0.3413	0.3438	0.3461	0.3485	0.3508	0.3531	0.3554	0.3577	0.3599	0.3621
1.1	0.3643	0.3665	0.3686	0.3708	0.3729	0.3749	0.3770	0.3790	0.3810	0.3830
1.2	0.3849	0.3869	0.3888	0.3907	0.3925	0.3944	0.3962	0.3980	0.3997	0.4015
1.3	0.4032	0.4049	0.4066	0.4082	0.4099	0.4115	0.4131	0.4147	0.4162	0.4177
1.4	0.4192	0.4207	0.4222	0.4236	0.4251	0.4265	0.4279	0.4292	0.4306	0.4319
1.5	0.4332	0.4345	0.4357	0.4370	0.4382	0.4394	0.4406	0.4418	0.4429	0.4441
1.6	0.4452	0.4463	0.4474	0.4484	0.4495	0.4505	0.4515	0.4525	0.4535	0.4545
1.7	0.4554	0.4564	0.4573	0.4582	0.4591	0.4599	0.4608	0.4616	0.4625	0.4633
1.8	0.4641	0.4649	0.4656	0.4664	0.4671	0.4678	0.4686	0.4693	0.4699	0.4706
1.9	0.4713	0.4719	0.4726	0.4732	0.4738	0.4744	0.4750	0.4756	0.4761	0.4767
2.0	0.4772	0.4778	0.4783	0.4788	0.4793	0.4798	0.4803	0.4808	0.4812	0.4817
2.1	0.4821	0.4826	0.4830	0.4834	0.4838	0.4842	0.4846	0.4850	0.4854	0.4857
2.2	0.4861	0.4864	0.4868	0.4871	0.4875	0.4878	0.4881	0.4884	0.4887	0.4890
2.3	0.4893	0.4896	0.4898	0.4901	0.4904	0.4906	0.4909	0.4911	0.4913	0.4916
2.4	0.4918	0.4920	0.4922	0.4925	0.4927	0.4929	0.4931	0.4932	0.4934	0.4936
2.5	0.4938	0.4940	0.4941	0.4943	0.4945	0.4946	0.4948	0.4949	0.4951	0.4952
2.6	0.4953	0.4955	0.4956	0.4957	0.4959	0.4960	0.4961	0.4962	0.4963	0.4964
2.7	0.4965	0.4966	0.4967	0.4968	0.4969	0.4970	0.4971	0.4972	04973	0.4974

z	0.00	0.01	0.02	0.03	0.04	0.05	0.06	0.07	0.08	0.09
2.8	0.4974	0.4975	0.4976	0.4977	0.4977	0.4978	0.4979	0.4979	0.4980	0.4981
2.9	0.4981	0.4982	0.4982	0.4983	0.4984	0.4984	0.4985	0.4985	0.4986	0.4986
3.0	0.49865	0.4987	0.4987	0.4988	0.4988	0.4989	0.4989	0.4989	0.4990	0.4990
4.0	0.49997									

From Hamburg, M., *Statistical Analysis for Decision Making*, 3rd ed. Copyright © 1983 by Wadsworth Publishing Co. With permission.

Appendix C

Normal deviates for statistical estimation and hypothesis testing

Critical Normal Deviates for Statistical Estimation

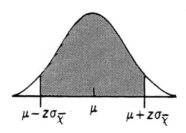

$$\mu - z\sigma_{\bar{X}} \qquad \mu \qquad \mu + z\sigma_{\bar{X}}$$

Reliability or confidence level	Normal deviate z
.80	1.28
.90	1.64
.95	1.96
.98	2.33
.99	2.57
.998	3.08
.999	3.27

Critical Normal Deviates for Hypothesis Testing

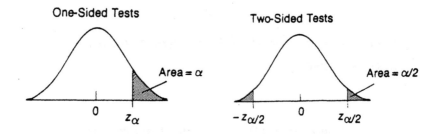

Single-Tail		Two-Tail	
Significance level α	Normal deviate z_α	Significance level α	Normal deviate $z_{\alpha/2}$
.10	1.28	.10	1.64
.05	1.64	.05	1.96
.025	1.96	.025	2.24
.01	2.33	.01	2.57
.005	2.57	.005	2.81
.001	3.08	.001	3.27

From Lapin, L., *Statistics for Modern Business Decision*, 4th ed. Wadsworth Publishing Co., Belmont, CA, 1987. With permission.

Appendix D

Random digits

85967	73152	14511	85285	36009	95892	36962	67835	63314	50162
07483	51453	11649	86348	76431	81594	95848	36738	25014	15460
96283	01898	61414	83525	04231	13604	75339	11730	85423	60698
49174	12074	98551	37895	93547	24769	09404	76548	05393	96700
97366	39941	21225	93629	19574	71565	33413	56087	40875	13351
90474	41469	16812	81542	81652	45554	27931	93994	22375	00953
28599	64109	09497	76235	41383	31555	12639	00619	22909	29563
25254	16210	89717	65997	82667	74624	36348	44018	64732	93589
28785	02760	24359	99410	77319	73408	58993	61098	04393	48245
84725	86576	86944	93296	10081	82454	76810	52975	10324	15457
41059	66456	47679	66810	15941	84602	14493	65515	19251	41642
67434	41045	82830	47617	36932	46728	71183	36345	41404	81110
72766	68816	37643	19959	57550	49620	98480	25640	67257	18671
92079	46784	66125	94932	64451	29275	57669	66658	30818	58353
29187	40350	62533	73603	34075	16451	42885	03448	37390	96328
74220	17612	65522	80607	19184	64164	66962	82310	18163	63495
03786	02407	06098	92917	40434	60602	82175	04470	78744	90775
75085	55558	15520	27038	25471	76107	90832	10819	56797	33751
09161	33015	19155	11715	00551	24909	31894	37774	37953	78837
75707	48992	64998	87080	39333	00767	45637	12538	67439	94914
21333	48660	31288	00086	79889	75532	28704	62844	92337	99695
65626	50061	42539	14812	48895	11196	34335	60492	70650	51108
84380	07389	87891	76255	89604	41372	10837	66992	93183	56920
46479	32072	80083	63868	70930	89654	05359	47196	12452	38234
59847	97197	55147	76639	76971	55928	36441	95141	42333	67483
31416	11231	27904	57383	31852	69137	96667	14315	01007	31929
82066	83436	67914	21465	99605	83114	97885	74440	99622	87912
01850	42782	39202	18582	46215	99228	79541	78298	75404	63648
32315	89276	89582	87138	16165	15984	21466	63830	30475	74729
59338	42703	55198	80380	67067	97155	34160	85019	03527	78140
58089	27632	50987	91373	07736	20436	96130	73483	85332	24384
61705	57285	30392	23660	75841	21931	04295	00875	09114	32101
18914	98982	60199	99275	41967	35208	30357	76772	92656	62318
11965	94089	34803	48941	69709	16784	44642	89761	66864	62803
85251	48111	80936	81781	93248	67877	16498	31924	51315	79921
66121	96986	84844	93873	46352	92183	51152	85878	30490	15974
53972	96642	24199	58080	35450	03482	66953	49521	63719	57615
14509	16594	78883	43222	23093	58645	60257	89250	63266	90858
37700	07688	65533	72126	23611	93993	01848	03910	38552	17472
85466	59392	72722	15473	73295	49759	56157	60477	83284	56367
52969	55863	42312	67842	05673	91878	82738	36563	79540	61935
42744	68315	17514	02878	97291	74851	42725	57894	81434	62041
26140	13336	67726	61876	29971	99294	96664	52817	90039	53211
95589	56319	14563	24071	06916	59555	18195	32280	79357	04224

39113	13217	59999	49952	83021	47709	53105	19295	88318	41626
41392	17622	18994	98283	07249	52289	24209	91139	30714	06604
54684	53645	79246	70183	87731	19185	08541	33519	07223	97413
89442	61001	36658	57444	95388	36682	38052	46719	09428	94012
36751	16778	54888	15357	68003	43564	90976	58904	40512	07725
98159	02564	21416	74944	53049	88749	02865	25772	89853	88714

Appendix E

Critical values of t

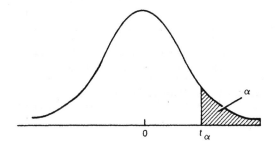

Degrees of freedom	Level of significance for one-tailed test					
	.10	.05	.025	.01	.005	.0005
	Level of significance for two-tailed test					
	.20	.10	.05	.02	.01	.001
1	3.078	6.314	12.706	31.821	63.657	636.619
2	1.886	2.920	4.303	6.965	9.925	31.598
3	1.638	2.353	3.182	4.541	5.841	12.941
4	1.533	2.132	2.776	3.747	4.604	8.610
5	1.476	2.015	2.571	3.365	4.032	6.859
6	1.440	1.943	2.447	3.143	3.707	5.959
7	1.415	1.895	2.365	2.998	3.499	5.405
8	1.397	1.860	2.306	2.896	3.355	5.041
9	1.383	1.833	2.262	2.821	3.250	4.781
10	1.372	1.812	2.228	2.764	3.169	4.587
11	1.363	1.796	2.201	2.718	3.106	4.437
12	1.356	1.782	2.179	2.681	3.055	4.318
13	1.350	1.771	2.160	2.650	3.012	4.221
14	1.345	1.761	2.145	2.624	2.977	4.140
15	1.341	1.753	2.131	2.602	2.947	4.073
16	1.337	1.746	2.120	2.583	2.921	4.015
17	1.333	1.740	2.110	2.567	2.898	3.965
18	1.330	1.734	2.101	2.552	2.878	3.922
19	1.328	1.729	2.093	2.539	2.861	3.883
20	1.325	1.725	2.086	2.528	2.845	3.850
21	1.323	1.721	2.080	2.518	2.831	3.819
22	1.321	1.717	2.074	2.508	2.819	3.792
23	1.319	1.714	2.069	2.500	2.807	3.767
24	1.318	1.711	2.064	2.492	2.797	3.745
25	1.316	1.708	2.060	2.485	2.787	3.725
26	1.315	1.706	2.056	2.479	2.779	3.707

Degrees of freedom	Level of significance for one-tailed test					
	.10	.05	.025	.01	.005	.0005
	Level of significance for two-tailed test					
	.20	.10	.05	.02	.01	.001
27	1.314	1.703	2.052	2.473	2.771	3.690
28	1.313	1.701	2.048	2.467	2.763	3.674
29	1.311	1.699	2.045	2.462	2.756	3.659
30	1.310	1.697	2.042	2.457	2.750	3.646
40	1.303	1.684	2.021	2.423	2.704	3.551
60	1.296	1.671	2.000	2.390	2.660	3.460
120	1.289	1.658	1.980	2.358	2.617	3.373
∞	1.282	1.645	1.960	2.326	2.576	3.291

Appendix F

F Distribution

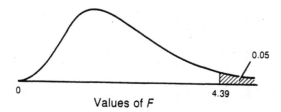

Example: In an F distribution with $v_1 = 5$ and $v_2 = 6$ degrees of freedom, the area to the right of an F value of 4.39 is 0.05. The value on the F scale to the right of which lies 0.05 of the area is in lightface type. The value on the F scale to the right of which lies 0.01 of the area is in boldface type. For the numerator, v_1 = number of degrees of freedom; for the denominator, v_2 = number of degrees of freedom.

F Distribution

α = .05 (roman type) and α = .01 boldface type)

v_1 Degrees of freedom

v_2		1	2	3	4	5	6	7	8	9	10	11	12	14	16	20	24	30	40	50	∞
1	.05	161.	200.	216.	225.	230.	234.	237.	239.	241.	242.	243.	244.	245.	246.	248.	249.	250.	251.	252.	254.
	.01	**4052.**	**4999.**	**5403.**	**5625.**	**5764.**	**5859.**	**5928.**	**5981.**	**6022.**	**6056.**	**6082.**	**6106.**	**6142.**	**6169.**	**6208.**	**6234.**	**6261.**	**6286.**	**6302.**	**6366.**
2	.05	18.51	19.00	19.16	19.25	19.30	19.33	19.36	19.37	19.38	19.39	19.40	19.41	19.42	19.43	19.44	19.45	19.46	19.47	19.47	19.50
	.01	**98.49**	**99.00**	**99.17**	**99.25**	**99.30**	**99.33**	**99.36**	**99.37**	**99.39**	**99.40**	**99.41**	**99.42**	**99.43**	**99.44**	**99.45**	**99.46**	**99.47**	**99.48**	**99.48**	**99.50**
3	.05	10.13	9.55	9.28	9.12	9.01	8.94	8.88	8.84	8.81	8.78	8.76	8.74	8.71	8.69	8.66	8.64	8.62	8.60	8.58	8.53
	.01	**34.12**	**30.82**	**29.46**	**28.71**	**28.24**	**27.91**	**27.67**	**27.49**	**27.34**	**27.23**	**27.13**	**27.05**	**26.92**	**26.83**	**26.69**	**26.60**	**26.50**	**26.41**	**26.35**	**26.12**
4	.05	7.71	6.94	6.59	6.39	6.26	6.16	6.09	6.04	6.00	5.96	5.93	5.91	5.87	5.84	5.80	5.77	5.74	5.71	5.70	5.63
	.01	**21.20**	**18.00**	**16.69**	**15.98**	**15.52**	**15.21**	**14.98**	**14.80**	**14.66**	**14.54**	**14.45**	**14.37**	**14.24**	**14.15**	**14.02**	**13.93**	**13.83**	**13.74**	**13.69**	**13.46**
5	.05	6.61	5.79	5.41	5.19	5.05	4.95	4.88	4.82	4.78	4.74	4.70	4.68	4.64	4.60	4.56	4.53	4.50	4.46	4.44	4.36
	.01	**16.26**	**13.27**	**12.06**	**11.39**	**10.97**	**10.67**	**10.45**	**10.29**	**10.15**	**10.05**	**9.96**	**9.89**	**9.77**	**9.68**	**9.55**	**9.47**	**9.38**	**9.29**	**9.24**	**9.02**
6	.05	5.99	5.14	4.76	4.53	4.39	4.28	4.21	4.15	4.10	4.06	4.03	4.00	3.96	3.92	3.87	3.84	3.81	3.77	3.75	3.67
	.01	**13.74**	**10.92**	**9.78**	**9.15**	**8.75**	**8.47**	**8.26**	**8.10**	**7.98**	**7.87**	**7.79**	**7.72**	**7.60**	**7.52**	**7.39**	**7.31**	**7.23**	**7.14**	**7.09**	**6.88**
7	.05	5.59	4.74	4.35	4.12	3.97	3.87	3.79	3.73	3.68	3.63	3.60	3.57	3.52	3.49	3.44	3.41	3.38	3.34	3.32	3.23
	.01	**12.25**	**9.55**	**8.45**	**7.85**	**7.46**	**7.19**	**7.00**	**6.84**	**6.71**	**6.62**	**6.54**	**6.47**	**6.35**	**6.27**	**6.15**	**6.07**	**5.98**	**5.90**	**5.85**	**5.65**
8	.05	5.32	4.46	4.07	3.84	3.69	3.58	3.50	3.44	3.39	3.34	3.31	3.28	3.23	3.20	3.15	3.12	3.08	3.05	3.03	2.93
	.01	**11.26**	**8.65**	**7.59**	**7.01**	**6.63**	**6.37**	**6.19**	**6.03**	**5.91**	**5.82**	**5.74**	**5.67**	**5.56**	**5.48**	**5.36**	**5.28**	**5.20**	**5.11**	**5.06**	**4.86**
9	.05	5.12	4.26	3.86	3.63	3.48	3.37	3.29	3.23	3.18	3.13	3.10	3.07	3.02	2.98	2.93	2.90	2.86	2.82	2.80	2.71
	.01	**10.56**	**8.02**	**6.99**	**6.42**	**6.06**	**5.80**	**5.62**	**5.47**	**5.35**	**5.26**	**5.18**	**5.11**	**5.00**	**4.92**	**4.80**	**4.73**	**4.64**	**4.56**	**4.51**	**4.31**
10	.05	4.96	4.10	3.71	3.48	3.33	3.22	3.14	3.07	3.02	2.97	2.94	2.91	2.86	2.82	2.77	2.74	2.70	2.67	2.64	2.54
	.01	**10.04**	**7.56**	**6.55**	**5.99**	**5.64**	**5.39**	**5.21**	**5.06**	**4.95**	**4.85**	**4.78**	**4.71**	**4.60**	**4.52**	**4.41**	**4.33**	**4.25**	**4.17**	**4.12**	**3.91**
11	.05	4.84	3.98	3.59	3.36	3.20	3.09	3.01	2.95	2.90	2.86	2.82	2.79	2.74	2.70	2.65	2.61	2.57	2.53	2.50	2.40
	.01	**9.65**	**7.20**	**6.22**	**5.67**	**5.32**	**5.07**	**4.88**	**4.74**	**4.63**	**4.54**	**4.46**	**4.40**	**4.29**	**4.21**	**4.10**	**4.02**	**3.94**	**3.86**	**3.80**	**3.60**
12	.05	4.75	3.88	3.49	3.26	3.11	3.00	2.92	2.85	2.80	2.76	2.72	2.69	2.64	2.60	2.54	2.50	2.46	2.42	2.40	2.30
	.01	**9.33**	**6.93**	**5.95**	**5.41**	**5.06**	**4.82**	**4.65**	**4.50**	**4.39**	**4.30**	**4.22**	**4.16**	**4.05**	**3.98**	**3.86**	**3.78**	**3.70**	**3.61**	**3.56**	**3.36**
13	.05	4.67	3.80	3.41	3.18	3.02	2.92	2.84	2.77	2.72	2.67	2.63	2.60	2.55	2.51	2.46	2.42	2.38	2.34	2.32	2.21
	.01	**9.07**	**6.70**	**5.74**	**5.20**	**4.86**	**4.62**	**4.44**	**4.30**	**4.19**	**4.10**	**4.02**	**3.96**	**3.85**	**3.78**	**3.67**	**3.59**	**3.51**	**3.42**	**3.37**	**3.16**
14	.05	4.60	3.74	3.34	3.11	2.96	2.85	2.77	2.70	2.65	2.60	2.56	2.53	2.48	2.44	2.39	2.35	2.31	2.27	2.24	2.13
	.01	**8.86**	**6.51**	**5.56**	**5.03**	**4.69**	**4.46**	**4.28**	**4.14**	**4.03**	**3.94**	**3.86**	**3.80**	**3.70**	**3.62**	**3.51**	**3.43**	**3.34**	**3.26**	**3.21**	**3.00**

F Distribution

α = .05 (roman type) and α = .01 boldface type)

v_1 Degrees of freedom

v_2	1	2	3	4	5	6	7	8	9	10	11	12	14	16	20	24	30	40	50	∞
15	4.54	3.68	3.29	3.06	2.90	2.79	2.70	2.64	2.59	2.55	2.51	2.48	2.43	2.39	2.33	2.29	2.25	2.21	2.18	2.07
	8.68	**6.36**	**5.42**	**4.89**	**4.56**	**4.32**	**4.14**	**4.00**	**3.89**	**3.80**	**3.73**	**3.67**	**3.56**	**3.48**	**3.36**	**3.29**	**3.20**	**3.12**	**3.07**	**2.87**
16	4.49	3.63	3.24	3.01	2.85	2.74	2.66	2.59	2.54	2.49	2.45	2.42	2.37	2.33	2.28	2.24	2.20	2.16	2.13	2.01
	8.53	**6.23**	**5.29**	**4.77**	**4.44**	**4.20**	**4.03**	**3.89**	**3.78**	**3.69**	**3.61**	**3.55**	**3.45**	**3.37**	**3.25**	**3.18**	**3.10**	**3.01**	**2.96**	**2.75**
17	4.45	3.59	3.20	2.96	2.81	2.70	2.62	2.55	2.50	2.45	2.41	2.38	2.33	2.29	2.23	2.19	2.15	2.11	2.08	1.96
	8.40	**6.11**	**5.18**	**4.67**	**4.34**	**4.10**	**3.93**	**3.79**	**3.68**	**3.59**	**3.52**	**3.45**	**3.35**	**3.27**	**3.16**	**3.08**	**3.00**	**2.92**	**2.86**	**2.65**
18	4.41	3.55	3.16	2.93	2.77	2.66	2.58	2.51	2.46	2.41	2.37	2.34	2.29	2.25	2.19	2.15	2.11	2.07	2.04	1.92
	8.28	**6.01**	**5.09**	**4.58**	**4.25**	**4.01**	**3.85**	**3.71**	**3.60**	**3.51**	**3.44**	**3.37**	**3.27**	**3.19**	**3.07**	**3.00**	**2.91**	**2.83**	**2.78**	**2.57**
19	4.38	3.52	3.13	2.90	2.74	2.63	2.55	2.48	2.43	2.38	2.34	2.31	2.26	2.21	2.15	2.11	2.07	2.02	2.00	1.88
	8.18	**5.93**	**5.01**	**4.50**	**4.17**	**3.94**	**3.77**	**3.63**	**3.52**	**3.43**	**3.36**	**3.30**	**3.19**	**3.12**	**3.00**	**2.92**	**2.84**	**2.76**	**2.70**	**2.49**
20	4.35	3.49	3.10	2.87	2.71	2.60	2.52	2.45	2.40	2.35	2.31	2.28	2.23	2.18	2.12	2.08	2.04	1.99	1.96	1.84
	8.10	**5.85**	**4.94**	**4.43**	**4.10**	**3.87**	**3.71**	**3.56**	**3.45**	**3.37**	**3.30**	**3.23**	**3.13**	**3.05**	**2.94**	**2.86**	**2.77**	**2.69**	**2.63**	**2.42**
21	4.32	3.47	3.07	2.84	2.68	2.57	2.49	2.42	2.37	2.32	2.28	2.25	2.20	2.15	2.09	2.05	2.00	1.96	1.93	1.81
	8.02	**5.78**	**4.87**	**4.37**	**4.04**	**3.81**	**3.65**	**3.51**	**3.40**	**3.31**	**3.24**	**3.17**	**3.07**	**2.99**	**2.88**	**2.80**	**2.72**	**2.63**	**2.58**	**2.36**
22	4.30	3.44	3.05	2.82	2.66	2.55	2.47	2.40	2.35	2.30	2.26	2.23	2.18	2.13	2.07	2.03	1.98	1.93	1.91	1.78
	7.94	**5.72**	**4.82**	**4.31**	**3.99**	**3.76**	**3.59**	**3.45**	**3.35**	**3.26**	**3.18**	**3.12**	**3.02**	**2.94**	**2.83**	**2.75**	**2.67**	**2.58**	**2.53**	**2.31**
23	4.28	3.42	3.03	2.80	2.64	2.53	2.45	2.38	2.32	2.28	2.24	2.20	2.14	2.10	2.04	2.00	1.96	1.91	1.88	1.76
	7.88	**5.66**	**4.76**	**4.26**	**3.94**	**3.71**	**3.54**	**3.41**	**3.30**	**3.21**	**3.14**	**3.07**	**2.97**	**2.89**	**2.78**	**2.70**	**2.62**	**2.53**	**2.48**	**2.26**
24	4.26	3.40	3.01	2.78	2.62	2.51	2.43	2.36	2.30	2.26	2.22	2.18	2.13	2.09	2.02	1.98	1.94	1.89	1.86	1.73
	7.82	**5.61**	**4.72**	**4.22**	**3.90**	**3.67**	**3.50**	**3.36**	**3.25**	**3.17**	**3.09**	**3.03**	**2.93**	**2.85**	**2.74**	**2.66**	**2.58**	**2.49**	**2.44**	**2.21**
25	4.24	3.38	2.99	2.76	2.60	2.49	2.41	2.34	2.28	2.24	2.20	2.16	2.11	2.06	2.00	1.96	1.92	1.87	1.84	1.71
	7.77	**5.57**	**4.68**	**4.18**	**3.86**	**3.63**	**3.46**	**3.32**	**3.21**	**3.13**	**3.05**	**2.99**	**2.89**	**2.81**	**2.70**	**2.62**	**2.54**	**2.45**	**2.40**	**2.17**
26	4.22	3.37	2.98	2.74	2.59	2.47	2.39	2.32	2.27	2.22	2.18	2.15	2.10	2.05	1.99	1.95	1.90	1.85	1.82	1.69
	7.72	**5.53**	**4.64**	**4.14**	**3.82**	**3.59**	**3.42**	**3.29**	**3.17**	**3.09**	**3.02**	**2.96**	**2.86**	**2.77**	**2.66**	**2.58**	**2.50**	**2.41**	**2.36**	**2.13**
27	4.21	3.35	2.96	2.73	2.57	2.46	2.37	2.30	2.25	2.20	2.16	2.13	2.08	2.03	1.97	1.93	1.88	1.84	1.80	1.67
	7.68	**5.49**	**4.60**	**4.11**	**3.79**	**3.56**	**3.39**	**3.26**	**3.14**	**3.06**	**2.98**	**2.93**	**2.83**	**2.74**	**2.63**	**2.55**	**2.47**	**2.38**	**2.33**	**2.10**
28	4.20	3.34	2.95	2.71	2.56	2.44	2.36	2.29	2.24	2.19	2.15	2.12	2.06	2.02	1.96	1.91	1.87	1.81	1.78	1.65
	7.64	**5.45**	**4.57**	**4.07**	**3.76**	**3.53**	**3.36**	**3.23**	**3.11**	**3.03**	**2.95**	**2.90**	**2.80**	**2.71**	**2.60**	**2.52**	**2.44**	**2.35**	**2.30**	**2.06**

F Distribution

$\alpha = .05$ (roman type) and $\alpha = .01$ boldface type)

v_1 Degrees of freedom

v_2	1	2	3	4	5	6	7	8	9	10	11	12	14	16	20	24	30	40	50	∞
29	4.18	3.33	2.93	2.70	2.54	2.43	2.35	2.28	2.22	2.18	2.14	2.10	2.05	2.00	1.94	1.90	1.85	1.80	1.77	1.64
	7.60	**5.42**	**4.54**	**4.04**	**3.73**	**3.50**	**3.33**	**3.20**	**3.08**	**3.00**	**2.92**	**2.87**	**2.77**	**2.68**	**2.57**	**2.49**	**2.41**	**2.32**	**2.27**	**2.03**
30	4.17	3.32	2.92	2.69	2.53	2.42	2.34	2.27	2.21	2.16	2.12	2.09	2.04	1.99	1.93	1.89	1.84	1.79	1.76	1.62
	7.56	**5.39**	**4.51**	**4.02**	**3.70**	**3.47**	**3.30**	**3.17**	**3.06**	**2.98**	**2.90**	**2.84**	**2.74**	**2.66**	**2.55**	**2.47**	**2.38**	**2.29**	**2.24**	**2.01**
32	4.15	3.30	2.90	2.67	2.51	2.40	2.32	2.25	2.19	2.14	2.10	2.07	2.02	1.97	1.91	1.86	1.82	1.76	1.74	1.59
	7.50	**5.34**	**4.46**	**3.97**	**3.66**	**3.42**	**3.25**	**3.12**	**3.01**	**2.94**	**2.86**	**2.80**	**2.70**	**2.62**	**2.51**	**2.42**	**2.34**	**2.25**	**2.20**	**1.96**
34	4.13	3.28	2.88	2.65	2.49	2.38	2.30	2.23	2.17	2.12	2.08	2.05	2.00	1.95	1.89	1.84	1.80	1.74	1.71	1.57
	7.44	**5.29**	**4.42**	**3.93**	**3.61**	**3.38**	**3.21**	**3.08**	**2.97**	**2.89**	**2.82**	**2.76**	**2.66**	**2.58**	**2.47**	**2.38**	**2.30**	**2.21**	**2.15**	**1.91**
36	4.11	3.26	2.86	2.63	2.48	2.36	2.28	2.21	2.15	2.10	2.06	2.03	1.98	1.93	1.87	1.82	1.78	1.72	1.69	1.55
	7.39	**5.25**	**4.38**	**3.89**	**3.58**	**3.35**	**3.18**	**3.04**	**2.94**	**2.86**	**2.78**	**2.72**	**2.62**	**2.54**	**2.43**	**2.35**	**2.26**	**2.17**	**2.12**	**1.87**
38	4.10	3.25	2.85	2.62	2.46	2.35	2.26	2.19	2.14	2.09	2.05	2.02	1.96	1.92	1.85	1.80	1.76	1.71	1.67	1.53
	7.35	**5.21**	**4.34**	**3.86**	**3.54**	**3.32**	**3.15**	**3.02**	**2.91**	**2.82**	**2.75**	**2.69**	**2.59**	**2.51**	**2.40**	**2.32**	**2.22**	**2.14**	**2.08**	**1.84**
40	4.08	3.23	2.84	2.61	2.45	2.34	2.25	2.18	2.12	2.07	2.04	2.00	1.95	1.90	1.84	1.79	1.74	1.69	1.66	1.51
	7.31	**5.18**	**4.31**	**3.83**	**3.51**	**3.29**	**3.12**	**2.99**	**2.88**	**2.80**	**2.73**	**2.66**	**2.56**	**2.49**	**2.37**	**2.29**	**2.20**	**2.11**	**2.05**	**1.81**
42	4.07	3.22	2.83	2.59	2.44	2.32	2.24	2.17	2.11	2.06	2.02	1.99	1.94	1.89	1.82	1.78	1.73	1.68	1.64	1.49
	7.27	**5.15**	**4.29**	**3.80**	**3.49**	**3.26**	**3.10**	**2.96**	**2.86**	**2.77**	**2.70**	**2.64**	**2.54**	**2.46**	**2.35**	**2.26**	**2.17**	**2.08**	**2.02**	**1.78**
44	4.06	3.21	2.82	2.58	2.43	2.31	2.23	2.16	2.10	2.05	2.01	1.98	1.92	1.88	1.81	1.76	1.72	1.66	1.63	1.48
	7.24	**5.12**	**4.26**	**3.78**	**3.46**	**3.24**	**3.07**	**2.94**	**2.84**	**2.75**	**2.68**	**2.62**	**2.52**	**2.44**	**2.32**	**2.24**	**2.15**	**2.06**	**2.00**	**1.75**
46	4.05	3.20	2.81	2.57	2.42	2.30	2.22	2.14	2.09	2.04	2.00	1.97	1.91	1.87	1.80	1.75	1.71	1.65	1.62	1.46
	7.21	**5.10**	**4.24**	**3.76**	**3.44**	**3.22**	**3.05**	**2.92**	**2.82**	**2.73**	**2.66**	**2.60**	**2.50**	**2.42**	**2.30**	**2.22**	**2.13**	**2.04**	**1.98**	**1.72**
48	4.04	3.19	2.80	2.56	2.41	2.30	2.21	2.14	2.08	2.03	1.99	1.96	1.90	1.86	1.79	1.74	1.70	1.64	1.61	1.45
	7.19	**5.08**	**4.22**	**3.74**	**3.42**	**3.20**	**3.04**	**2.90**	**2.80**	**2.71**	**2.64**	**2.58**	**2.48**	**2.40**	**2.28**	**2.20**	**2.11**	**2.02**	**1.96**	**1.70**
50	4.03	3.18	2.79	2.56	2.40	2.29	2.20	2.13	2.07	2.02	1.98	1.95	1.90	1.85	1.78	1.74	1.69	1.63	1.60	1.44
	7.17	**5.06**	**4.20**	**3.72**	**3.41**	**3.18**	**3.02**	**2.88**	**2.78**	**2.70**	**2.62**	**2.56**	**2.46**	**2.39**	**2.26**	**2.18**	**2.10**	**2.00**	**1.94**	**1.68**
55	4.02	3.17	2.78	2.54	2.38	2.27	2.18	2.11	2.05	2.00	1.97	1.93	1.88	1.83	1.76	1.72	1.67	1.61	1.58	1.41
	7.12	**5.01**	**4.16**	**3.68**	**3.37**	**3.15**	**2.98**	**2.85**	**2.75**	**2.66**	**2.59**	**2.53**	**2.43**	**2.35**	**2.23**	**2.15**	**2.06**	**1.96**	**1.90**	**1.64**
60	4.00	3.15	2.76	2.52	2.37	2.25	2.17	2.10	2.04	1.99	1.95	1.92	1.86	1.81	1.75	1.70	1.65	1.59	1.56	1.39
	7.08	**4.98**	**4.13**	**3.65**	**3.34**	**3.12**	**2.95**	**2.82**	**2.72**	**2.63**	**2.56**	**2.50**	**2.40**	**2.32**	**2.20**	**2.12**	**2.03**	**1.93**	**1.87**	**1.60**

F Distribution
$\alpha = .05$ (roman type) and $\alpha = .01$ (boldface type)

v_1 Degrees of freedom

v_2	1	2	3	4	5	6	7	8	9	10	11	12	14	16	20	24	30	40	50	∞
65	3.99	3.14	2.75	2.51	2.36	2.24	2.15	2.08	2.02	1.98	1.94	1.90	1.85	1.80	1.73	1.68	1.63	1.57	1.54	1.37
	7.04	**4.95**	**4.10**	**3.62**	**3.31**	**3.09**	**2.93**	**2.79**	**2.70**	**2.61**	**2.54**	**2.47**	**2.37**	**2.30**	**2.18**	**2.09**	**2.00**	**1.90**	**1.84**	**1.56**
70	3.98	3.13	2.74	2.50	2.35	2.23	2.14	2.07	2.01	1.97	1.93	1.89	1.84	1.79	1.72	1.67	1.62	1.56	1.53	1.35
	7.01	**4.92**	**4.08**	**3.60**	**3.29**	**3.07**	**2.91**	**2.77**	**2.67**	**2.59**	**2.51**	**2.45**	**2.35**	**2.28**	**2.15**	**2.07**	**1.98**	**1.88**	**1.82**	**1.53**
80	3.96	3.11	2.72	2.48	2.33	2.21	2.12	2.05	1.99	1.95	1.91	1.88	1.82	1.77	1.70	1.65	1.60	1.54	1.51	1.32
	6.96	**4.88**	**4.04**	**3.56**	**3.25**	**3.04**	**2.87**	**2.74**	**2.64**	**2.55**	**2.48**	**2.41**	**2.32**	**2.24**	**2.11**	**2.03**	**1.94**	**1.84**	**1.78**	**1.49**
100	3.94	3.09	2.70	2.46	2.30	2.19	2.10	2.03	1.97	1.92	1.88	1.85	1.79	1.75	1.68	1.63	1.57	1.51	1.48	1.28
	6.90	**4.82**	**3.98**	**3.51**	**3.20**	**2.99**	**2.82**	**2.69**	**2.59**	**2.51**	**2.43**	**2.36**	**2.26**	**2.19**	**2.06**	**1.98**	**1.89**	**1.79**	**1.73**	**1.43**
125	3.92	3.07	2.68	2.44	2.29	2.17	2.08	2.01	1.95	1.90	1.86	1.83	1.77	1.72	1.65	1.60	1.55	1.49	1.45	1.25
	6.84	**4.78**	**3.94**	**3.47**	**3.17**	**2.95**	**2.79**	**2.65**	**2.56**	**2.47**	**2.40**	**2.33**	**2.23**	**2.15**	**2.03**	**1.94**	**1.85**	**1.75**	**1.68**	**1.37**
150	3.91	3.06	2.67	2.43	2.27	2.16	2.07	2.00	1.94	1.89	1.85	1.82	1.76	1.71	1.64	1.59	1.54	1.47	1.44	1.22
	6.81	**4.75**	**3.91**	**3.44**	**3.14**	**2.92**	**2.76**	**2.62**	**2.53**	**2.44**	**2.37**	**2.30**	**2.20**	**2.12**	**2.00**	**1.91**	**1.83**	**1.72**	**1.66**	**1.33**
200	3.89	3.04	2.65	2.41	2.26	2.14	2.05	1.98	1.92	1.87	1.83	1.80	1.74	1.69	1.62	1.57	1.52	1.45	1.42	1.19
	6.76	**4.71**	**3.88**	**3.41**	**3.11**	**2.90**	**2.73**	**2.60**	**2.50**	**2.41**	**2.34**	**2.28**	**2.17**	**2.09**	**1.97**	**1.88**	**1.79**	**1.69**	**1.62**	**1.28**
400	3.86	3.02	2.62	2.39	2.23	2.12	2.03	1.96	1.90	1.85	1.81	1.78	1.72	1.67	1.60	1.54	1.49	1.42	1.38	1.13
	6.70	**4.66**	**3.83**	**3.36**	**3.06**	**2.85**	**2.69**	**2.55**	**2.46**	**2.37**	**2.29**	**2.23**	**2.12**	**2.04**	**1.92**	**1.84**	**1.74**	**1.64**	**1.57**	**1.19**
1000	3.85	3.00	2.61	2.38	2.22	2.10	2.02	1.95	1.89	1.84	1.80	1.76	1.70	1.65	1.58	1.53	1.47	1.41	1.36	1.08
	6.66	**4.62**	**3.80**	**3.34**	**3.04**	**2.82**	**2.66**	**2.53**	**2.43**	**2.34**	**2.26**	**2.20**	**2.09**	**2.01**	**1.89**	**1.81**	**1.71**	**1.61**	**1.54**	**1.11**
∞	3.84	2.99	2.60	2.37	2.21	2.09	2.01	1.94	1.88	1.83	1.79	1.75	1.69	1.64	1.57	1.52	1.46	1.40	1.35	1.00
	6.63	**4.60**	**3.78**	**3.32**	**3.02**	**2.80**	**2.64**	**2.51**	**2.41**	**2.32**	**2.24**	**2.18**	**2.07**	**1.99**	**1.87**	**1.79**	**1.69**	**1.59**	**1.52**	**1.00**

From Snedecor, G. W. and Cochran, W. G., *Statistical Methods*, 7th ed., Iowa State University Press, Ames, IA, 1980. With permission.

Appendix G

Chi-square (χ^2) distribution

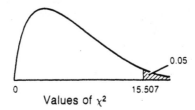

Values of χ^2

Example: In a χ^2 distribution with $v = 8$ degrees of freedom ($d.f.$) the area to the right of a χ^2 value of 15.507 is 0.05.

Degrees of freedom v	Area in right tail				
	0.20	.010	0.05	0.02	0.01
1	1.642	2.706	3.841	5.412	6.635
2	3.219	4.605	5.991	7.824	9.210
3	4.642	6.251	7.815	9.837	11.345
4	5.989	7.779	9.488	11.668	13.277
5	7.289	9.236	11.070	13.388	15.086
6	8.558	10.645	12.592	15.033	16.812
7	9.803	12.017	14.067	16.622	18.475
8	11.030	13.362	15.507	18.168	20.090
9	12.242	14.684	16.919	19.679	21.666
10	13.442	15.987	18.307	21.161	23.209
11	14.631	17.275	19.675	22.618	24.725
12	15.812	18.549	21.026	24.054	26.217
13	16.985	19.812	22.362	25.472	27.688
14	18.151	21.064	23.685	26.873	29.141
15	19.311	22.307	24.996	28.259	30.578
16	20.465	23.542	26.296	29.633	32.000
17	21.615	24.769	27.587	30.995	33.409
18	22.760	25.989	28.869	32.346	34.805
19	23.900	27.204	30.144	33.687	36.191
20	25.038	28.412	31.410	35.020	37.566
21	26.171	29.615	32.671	36.343	38.932
22	27.301	30.813	33.924	37.659	40.289
23	28.429	32.007	35.172	38.968	41.638
24	29.553	33.196	36.415	40.270	42.980
25	30.675	34.382	37.652	41.566	44.314
26	31.795	35.563	38.885	42.856	45.642
27	32.912	36.741	40.113	44.140	46.963

Degrees of freedom ν	Area in right tail				
	0.20	.010	0.05	0.02	0.01
28	34.027	37.916	41.337	45.419	48.278
29	35.139	39.087	42.557	46.693	49.588
30	36.250	40.256	43.773	47.962	50.892

Appendix H

Wilcoxon T values

Critical Values of T, the Wilcoxon Signed Rank Statistic, Where T is
the Largest Integer such That $Pr(T \leq t/N) \leq \alpha$ the Cumulative
One-Tailed Probability

N	$2\alpha.15$ $\alpha.075$	.10 .050	.05 .025	.04 .020	.03 .015	.02 .010	.01 .005
4	0						
5	1	0					
6	2	2	0	0			
7	4	3	2	1	0	0	
8	7	5	3	3	2	1	0
9	9	8	5	5	4	3	1
10	12	10	8	7	6	5	3
11	16	13	10	9	8	7	5
12	19	17	13	12	11	9	7
13	24	21	17	16	14	12	9
14	28	25	21	19	18	15	12
15	33	30	25	23	21	19	15
16	39	35	29	28	26	23	19
17	45	41	34	33	30	27	23
18	51	47	40	38	35	32	27
19	58	53	46	43	41	37	32
20	65	60	52	50	47	43	37
21	73	67	58	56	53	49	42
22	81	75	65	63	59	55	48
23	89	83	73	70	66	62	54
24	98	91	81	78	74	69	61
25	108	100	89	86	82	76	68
26	118	110	98	94	90	84	75
27	128	119	107	103	99	92	83
28	138	130	116	112	108	101	91
29	150	140	126	122	117	110	100
30	161	151	137	132	127	120	109
31	173	163	147	143	137	130	118
32	186	175	159	154	148	140	128
33	199	187	170	165	159	151	138
34	212	200	182	177	171	162	148
35	226	213	195	189	182	173	159
40	302	286	264	257	249	238	220
50	487	466	434	425	413	397	373
60	718	690	648	636	620	600	567
70	995	960	907	891	872	846	805
80	1318	1276	1211	1192	1168	1136	1086
90	1688	1638	1560	1537	1509	1471	1410
100	2105	2045	1955	1928	1894	1850	1779

Adapted from Verdoorn, L. R., *Biometrika*, Vol. 50, *Extended Table of Critical Values for Wilcoxon's*, 1963. Reproduced by permission of the Biometrika Trustees.

Appendix I

Critical values of U in the Mann–Whitney test

In the First Table the Entries are the Critical Values of U for a One-Tailed Test at .025 or for a Two-Tailed Test at .05; in the Second, for a One-Tailed Test at .05 or for a Two-Tailed Test at .10

$n_2 \backslash n_1$	1	2	3	4	5	6	7	8	9	10	11	12	13	14	15	16	17	18	19	20
1																				
2								0	0	0	0	1	1	1	1	1	2	2	2	2
3					0	1	1	2	2	3	3	4	4	5	5	6	6	7	7	8
4				0	1	2	3	4	4	5	6	7	8	9	10	11	11	12	13	13
5			0	1	2	3	5	6	7	8	9	11	12	13	14	15	17	18	19	20
6			1	2	3	5	6	8	10	11	13	14	16	17	19	21	22	24	25	27
7			1	3	5	6	8	10	12	14	16	18	20	22	24	26	28	30	32	34
8		0	2	4	6	8	10	13	15	17	19	22	24	26	29	31	34	36	38	41
9		0	2	4	7	10	12	15	17	20	23	26	28	31	34	37	39	42	45	48
10		0	3	5	8	11	14	17	20	23	26	29	33	36	39	42	45	48	52	55
11		0	3	6	9	13	16	19	23	26	30	33	37	40	44	47	51	55	58	62
12		1	4	7	11	14	18	22	26	29	33	37	41	45	49	53	57	61	65	69
13		1	4	8	12	16	20	24	28	33	37	41	45	50	54	59	63	67	72	76
14		1	5	9	13	17	22	26	31	36	40	45	50	55	59	64	67	74	78	83
15		1	5	10	14	19	24	29	34	39	44	49	54	59	64	70	75	80	85	90
16		1	6	11	15	21	26	31	37	42	47	53	59	64	70	75	81	86	92	98
17		2	6	11	17	22	28	34	39	45	51	57	63	67	75	81	87	93	99	105
18		2	7	12	18	24	30	36	42	48	55	61	67	74	80	86	93	99	106	112
19		2	7	13	19	25	32	38	45	52	58	65	72	78	85	92	99	106	113	119
20		2	8	13	20	27	34	41	48	55	62	69	76	83	90	98	105	112	119	127

$n_2 \backslash n_1$	1	2	3	4	5	6	7	8	9	10	11	12	13	14	15	16	17	18	19	20
1																			0	0
2					0	0	0	1	1	1	1	2	2	2	3	3	3	4	4	4
3			0	0	1	2	2	3	3	4	5	5	6	7	7	8	9	9	10	11
4			0	1	2	3	4	5	6	7	8	9	10	11	12	14	15	16	17	18
5		0	1	2	4	5	6	8	9	11	12	13	15	16	18	19	20	22	23	25
6		0	2	3	5	7	8	10	12	14	16	17	19	21	23	25	26	28	30	32
7		0	2	4	6	8	11	13	15	17	19	21	24	26	28	30	33	35	37	39
8		1	3	5	8	10	13	15	18	20	23	26	28	31	33	36	39	41	44	47
9		1	3	6	9	12	15	18	21	24	27	30	33	36	39	42	45	48	51	54
10		1	4	7	11	14	17	20	24	27	31	34	37	41	44	48	51	55	58	62
11		1	5	8	12	16	19	23	27	31	34	38	42	46	50	54	57	61	65	69
12		2	5	9	13	17	21	26	30	34	38	42	47	51	55	60	64	68	72	77
13		2	6	10	15	19	24	28	33	37	42	47	51	56	61	65	70	75	80	84
14		2	7	11	16	21	26	31	36	41	46	51	56	61	66	71	77	82	87	92
15		3	7	12	18	23	28	33	39	44	50	55	61	66	72	77	83	88	94	100
16		3	8	14	19	25	30	36	42	48	54	60	65	71	77	83	89	95	101	107
17		3	9	15	20	26	33	39	45	51	57	64	70	77	83	89	96	102	109	115
18		4	9	16	22	28	35	41	48	55	61	68	75	82	88	95	102	109	116	123
19	0	4	10	17	23	30	37	44	51	58	65	72	80	87	94	101	109	116	123	130
20	0	4	11	18	25	32	39	47	54	62	69	77	84	92	100	107	115	123	130	138

Appendix J

Values of d_L and d_U for the Durbin–Watson Test

For $\alpha = .05$

	k = 1		k = 2		k = 3		k = 4		k = 5	
n	d_L	d_U	d_L	d_U	d_L	d_U	d_L	d_U	d_L	d_U
15	1.08	1.36	0.95	1.54	0.82	1.75	0.69	1.97	0.56	2.21
16	1.10	1.37	0.98	1.54	0.86	1.73	0.74	1.93	0.62	2.15
17	1.13	1.38	1.02	1.54	0.90	1.71	0.78	1.90	0.67	2.10
18	1.16	1.39	1.05	1.53	0.93	1.69	0.82	1.87	0.71	2.06
19	1.18	1.40	1.08	1.53	0.97	1.68	0.86	1.85	0.75	2.02
20	1.20	1.41	1.10	1.54	1.00	1.68	0.90	1.83	0.79	1.99
21	1.22	1.42	1.13	1.54	1.03	1.67	0.93	1.81	0.83	1.96
22	1.24	1.43	1.15	1.54	1.05	1.66	0.96	1.80	0.86	1.94
23	1.26	1.44	1.17	1.54	1.08	1.66	0.99	1.79	0.90	1.92
24	1.27	1.45	1.19	1.55	1.10	1.66	1.01	1.78	0.93	1.90
25	1.29	1.45	1.21	1.55	1.12	1.66	1.04	1.77	0.95	1.89
26	1.30	1.46	1.22	1.55	1.14	1.65	1.06	1.76	0.98	1.88
27	1.32	1.47	1.24	1.56	1.16	1.65	1.08	1.76	1.01	1.86
28	1.33	1.48	1.26	1.56	1.18	1.65	1.10	1.75	1.03	1.85
29	1.34	1.48	1.27	1.56	1.20	1.65	1.12	1.74	1.05	1.84
30	1.35	1.49	1.28	1.57	1.21	1.65	1.14	1.74	1.07	1.83
31	1.36	1.50	1.30	1.57	1.23	1.65	1.16	1.74	1.09	1.83
32	1.37	1.50	1.31	1.57	1.24	1.65	1.18	1.73	1.11	1.82
33	1.38	1.51	1.32	1.58	1.26	1.65	1.19	1.73	1.13	1.81
34	1.39	1.51	1.33	1.58	1.27	1.65	1.21	1.73	1.15	1.81
35	1.40	1.52	1.34	1.58	1.28	1.65	1.22	1.73	1.16	1.80
36	1.41	1.52	1.35	1.59	1.29	1.65	1.24	1.73	1.18	1.80
37	1.42	1.53	1.36	1.59	1.31	1.66	1.25	1.72	1.19	1.80
38	1.43	1.54	1.37	1.59	1.32	1.66	1.26	1.72	1.21	1.79
39	1.43	1.54	1.38	1.60	1.33	1.66	1.27	1.72	1.22	1.79
40	1.44	1.54	1.39	1.60	1.34	1.66	1.29	1.72	1.23	1.79
45	1.48	1.57	1.43	1.62	1.38	1.67	1.34	1.72	1.29	1.78
50	1.50	1.59	1.46	1.63	1.42	1.67	1.38	1.72	1.34	1.77
55	1.53	1.60	1.49	1.64	1.45	1.68	1.41	1.72	1.38	1.77
60	1.55	1.62	1.51	1.65	1.48	1.69	1.44	1.73	1.41	1.77
65	1.57	1.63	1.54	1.66	1.50	1.70	1.47	1.73	1.44	1.77
70	1.58	1.64	1.55	1.67	1.52	1.70	1.49	1.74	1.46	1.77
75	1.60	1.65	1.57	1.68	1.54	1.71	1.51	1.74	1.49	1.77
80	1.61	1.66	1.59	1.69	1.56	1.72	1.53	1.74	1.51	1.77
85	1.62	1.67	1.60	1.70	1.57	1.72	1.55	1.75	1.52	1.77
90	1.63	1.68	1.61	1.70	1.59	1.73	1.57	1.75	1.54	1.78
95	1.64	1.69	1.62	1.71	1.60	1.73	1.58	1.75	1.56	1.78
100	1.65	1.69	1.63	1.72	1.61	1.74	1.59	1.76	1.57	1.78

Note: n = Number of observations; k = number of independent variables.

From Durbin, J. and Watson, G.S., Testing for serial correlation in least squares regression, *Biometrika*, 38, June 1951. With permission.

For α = .01

n	k = 1		k = 2		k = 3		k = 4		k = 5	
	d_L	d_U	d_L	d_U	d_L	d_U	d_L	d_U	d_L	d_U
15	0.81	1.07	0.70	1.25	0.59	1.46	0.49	1.70	0.39	1.96
16	0.84	1.09	0.74	1.25	0.63	1.44	0.53	1.66	0.44	1.90
17	0.87	1.10	0.77	1.25	0.67	1.43	0.57	1.63	0.48	1.85
18	0.90	1.12	0.80	1.26	0.71	1.42	0.61	1.60	0.52	1.80
19	0.93	1.13	0.83	1.26	0.74	1.41	0.65	1.58	0.56	1.77
20	0.95	1.15	0.86	1.27	0.77	1.41	0.68	1.57	0.60	1.74
21	0.97	1.16	0.89	1.27	0.80	1.41	0.72	1.55	0.63	1.71
22	1.00	1.17	0.91	1.28	0.83	1.40	0.75	1.54	0.66	1.69
23	1.02	1.19	0.94	1.29	0.86	1.40	0.77	1.53	0.70	1.67
24	1.04	1.20	0.96	1.30	0.88	1.41	0.80	1.53	0.72	1.66
25	1.05	1.21	0.98	1.30	0.90	1.41	0.83	1.52	0.75	1.65
26	1.07	1.22	1.00	1.31	0.93	1.41	0.85	1.52	0.78	1.64
27	1.09	1.23	1.02	1.32	0.95	1.41	0.88	1.51	0.81	1.63
28	1.10	1.24	1.04	1.32	0.97	1.41	0.90	1.51	0.83	1.62
29	1.12	1.25	1.05	1.33	0.99	1.42	0.92	1.51	0.85	1.61
30	1.13	1.26	1.07	1.34	1.01	1.42	0.94	1.51	0.88	1.61
31	1.15	1.27	1.08	1.34	1.02	1.42	0.96	1.51	0.90	1.60
32	1.16	1.28	1.10	1.35	1.04	1.43	0.98	1.51	0.92	1.60
33	1.17	1.29	1.11	1.36	1.05	1.43	1.00	1.51	0.94	1.59
34	1.18	1.30	1.13	1.36	1.07	1.43	1.01	1.51	0.95	1.59
35	1.19	1.31	1.14	1.37	1.08	1.44	1.03	1.51	0.97	1.59
36	1.21	1.32	1.15	1.38	1.10	1.44	1.04	1.51	0.99	1.59
37	1.22	1.32	1.16	1.38	1.11	1.45	1.06	1.51	1.00	1.59
38	1.23	1.33	1.18	1.39	1.12	1.45	1.07	1.52	1.02	1.58
39	1.24	1.34	1.19	1.39	1.14	1.45	1.09	1.52	1.03	1.58
40	1.25	1.34	1.20	1.40	1.15	1.46	1.10	1.52	1.05	1.58
45	1.29	1.38	1.24	1.42	1.20	1.48	1.16	1.53	1.11	1.58
50	1.32	1.40	1.28	1.45	1.24	1.49	1.20	1.54	1.16	1.59
55	1.36	1.43	1.32	1.47	1.28	1.51	1.25	1.55	1.21	1.59
60	1.38	1.45	1.35	1.48	1.32	1.52	1.28	1.56	1.25	1.60
65	1.41	1.47	1.38	1.50	1.35	1.53	1.31	1.57	1.28	1.61
70	1.43	1.49	1.40	1.52	1.37	1.55	1.34	1.58	1.31	1.61
75	1.45	1.50	1.42	1.53	1.39	1.56	1.37	1.59	1.34	1.62
80	1.47	1.52	1.44	1.54	1.42	1.57	1.39	1.60	1.36	1.62
85	1.48	1.53	1.46	1.55	1.43	1.58	1.41	1.60	1.39	1.63
90	1.50	1.54	1.47	1.56	1.45	1.59	1.43	1.61	1.41	1.64
95	1.51	1.55	1.49	1.57	1.47	1.60	1.45	1.62	1.42	1.64
100	1.52	1.56	1.50	1.58	1.48	1.60	1.46	1.63	1.44	1.65

Index

Milton Keynes UK
Ingram Content Group UK Ltd.
UKHW021854071024
449327UK00021B/1570